Y0-BDD-109

Immunoinformatics

METHODS IN MOLECULAR BIOLOGY™

John M. Walker, SERIES EDITOR

METHODS IN MOLECULAR BIOLOGY™

Immunoinformatics

Predicting Immunogenicity In Silico

Edited by

Darren R. Flower

The Jenner Institute, University of Oxford, Berkshire, UK

HUMANA PRESS ✱ TOTOWA, NEW JERSEY

999 Riverview Drive, Suite 208
Totowa, New Jersey 07512

www.humanapress.com

This publication is printed on acid-free paper. ∞
ANSI Z39.48-1984 (American Standards Institute) Permanence of Paper for Printed Library Materials

Cover illustration: derived from Fig. 1, Chapter 13; *see* complete figure and legend on p. 191.

For additional copies, pricing for bulk purchases, and/or information about other Humana titles, contact Humana at the above address or at any of the following numbers: Tel.: 973-256-1699; Fax: 973-256-8341; E-mail: humana@humanapr.com; or visit our Website: www.humanapress.com

eISBN 978-1-60327-118-9

Printed in the United States of America. 10 9 8 7 6 5 4 3 2 1

Library of Congress Control Number: 2007927138

Preface

This is a book about immunoinformatics and its developing role in the computational prediction of immunogenicity. Immunogenicity is the ability to induce a specific immune response when a pathogen is exposed to initial surveillance by the immune system. Immunoinformatics is, as well seen time and time again, the application of informatics techniques drawn from computer science to molecules of the immune system and their interactions. In recent years, immunoinformatics has grown significantly in scientific stature and ready applicability and is now able to contribute in a genuine manner to all kinds of problems in immunology, not least the prediction of immunogenicity. The practical utility of such approaches in the discovery and development of vaccines is a question which remains open for many, but for those able to grasp and capitalize on its potential, immunoinformatics is set to become a tool of incomparable value. This book is thus a primer for those keen to come to grips with this emerging technology. Albeit not pretending to be completely comprehensive, *Immunoinformatics: Predicting Immunogenicity In Silico* nonetheless sets out to sample the major areas in immunoinformatics. It seeks to equip the reader with a grasp of where the field is and where it is going. Hopefully, it will both engage the reader and provide a sound background for the use of immunoinformatics in immunology and vaccinology. As high-throughput systems biology begins to gather speed and threatens to sweep all before it, the future of biological science, of which immunology is such a profound part, will rightly belong to those able to combine seamlessly the experimental and the theoretical aspects of bioscience, merging without effort or obvious discontinuity the skill sets of the lab-based and the computer-based science professional. It may take some time for the full ascendancy of this dynamic hybrid to properly assert itself, but the day will come when both the atavistic, pipette-wielding Luddite in the white coat and the socially inept, geeky, nerdy weirdo staring into the computer screen will become stereotypes as outmoded and redundant as the most extinct of Dodos. As a community, science should engage this change as wholeheartedly as it can. Such an eventuality can be avoided but only for so long. So begin by reading this *Immunoinformatics: Predicting Immunogenicity In Silico*: garner its wisdom should you find some, savour its gems, gather up its insight, and forgive its foibles, its inconsistencies, its shortcomings, and its many omissions, yet above all learn from it and even try to enjoy it.

I wish to thank all the authors for their worthy contributions to the book. It goes without saying that without the chapters that they contributed the book would not exist. Having said that, the quality that their work evinces is nonetheless outstanding. I should also like to thank Prof. John Walker, Editor-in-Chief of the Methods in Molecular Biology series, whose help and encouragement has been steadfast and has greatly eased the passage from inception to publication. Likewise, my thanks go to all the staff at Humana for their inestimable contributions in administration and book production, which have complemented marvelously the work of all the authors. Though I am deeply indebted to all contributors for all their help and advice, I must take on myself blame for any mistakes and omissions you find herein.

Finally, some local thanks. In particular, I should also like to thank members of my research group for helping to make all this possible: Dr. Irini Doytchinova, Valerie Walshe, Martin Blythe, Christianna Zygouri, Debra Clayton (née Taylor), Shelley Hemsley, Christopher Toseland, Kelly Paine, Dr. Pingping Guan, Dr. Paul Taylor, Dr. Helen McSparron, Dr. Matthew N. Davies, Dr. Channa Hattotuwagama, and Dr. Shunzhou Wan. I should also like to thank other staff members for their help and for stimulating discussions: Prof. Peter Beverley, Dr. Persephone Borrow, Dr. Shirley Ellis, Dr. Simon Wong, Dr. Helen Bodmer, Dr. Sam Hou, Dr. Lisa Hyland, Dr. David Tough, Dr. Elma Tchillian, and Dr. Josef Walker. I should also like to thank my colleagues and co-workers at the EJIVR and the Institute for Animal Health (IAH), Compton for their close and supportive collaboration. Finally, I thank a number of others: Prof. Terri Attwood, Prof. Peter Coveney, Prof. Vladimir Brusic, and Dr. Anne DeGroot.

Darren R. Flower

Contents

Contributors

MANOJ K. BHASIN • *Institute of Microbial Technology, Chandigarh, India, and Dana-Farber Cancer Institute, Harvard Medical School, Boston, MA*
RAINER BLASCZYK • *Institute for Transfusion Medicine, Hannover Medical School, Hannover, Germany*
FRANK R. BURDEN • *SciMetrics, Harrow Enterprises Pty. Ltd., Fitzroy, Victoria, Australia*
PETER V. COVENEY • *Centre for Computational Science, Chemistry Department, University College of London, London, UK*
YANG DAI • *Bioengineering Bioinformatics, University of Illinois at Chicago, Chicago, IL*
MATTHEW N. DAVIES • *The Jenner Institute, University of Oxford, Berkshire, UK*
DAVID S. DELUCA • *Institute for Transfusion Medicine, Hannover Medical School, Hannover, Germany*
PIERRE DÖNNES • *Division for Simulation of Biological Systems, Eberhard Karls University Tübingen, Tübingen, Germany*
IRINI A. DOYTCHINOVA • *The Jenner Institute, University of Oxford, Berkshire, UK*
DARREN R. FLOWER • *The Jenner Institute, University of Oxford, Berkshire, UK*
PINGPING GUAN • *Computational Biology Group, John Innes Centre, Norwich, UK*
CHANNA K. HATTOTUWAGAMA • *The Jenner Institute, University of Oxford, Berkshire, UK*
LEI HUANG • *Bioengineering Bioinformatics, University of Illinois at Chicago, Chicago, IL*
PANDJASSARAME KANGUEANE • *School of Mechanical and Aerospace Engineering, NANYANG Technological University, Singapore*
OLEKSIY KARPENKO • *Bioengineering Bioinformatics, University of Illinois at Chicago, Chicago, IL*
SNEH LATA • *Institute of Microbial Technology, Chandigarh, India*

MARIE-PAULE LEFRANC • *Institut Universitaire de France, Laboratoire d'ImmunoGénétique Moléculaire, LIGM, Université Montpellier II, UPR CNRS 1142, Institut de Génétique Humaine, Montpellier Cedex, France*
TONGBIN LI • *Department of Neuroscience, University of Minnesota, Minneapolis, MN*
THY-HOU LIN • *Institute of Molecular Medicine and Department of Life Science, National Tsing Hua University, Hsinchu, Taiwan*
WEN LIU • *Department of Neuroscience, University of Minnesota, Minneapolis, MN*
RONNA REUBEN MALLIOS • *Grants and Research Office, Fresno, CA*
STEVEN G. E. MARSH • *Department of Haematology, Anthony Nolan Research Institute, Royal Free Hospital, Hampstead, London, UK*
XIANGSHAN MENG • *Department of Neuroscience, University of Minnesota, Minneapolis, MN*
NAVEEN MURUGAN • *Bioengineering Bioinformatics, University of Illinois at Chicago, Chicago, IL*
MARIA-DOROTHEA NASTKE • *Department of Immunology, Institute for Cell Biology, University of Tübingen, Tübingen, Germany*
GAJENDRA P. S. RAGHAVA • *Institute of Microbial Technology, Chandigarh, India; and UAMS, BRCII, Little Rock, AR*
SHOBA RANGANATHAN • *Department of Chemistry and Biomolecular Sciences & Biotechnology Research Institute, Macquarie University, New South Wales, Australia; and Department of Biochemistry, Yong Loo Lin School of Medicine, National University of Singapore, Singapore*
PEDRO A. RECHE • *Laboratory of Immunobiology and Department of Medical Oncology, Dana-Farber Cancer Institute; and Department of Medicine, Harvard Medical School, Boston, MA*
ELLIS L. REINHERZ • *Dana-Farber Cancer Institute, Harvard Medical School, Boston, MA*
JAMES ROBINSON • *Anthony Nolan Research Institute, Royal Free Hospital, Hampstead, London, UK*
SUDIPTO SAHA • *Institute of Microbial Technology, Chandigarh, India*
MEENA KISHORE SAKHARKAR • *School of Mechanical and Aerospace Engineering, NANYANG Technological University, Singapore*
MATHIAS M. SCHULER • *Department of Immunology, Institute for Cell Biology, University of Tübingen, Tübingen, Germany*
RICHARD SIMON • *Biometric Research Branch, National Cancer Institute, National Institutes of Health, Rockville, MD*

MAHENDER KUMAR SINGH • *Institute of Microbial Technology, Chandigarh, India*
SHILPY SRIVASTAVA • *Institute of Microbial Technology, Chandigarh, India*
STEFAN STEVANOVIĆ • *Department of Immunology, Institute for Cell Biology, University of Tübingen, Tübingen, Germany*
MYONG-HEE SUNG • *Laboratory of Receptor Biology and Gene Expression, Staff Scientist National Cancer Institute, Bethesda, MD*
JOO CHUAN TONG • *Department of Biochemistry, National University of Singapore, Singapore, and Institute for Infocomm Research, Singapore*
GRISH C. VARSHNEY • *Institute of Microbial Technology, Chandigarh, India*
JI WAN • *Department of Neuroscience, University of Minnesota, Minneapolis, MN*
SHUNZHOU WAN • *Centre for Computational Science, Chemistry Department, University College of London, London, UK*
DAVID A. WINKLER • *Centre for Complexity in Drug Discovery, CSIRO Molecular and Health Technologies, Clayton, Australia*
YINGDONG ZHAO • *National Cancer Institute, National Institutes of Health, Rockville, MD*

Color Plates

Color plates follow p. 32.

COLOR PLATE 1 *IMGT* Collier de Perles of a V-DOMAIN. The *IMGT* Collier de Perles of V-DOMAIN is based on the *IMGT* unique numbering for V-DOMAIN and V-LIKE-DOMAIN (***10***). (Chapter 2, Fig. 1; *see* full caption on p. 25 and discussion on p. 24.)

COLOR PLATE 2 Three-dimensional structures and *IMGT* Collier de Perles of a V-DOMAIN, a C-DOMAIN and G-DOMAINs. (Chapter 2, Fig. 2; *see* full caption on p. 27 and discussion on p. 24.)

COLOR PLATE 3 *IMGT* pMHC contact sites of human HLA-A*0201 MHC-I and a 9-amino acid peptide side chains (*IMGT/3Dstructure-DB*: 1im3). (Chapter 2, Fig. 4; *see* full caption on p. 36 and discussion on p. 35.)

COLOR PLATE 4 *IMGT* pMHC contact sites of human HLA-DRA*0101 and HLA-DRB5*0101 MHC-II and the peptide side chains (9 amino acids located in the groove) (*IMGT/3Dstructure-DB*: 1fv1). (Chapter 2, Fig. 5; *see* full caption on p. 37 and discussion on p. 35.)

1

Immunoinformatics and the In Silico Prediction of Immunogenicity

An Introduction

Darren R. Flower

Summary

Immunoinformatics is the application of informatics techniques to molecules of the immune system. One of its principal goals is the effective prediction of immunogenicity, be that at the level of epitope, subunit vaccine, or attenuated pathogen. Immunogenicity is the ability of a pathogen or component thereof to induce a specific immune response when first exposed to surveillance by the immune system, whereas antigenicity is the capacity for recognition by the extant machinery of the adaptive immune response in a recall response. In thisbook, we introduce these subjects and explore the current state of play in immunoinformatics and the in silico prediction of immunogenicity.

Key Words: Antigen presentation; bioinformatics; computational chemistry; computational vaccinology; immunoinformatics; MHC binding; vaccine design

1. Introduction

Immunology is important because the domain of infectious disease is the domain of immunology. For immunology is, amongst many other studies, the study of how the body is able to defend itself against infection; from the standpoint of human disease, an accurate appreciation of adaptive and, increasingly, innate immunity is unequivocally fundamental to our continuing assault on contagious disease, the greatest source of preventable human mortality and morbidity. Its societal importance is unquestionable, for immunology deals with the physiological function of the immune system in both health and disease.

From: *Methods in Molecular Biology, vol. 409: Immunoinformatics: Predicting Immunogenicity In Silico*
Edited by: D. R. Flower © Humana Press Inc., Totowa, NJ

Our knowledge concerning the varied molecular and cellular mechanisms that underlie the macroscopic manifestation of immunity at the level of the whole organism has facilitated and fomented the development of new clinical and non-clinical technologies. Likewise, the inexorable move towards automation and high-throughput science is having an important effect on immunology: after a 100 years of empirical research, immunology is hovering on the brink of reinventing itself as a quantitative, genome-based science. Immunology is thus poised at a turning point in its long and distinguished history, whether or not the multitude of practitioners of immunology wish to acknowledge it. Like most biological sciences, immunology needs to make the most of the potentially overwhelming cascade of new information delivered by high-throughput technologies.

As much of its focus is strongly anthrocentric, being centered primarily on the adaptive immune system of vertebrates, immunology is rightly viewed as an important—even a paramount—science. Immunologists are sometimes viewed—rightly perhaps—as a discipline apart. Immunology has a high standing in the wider scientific community: its journals have high-impact factors and it is a large and, generally speaking, a well-funded discipline. Thus, the realm of immunology is indeed broad, encompassing, as it does, the malfunctioning of immunity in immunological disorders, including autoimmune diseases, allograph rejection, and immune deficiency, as well as the in vivo, in vitro, and in situ, physico-chemical and functional properties of immunological components of the immune system.

Although the use of computers to combat infection and other disease states may seem far-fetched to some, computational approaches have nonetheless long been used to design small-molecule drugs, with all the implications for human health that they entail. We are now beginning to see the way in which the use of computers is also impacting on the discovery of immunotherapeutics and prophylactic vaccines.

The synergy of the in silico and in vitro is made manifest through the discipline of immunoinformatics. Immunoinformatics, a profound new branch of computational science that has the potential to greatly accelerate the celerity and effectiveness of the search for new immunotherapeutics, has recently emerged as a buoyant subdiscipline within bioinformatics. Immunoinformatics is thus the application of bioinformatic methods to the unique problems of immunology and vaccinology. Immunoinformatics, as a principal component of incipient immunomic technologies, is also beginning to catalyse key alterations in the way that immunology is done. Immunology is finally coming to grips with the egregious implications of the post-genomic revolution and has begun to release itself from the empirical straight jacket that has constrained its development hitherto. It is clear that such high-throughput approaches will engender a paradigm shift from hypothesis to data-driven research, with new understanding emerging from the analysis of data sets which initially seem both complicated and confusing.

In response to such pressures, there has been much interest recently in the effective deployment of informatics tools, which can analyze data arising from immunological research. In turn, this has caused two kinds of immunological computer support to grow. The first is straightforward bioinformatics support that is technically indistinguishable from support for other areas of biology; this includes, for example, the annotation of both human and microbial genomes. The other type of support is the more focussed and specialized strand of immunoinformatics. It is a discipline firmly grounded in computer science, but one that increasingly integrates a whole range of interdisciplinary techniques from physical biochemistry, biophysics, computational and medicinal and analytical chemistry, structural biology and protein homology modeling, as well as many others. The principal task hitherto of this exciting and dynamic specialism has been the accurate prediction of immunogenicity, be that manifest as the identification of epitopes or the prediction of whole protein immunogens. This endeavor is the focus of this book.

It is a well-known truism that the immune system is both complex and hierarchical, exhibiting startling emergent behavior at all levels. The complexity exhibited by the immune systems is undoubtedly confounding, and, although there are many who deny it, our ignorance of fundamental immunology remains unexpectedly profound. Yet, at its heart there lie straightforward and unequivocally explicable molecular recognition events: the coming together of two or more molecules to form stable complexes of measurable duration. The binding of an epitope to a major histocompatibility complex (MHC) protein, or T-cell receptor (TCR) to a peptide–MHC complex, is in terms of underlying physicochemical phenomena, identical to any other molecular interaction in any other area of biological science. It is only at higher levels—when thousands or millions of different molecules work synergistically together—that the immune system displays, in space and time, emergent properties. Immunogenicity is such an emergent property.

In seeking to address the prediction of immunogenicity, immunoinformatics exploits the observation that immunogenicity is, ultimately, based on simple and understandable molecular events. Immunogenicity is the property of a molecular, or supramolecular, moiety that allows it to induce a significant response from the immune system. Here a molecular moiety may be a protein, lipid, carbohydrate, or some combination thereof. A supramolecular moiety may be a virus, bacteria, or protozoan parasite. An immunogen—a moiety exhibiting immunogenicity—is a substance which can elicit a specific immune response, whereas an antigen—a moiety exhibiting antigenicity—is a substance recognized, in a recall response, by the extant machinery of the adaptive immune response, such as T cells or antibodies. Thus, antigenicity is the capacity, exhibited by an antigen, for recognition by one or several parts of the antibody or TCR immune repertoire. Immunogenicity, on the other hand, is the ability of an immunogen to induce a specific immune response when it is exposed to initial surveillance by the immune system. These two properties are clearly coupled, but properly understanding how they are interrelated is by no means facile.

Predicting actual antigenicity and/or immunogenicity of a complex protein remains problematic. It depends simultaneously upon the context in which it is presented and the nature of the immune repertoire that recognizes it. Either or both of these components may be critical. For example, the immune response in many immunogens or antigens is focussed on a handful of immunodominant structures, while much of the rest of the molecule may be unable to mount a response. In seeking to assess immunogenicity, we must consider properties of the host and the pathogenic organism of origin, and not just the intrinsic properties of the antigen itself. The composition of the available immune repertoire will affect its response to a given epitope and alter its recognition of a particular target. When mounting a response in vivo, those elements of an immune repertoire capable of participating, in a given response, might have been deleted through their cross-reactivity with host antigens. Moreover, fundamental restrictions on the antibody repertoire, for example, as imposed by the limited number of genes that encode the antigen-binding site of the antibody, may also curtail possible responses. Overall, it is clear that antigenicity and immunogenicity have many interlinked causes. The induction of immune responses requires critical interaction between innate parts of the immune system, which respond rapidly and in a relatively non-specific manner, and other, more specific, components, which recognize individual epitopes.

In order to protect the host against infectious disease, the immune system must recognize a variety of microbial pathogens (bacteria, fungi, parasites, and viruses), principally through the recognition of biological macromolecules, typically whole, or degraded, proteins. However, epitopes do not need to be proteinacious; carbohydrates, lipid, and even nucleic acid can act as an epitope, either alone or in combination with peptide. Glycosylation, as a particular kind of posttranslational modification, is a common event that contributes to protein immunogenicity, whether mediated through humoral or cellular immunity.

Glycosylated peptide epitopes can be bound by antibodies and can be presented by both class I or class II MHC molecules and then be bound by glycopeptide-specific T-cell clones. Thus, immunogenicity can, in general, manifest itself through both arms of the adaptive immune response: humoral (mediated through the binding of whole protein antigens by antibodies) and cellular immunology (mediated by the recognition of proteolytically cleaved peptides by T cells).

Humoral immunogenicity, as mediated by soluble or membrane-bound cell-surface antibodies, can be measured in several ways. Methods such as enzyme-linked immunosorbent assay (ELISA) or competitive inhibition assays yield values for the antibody titer, the concentration at which the ability of antibodies in the blood to bind an antigen has reached its half maximal value. One can also measure directly the affinity of antibody and antigen, using, say, equilibrium dialysis. Measurements of cellular immunity have become legion. For class I presentation, arguably the most direct approach is to measure T-cell killing. Cytotoxic T lymphocytes (CTLs) can lyse target cells. This can be measured using a chromium radioisotope, which is taken up by target cells and released during lysis. For class II presentation, the proliferative response of CD4+ T cells, which, in turn, activates macrophages or B cells, is measurable through tritiated thymidine incorporation into T-cell DNA. One can also measure cytokine production by class I and/or class II T cells. Recently, attention has moved towards tetramers as tools for the detection of T-cell responses ***(1)***.

Much of immunogenicity is determined by the presence of epitopes, the principal chemical moieties recognized by the immune system. Consequently, the accurate prediction of B-cell and T-cell epitopes is the pivotal challenge for immunoinformatics. Epitope prediction can be fairly described as both the high frontier of immunoinformatic investigation and a grand scientific challenge: it is difficult, yet exciting, and, as a central tool in the drive to develop improved vaccines and diagnostics, is also of true practical value.

Despite a growing appreciation of the role played by non-peptide epitopes, such as carbohydrates and lipids, peptidic B-cell and T-cell epitopes (as mediated by the humoral or cellular immune systems, respectively) remain the principal tools by which the intricacy of immune responses can be surveyed and manipulated, as it is the recognition of epitopes by T cells, B cells, and soluble antibodies that lies at the heart of the adaptive immune response. Such initial responses lead, in turn, to the activation of the cellular and humoral immune systems and, ultimately, to the effective destruction of pathogenic organisms.

The word *epitope* is widely used amongst biological scientists. Etymologically speaking, its roots are Greek, and, like most words, its meanings are diverse and in a state of constant flux. It is most often used to refer to any region of a biomacromolecule which is recognized, or bound, by another biomacromolecule. For an immunologist, the meaning is more restricted and refers to particular structures recognized by the immune system in particular ways. B cell epitopes are regions of a protein recognized by antibody molecules. T-cell epitopes are short peptides which are bound by MHCs and subsequently recognized by T cells.

The region on a macromolecule, which undertakes the recognition of an epitope, is called a paratope. In terms of the physical chemistry of binding, then we need think only of equal partners in a binding reaction. However, viewed within the context of protein and organismal function, itself strictly a teleonomic, or even anthropomorphic, construct of limited explicit veracity, then the distinction between epitope and paratope, with all its intentionality of meaning, gains some epistemological authenticity, albeit more operational than actual.

A B-cell epitope is a region of a protein, or other biomacromolecule, recognized by soluble or membrane-bound antibodies. B-cell epitopes are classified as either linear or discontinuous epitopes. Linear epitopes comprise a single continuous stretch of amino acids within a protein sequence, whereas an epitope whose residues are distantly separated in the sequence and are brought into physical proximity by protein folding is called a discontinuous epitope. Although most epitopes are, in all likelihood, discontinuous, experimental epitope detection has focussed on linear epitopes. Linear epitopes are believed to be able to elicit antibodies that can subsequently cross-react with its parent protein. Chapter 29 addresses the prediction of B-cell epitopes.

A T-cell epitope is a short peptide bound, in turn, by MHC and TCR, to form a ternary complex. The formation of such a complex is the primary, but not sole, molecular recognition event in the activation of T cells. Many other co-receptors and accessory molecules, in addition to CD4 and CD8 molecules, are also involved in T-cell recognition. The recognition process is not simple and remains poorly understood. However, it has emerged that the process involves the creation of the immunological synapse, a highly organized, spatio-temporal arrangement of receptors and accessory molecules of many types. The involvement of these accessory molecules, although essential, is not properly understood, at least from a quantitative perspective. Ultimately, the accurate modeling of all these complex processes will be required to gain full and complete insight into the process of epitope presentation.

While the accurate and reliable prediction of B-cell epitopes remains at an early stage, a large number of sophisticated, and successful, methods for the prediction of T-cell epitopes have been developed. These began with early motif methods and have grown to exploit both qualitative and semiquantitative approaches, typified by neural network classification methods and a variety of more quantitative approaches. Most modern methods for T-cell-epitope prediction rely on predicting the affinity of peptides binding to MHCs.

As everyone knows, MHCs bind peptides. These are themselves derived through the degradation, by proteolytic enzymes, of foreign or self-proteins. Foreign epitopes originate from benign or pathogenic microbes, such as viruses and bacteria. Self-epitopes originate from host proteins that find their way into the degradation pathway as part of the cell's intrinsic quality control procedures. The proteolytic pathway by which peptides become available to MHCs is very complex and many important details and components remain to be elucidated. Yet, it is the complexity of the T-cell presentation pathway that allows peptides with diverse posttranslational modifications, such as phosphorylation or glycosylation, to form peptide-MHC complexes (pMHC), and thus, ultimately, to be recognized by TCRs. Moreover, MHCs are very catholic in terms of the molecules they bind and are not restricted to peptides. Chemically modified peptides and peptidomimetics are also bound by MHCs. It is also well known that many drug-like molecules bind to MHCs *(2)*.

There are several alternative processing pathways, but the principal ones seem linked to the two major types of MHC: class I and class II. Class I MHCs are expressed by almost all cells in the body. They are recognized by T cells whose surfaces are rich in CD8 co-receptor protein. Class II MHCs are only expressed on so-called professional antigen-presenting cells and are recognized by T cells whose surfaces are rich in CD4 co-receptors. MHCs are polymorphic. Generally, most humans have six classic MHCs—3 class I [human leukocyte antigen (HLA)-A, HLA-B, and HLA-C] and 3 class II (HLA-DR, HLA-DP, and HLA-DQ); these proteins will have different sequences, or different HLA alleles, in different individuals. Different MHC alleles, both class I and class II, have different peptide specificities. A simple way to look at this phenomenon is to say that MHCs bind peptides that exhibit certain particular sequence patterns and not others. Within the human population, there are a large number of different, possible variant, genes coding for MHC proteins, each exhibiting different peptide-binding sequence selectivities. TCRs, in their turn, also exhibit different and typically weaker affinities for different peptide–MHC complexes. The combination of MHC and TCR selectivities thus determines the power of peptide recognition in the immune system and thus the recognition of foreign proteins and pathogens. This will be discussed more thoroughly in accompanying chapters.

Whatever dyed-in-the-wool immunologists may say, such interactions form the quintessential nucleus of immune recognition, and thus the principal point of intervention by immunotherapeutics.

The peptides presented by class I and class II MHCs differ principally in terms of their length. Class I peptides are primarily derived from intracellular proteins, such as viruses. These proteins are targeted to the proteasome, which cuts them into short peptides. Subsidiary enzymes also cleave these peptides, producing a range of peptide lengths, of which the distribution used to be believed to fall neatly into the range 8–11 amino acids. More recently, however, this has been shown that much longer peptides, currently up to 15 amino acids, can also be bound by MHCs and recognized by TCRs *(3)*. For class II, the receptor-mediated intake of extracellular protein derived from a pathogen is targeted to an endosomal compartment, where such proteins are cleaved by cathepsins, a particular class of protease, to produce peptides that are typically somewhat longer than class I. These, again, exhibit a considerable distribution of lengths, centered on a range of 15–20 amino acids. However, longer and shorter peptides can also be presented, through class II MHCs, to immune surveillance.

There are many other aspects of immunogenicity which have yet to be properly explored experimentally. Although the anecdotal evidence is suggestive, it is not yet easily amendable to predictive methods. However, simple observations, for example that the larger and more chemically complex a protein and the more distant its sequence is from those of self-proteins the more likely it is to be immunogenic, seem almost self-evident. While the observation that particulate or aggregated protein is more likely to evoke a response does not afford so obvious an explanation. Other factors, such as the affinity of antigens for the apparatus of the endocytic pathway, while clearly germane to the issue of immunogenicity, are as yet poorly understood, if at all. Moreover, an understanding of the pathogen as well as the host is important. While it is clear that bacterial pathogens have developed successfully many inventive ways to attack the human host, it is also clear that many seemingly diverse pathogens share virulence traits. Aspects of this are addressed in Chapter 31. Properties of microbial protein, including those independent of the host, such as subcellular location and expression levels, are also important. Immunogenicity is neither a property conferred solely by the host nor solely by the pathogen, but one that arises from a synergistic combination of both. Thus, ultimately, we will need to address the problem using a holistic, integrative approach, drawing inference from a variety of sources ranging from the molecular through to the fully organismal.

It is now generally accepted that only peptides that bind to MHC at an affinity above a certain threshold will act as T-cell epitopes and that, to some extent at least, peptide affinity for the MHC correlates with T-cell response. This particular issue is somewhat complicated and obscured by hearsay and dogma: as with many questions important to immunoinformatics, the key, systematic studies remain to be done. Deprecating counterexamples, the behavior of heteroclitic peptides, where synthetic enhancements to binding affinity are often reflected in enhanced T-cell reactivity, seems compelling evidence. However, and whatever people may say, affinity of binding is an important component of recognition and of the overall immune response, not only, or, necessarily, the most important part, but an essential component. Its importance is debated, particularly by people who are vocal in their criticism of the immunoinformatic endeavor. Nevertheless, its utility is clear. Experimental immunologists and vaccinologists are constantly using nascent immunoinformatic approaches to select, filter, or prune candidate peptides in order to identify functional epitopes.

Many questions relating to immunogenicity remain decidedly open. What underlying molecular mechanisms give rise to immunogenicity? For cellular immunology, is it, as many believe, the lifetime—the off rate essentially—of the MHC–peptide complex that determines how immunogenic an epitope is? Or is it, as many continue to assume, the affinity of peptide for MHC? Or is immunogenicty related to the total population of peptide-bound MHCs on the surface of an antigen-presenting cell? Or is it some combination of these phenomena leading, say, to an appropriate duration of the immunological synapse? However, it is vital to emphasize once more that immunogenicity is not an isolated function of peptide binding to MHC molecules, but a phenomenon that arises from an organism recognizing a variety of signals, of which recognition of a particular bound peptide is just one. Peptide binding is a necessary, but not sufficient, condition for immunogenicity.

For some time, the database has been the *lingua franca* or, more prosaically, the common language of bioinformatics. Although the specific dialect—the type of data archived—may change, the use, the creation, and the manipulation of databases, which contain biologically relevant information, are the most critical feature of contemporary bioinformatics. The same is broadly true also of immunoinformatics. This is manifest through its support for post-genomic bioscience and as a discipline in its own right. Functional data, as housed in databases, will rapidly become the principal currency in the dynamic information economy of twenty-first century immunology. Having said all that, there is nothing particularly new about immunological databases, at least in the sense that they do no more than apply standard data warehousing techniques in an immunological context. Nonetheless, the continuing development of an expanding variety of immunoinformatic database systems indicates that the application of bioinformatics to immunology is beginning to broaden and mature.

For example, the IMGT initiative (described in Chapters 2, 3, and 4) has made the sequence analysis of important immunological macromolecules its focus for many years. Functional, or epitope-orientated, databases are somewhat newer, but their provenance is now well established. Generally speaking, such databases record data on T-cell epitopes or peptide–MHC binding affinity. Some would say that the best database available currently is the HIV Molecular Immunology database *(4)*, which focuses on the properties of a single virus. The scope of the database is, in terms of the kinds of data it archives, broader than many, with information on both cellular (T-cell epitopes and MHC-binding motifs) and humoral immunology (linear and conformational B-cell epitopes). Another widely used database is SYFPEITHI (described in Chapter 5), a high-quality development, which contains an up-to-date compendium of T-cell epitopes. SYFPEITHI also contains much data on MHC peptide ligands, peptides isolated from cell-surface MHC proteins ex vivo, but excludes data on synthetic peptide 'binders'.

MHCPEP *(5)*, a now defunct database, pooled both T-cell epitope and MHC-binding data in a flat file, introducing a widely used conceptual simplification, which combines together the bewildering variety of binding measures, reclassifying peptides as either 'binders' or 'non-binders'. Binders are further subdivided as high binders, medium binders, and low binders. More recently, Brusic and coworkers have developed a much more complex and sophisticated database: FIMM *(6)*. This system integrates a variety of data on MHC–peptide interactions: in addition to T-cell epitopes and MHC-peptide binding data, it also archives a wide variety of other data, including sequence data on MHCs themselves together with data on the disease associations of particular MHC alleles.

Databases have begun to diversify and now address a wider and more varied tranche of immunological data. AntiJen, formerly known as JenPep *(7,8,9)*, is a database developed recently, which brings together a variety of kinetic, thermodynamic, functional, and cellular data within immunobiology. While it retains a focus on both T-cell and B-cell epitopes, AntiJen is the first functional database in immunology to contain continuous quantitative binding data on a variety of immunological molecular interactions, rather than the kind of subjective classifications described above. Data archived includes thermodynamic and kinetic measures of peptide binding to transporter associated with antigen processing (TAP) and MHC, peptide–MHC complex binding to TCRs, and general immunological protein–protein interactions, such as the interaction of co-receptors and interactions with superantigens. Although the nature of the data within AntiJen sets it apart from other immunology databases, there is, nonetheless, considerable overlap between other systems and our database. In Chapters 6, 7, and 8, databases originating in Raghava's group address three areas: cellular immunology (MHCBN, Chapter 6), B-cell-mediated immunology (BCIPEP, Chapter 7), and small-molecule Haptens (HAPTENDB, Chapter 8). Clearly, the ability of all these databases, each with its own particular focus, properly to address the increasing needs of present-day immunoinformaticians is greatest when they are combined synergistically.

Polymorphism confounds effective study of MHC-mediated peptide specificity. Indeed, and for a number of diverse reasons, HLA is the best and most extensively studied of all human proteins with regard to polymorphism. As perusal of Chapter 3 will ably demonstrate the IMGT/HLA database stores literally thousands of distinct HLA class I and class II allele sequences. Such allelic variants have come into being through a process of random mutation, albeit filtered and constrained by evolutionary processes operating with an environment characterized by morbid host–pathogen interactions, themselves constrained by time and geography. Because MHCs exhibit such extensive polymorphic amino acid variation, small alterations in the identity of binding site residues should give rise to differences in peptide selectivity exhibited during peptide binding. Many HLA alleles have been demonstrated to bind peptides with a similar specificity. This has led to the concept of MHC supertypes and the idea that sequence distinct MHCs can be clustered or grouped into distinct classes that exhibit equivalent, if not necessarily identical, peptide specificities. The pace and verity of vaccine discovery would be greatly enhanced if one could delineate effective rules able to group together HLA alleles with similar specificities. Such a classification, if accurate and sufficiently extensive, would greatly reduce experimental work as it would no longer be necessary to study every allele, thus making the discovery of epitope-based vaccines targeted at multiple alleles more efficient. Some have sought insights into MHC supertypes from a sequence perspective, others from structural data. A number of such approaches are discussed in Chapters 9, 10, 11, and 12.

A useful simplification of biological computation is to divide methods between the areas of simulation and data mining, although, in truth, there is a continuous spectrum of techniques stretching from one extreme to the other. In the minds of many, data mining is synonymous with text mining: the unsupervised extraction of data and information directly from the bioscience literature. The distinction between data and information is an important one. Data may include the sequences of peptides or numerical values, such as an IC_{50}. These can be identified readily on the basis of case or the unequivocal association of an unambiguous symbol with a fixed-format number. Information, on the other hand, is highly context dependent and might include things as elusive as the unwritten implications of an observation or set of observations.

Within immunology, the principal example of data mining has, over the years, been the identification of peptide-binding motifs, which seeks to characterize the peptide specificity of different MHC alleles in terms of dominant anchor positions with a strong preference for certain amino acids ***(10)***. Because such motifs are simple to understand and simple to implement either visually or computationally, they have enjoyed considerable popularity amongst immunologists. Human class I allele HLA-A*0201, probably the best-studied allele, has anchor residues at peptide positions P2 and P9. At P2, acceptable amino acids would be L and M, and V and L at position P9. Secondary anchors, which are residues that are favorable, but not essential, for binding, may also be present. A vast tranche of papers have, over the last 15 years, successfully extended the list of known motifs to include the specificity patterns of numerous alleles, from humans, mice, and many other animals. However, despite such apparent success, there are still many fundamental problems associated with the motif approach to the characterization of peptide specificity. The most significant of these problems is that the method is deterministic: a peptide either is or is not a binder. A brief reading of the literature however shows that motif matches produce many false positives and probably also produces many false negatives, although predicted non-binders are seldom, if ever, assayed. Thus being motif-positive, as the jargon sometimes puts it, is neither necessary nor sufficient for a peptide to possess MHC affinity. Although it is clear that so-called primary anchors do often dominate binding, we have shown unequivocally that binding motifs as descriptions of the process, are fundamentally flawed. Not hopeless, not useless, but incomplete, partial, and inadequate for purpose. In the sense that motifs are widely used and widely understood, they are indeed most useful, but as accurate predictors of binding they leave much to be desired.

Shortcomings in motifs have led many to seek alternative data-mining solutions to the peptide–MHC affinity problem. The development of data-driven predictive methods in immunoinformatics is now several decades old. Early methods attempted to predict epitopes directly, and in the absence of knowledge of the peptide preferences of MHC restriction, were not very successful ***(11)***. Several groups have used techniques from artificial intelligence research, such as artificial neural networks (ANNs) and hidden Markov models (HMMs), to tackle the problem of predicting peptide–MHC affinity. ANNs and HMMs are, for most applications and for most bioinformaticians, techniques of choice for building predictive models. The successful application of ANNs is often complicated by the presence of several adjustable, and impenetrable, factors: over-fitting, over-training (or memorization), and interpretation. While over-fitting and over-training have been largely overcome, interpretation remains essentially intractable. Of course, many other methods—indeed, in all probability, all methods—suffer similar or equivalent problems. Over-fitting is the curse of all data-driven methods. Support vector machines (SVM) are currently flavor-of-the-month, while we used Partial Least Squares (PLS) as its principal statistical inference engine. Whether either of these—or, indeed, any other method of which we can conceive—will ever escape the traps which have caught-out other techniques remains to be seen.

Section 3 details a whole variety of different cutting-edge approaches to the problem of predicting MHC binding and thus, ultimately, the prediction of immunogenicity. Chapters 13 through 16 describe general and/or well-established methods for the prediction of MHC–peptide binding. Later chapters, 17 through 20, discuss some more up-to-date approaches, including SVM, and SVM-based regressions approaches. Chapters 21 and 22 look at how structural modeling can be used to look at peptide binding, while Chapters 23 and 24 concentrate in more detail on the use of molecular dynamic simulations to address this problem. Finally, Chapters 25, 26, and 27 look at the more complex data mining problems associated with data-driven approaches to predicting the binding affinity of class II peptide–MHC interactions.

Currently, in vitro and in vivo testing is required to differentiate high-binding peptides from true epitopes. Sets of MHC binders identified by prediction methods can then be evaluated experimentally as potential epitopes. This will reduce the required experiment burden by several orders of magnitude: the alternative—assaying overlapping peptides from, say, each of the 4000 genes in a bacterial genome—equates to the experimental evaluation of literally tens of thousands of potential peptides. To evaluate the immunogenicity of these using hand assays would be prohibitively inefficient. Even if this could be done using high-throughput technology, the associated time and cost would still be excessive.

We have come to a turning point, where a number of technologies have obtained the necessary level of maturity: post-genomic strategies on the one hand and predictive computational methods on the other. Progress will occur in two ways. One will involve closer connections between immunoinformaticians and experimentalists seeking to discover new vaccines. In such a situation, work would progress through a cyclical process of using and refining models and experiments, at each stage moving closer towards a common goal of effective, cost-efficient vaccine development. The other way is the devolved model, where methods are made accessible and accessible remotely via the web.

There is also a clear and obvious need for experimental work to be conducted in support of the development of accurate in silico methods. Our ability to combine in vitro and in silico analysis allows us to improve both the scope and power of our predictions, in a way that would be impossible using only data from the literature. To ensure we produce useful, quality in silico models, rather than worthless and unusable methods, we need to value the prediction generated by immunoinformatics for themselves and conduct experiments appropriately.

Immunoinformatics is developing at an unprecedented pace, with many groups trying to improve databases and algorithms. There has also been a diversification in what is being done, as there always is as a new field grows and expands and matures. In this context, Section 4 addresses some different approaches to the prediction of characteristics germane to the complex and subtle property of immunogenicity. Chapter 28 looks at binding to TAP, an important checkpoint on the class I antigen-presentation pathway, while Chapter 29 looks at the prediction of B-cell epitopes. Chapter30 looks at the medically important area of histocompatibility prediction, and Chapter 31 looks at the prediction of bacterial virulence.

Despite the steady increase in studies reporting the real-world use of prediction algorithms, there remains a lingering feeling that truly convincing validations of the underlying approach are still required. In time, this will come, but only incrementally. Yet, we should feel confident that the great synergy arising from these disciplines will be of true benefit to immunology, with clear improvements in vaccine candidates, diagnostics, and laboratory reagents. Methods able to accurately predict immunogenicity will yet become pivotal tools for tomorrow's vaccinologist. Yet, for these improved methodologies to be ultimately effective, they must be used routinely by experimental immunologists. To do this requires two things. First, more and better algorithms and more user-friendly software are required. In spite of their increasing accuracy and reliability, most of these tools remain daunting for laboratory-based immunologists. This must be addressed. Second, it requires the confidence of experimentalists to exploit the methodology and to commit laboratory experimentation. The use of these methods should be routine. It is not only a matter of training and education, however. These methods must, ultimately, be made more accessible and robust. But equally well, experimentalists cannot hide behind arguments about usability forever, they must engage fully and completely with advancing technology. Hopefully, this book will encourage them to do so.

References

1. Doherty, PC, Riberdy, JM, Belz, GT. (2000). Quantitative analysis of the CD8+ T-cell response to readily eliminated and persistent viruses. Philos Trans R Soc Lond B Biol Sci 355, 1093.
2. Pichler, WJ. (2002). Modes of presentation of chemical neoantigens to the immune system. Toxicology Dec 27, 181–182, 49–54.
3. Probst-Kepper, M, Hecht HJ, Herrmann H, Janke, V, Ocklenburg, F, Klempnauer, J, van den Eynde, BJ, Weiss, S. (2004). Conformational restraints and flexibility of 14-Meric peptides in complex with HLA-B*3501. J Immunol 173, 5610–5616.
4. Korber, BTM, Brander, C, Haynes, BF, Koup, R, Kuiken, C, Moore, JP, Walker, BD, Watkins, D. (2001). *HIV Molecular Immunology 2001*, Los Alamos National Laboratory: Theoretical Biology and Biophysics, Los Alamos, NM.
5. Brusic, V, Rudy, G, Harrison, LC. (1998). MHCPEP, a database of MHC-binding peptides: update 1997. Nucleic Acids Res 26, 368–371.
6. Schonbach, C, Koh. JL, Flower, DR, Brusic, V. (2005). An Update on the Functional Molecular Immunology (FIMM) Database. Appl Bioinformatics 4, 25–31.
7. Blythe, MJ, Doytchinova, IA, Flower, DR. (2002). JenPep, a database of quantitative functional peptide data for immunology. Bioinformatics 18, 434–439.
8. McSparron, H, Blythe, MJ, Zygouri, C, Doytchinova, IA, Flower, DR. (2003). JenPep: a novel computational information resource for immunobiology and vaccinology. J Chem Inf Comput Sci. 43, 1276–1287. Antigen Immunomics Research.
9. Toseland, CP, Clayton, DJ, McSparron, H, Hemsley, SL, Blythe, MJ, Paine, K, Doytchinova, IA, Guan, P, Hattotuwagama, CK, Flower, DR. AntiJen: a quantitative immunology database integrating functional, thermodynamic, kinetic, biophysical, and cellular data. Immunome Res 1, 4. Online.
10. Sette, A, Buus, S, Appella, E, Smith, JA, Chesnut, R, Miles, C, Colon, SM, Grey, HM. (1989). Prediction of major histocompatibility complex binding regions of protein antigens by sequence pattern analysis. Proc Natl Acad Sci USA 86, 3296–3300.
11. Deavin, AJ, Auton, TR, Greaney, PJ. (1996). Statistical comparison of established T-cell epitope predictors against a large database of human and murine antigens. Mol Immunol 33, 145–155.

I

Databases

2

IMGT®, the International ImmunoGeneTics Information System® for Immunoinformatics

Methods for Querying IMGT® Databases, Tools, and Web Resources in the Context of Immunoinformatics

Marie-Paule Lefranc

Summary

IMGT®, the international ImMunoGeneTics information system® (http://imgt.cines.fr), was created in 1989 by the Laboratoire d'ImmunoGénétique Moléculaire (LIGM) (Université Montpellier II and CNRS) at Montpellier, France, in order to standardize and manage the complexity of immunogenetics data. *IMGT*® is recognized as the international reference in immunogenetics and immunoinformatics. *IMGT*® is a high quality integrated knowledge resource, specialized in (i) the immunoglobulin (IG), T cell receptors (TR), major histocompatibility complex (MHC) of human and other vertebrates; (ii) proteins that belong to the immunoglobulin superfamily (IgSF) and to the MHC superfamily (MhcSF); and (iii) related proteins of the immune systems (RPI) of any species. *IMGT*® provides a common access to standardized data from genome, proteome, genetics, and three-dimensional (3D) structures for the IG, TR, MHC, IgSF, MhcSF, and RPI. *IMGT*® interactive on-line tools are provided for genome, sequence, and 3D structure analysis. *IMGT*® Web resources comprise 8,000 HTML pages of synthesis and knowledge (*IMGT* Scientific chart, *IMGT* Repertoire, *IMGT* Education, etc.) and external links (*IMGT* Bloc-notes and *IMGT* other accesses).

Key Words: IMGT; ontology; immunoglobulin; T cell receptor; MHC; IgSF; MhcSF

1. Introduction

The number of genomics, genetics, three-dimensional (3D), and functional data published in the immunogenetics field is growing exponentially and involves fundamental, clinical, veterinary, and pharmaceutical research. The

From: *Methods in Molecular Biology, vol. 409: Immunoinformatics: Predicting Immunogenicity In Silico*
Edited by: D. R. Flower © Humana Press Inc., Totowa, NJ

number of potential protein forms of the antigen receptors, immunoglobulins (IG), and T cell receptors (TR) is almost unlimited. The potential repertoire of each individual is estimated to comprise about 10^{12} different IG (or antibodies) and TR, and the limiting factor is only the number of B and T cells that an organism is genetically programmed to produce. This huge diversity is inherent to the particularly complex and unique molecular synthesis and genetics of the antigen receptor chains. This includes biological mechanisms such as DNA molecular rearrangements in multiple loci (three for IG and four for TR in humans) located on different chromosomes (four in humans), nucleotide deletions and insertions at the rearrangement junctions (or N-diversity), and somatic hypermutations in the IG loci (for review, see ref. ***1**,**2***).

IMGT®, the international ImMunoGeneTics information system® (http://imgt.cines.fr) ***(3,4)***, was created in 1989 by the Laboratoire d'ImmunoGénétique Moléculaire (LIGM) (Université Montpellier II and CNRS) at Montpellier, France, in order to standardize and manage the complexity of the immunogenetics data. *IMGT®* is as the international reference in immunogenetics and immunoinformatics. *IMGT®* is a high quality integrated knowledge resource, specialized in (i) the IG, TR, major histocompatibility complex (MHC) of human and other vertebrates, (ii) proteins that belong to the immunoglobulin superfamily (IgSF) and to the MHC superfamily (MhcSF), and (iii) related proteins of the immune systems (RPI) of any species. *IMGT®* provides a common access to standardized data from genome, proteome, genetics and 3D structures for the IG, TR, MHC, IgSF, MhcSF and RPI ***(3,4)***.

The *IMGT®* information system consists of databases, tools, and Web resources ***(3)***. *IMGT®* databases include one genome database, three sequence databases, and one 3D structure database. *IMGT®* interactive on-line tools are provided for genome, sequence, and 3D structure analysis. *IMGT®* Web resources comprise 8,000 HTML pages of synthesis and knowledge (*IMGT* Scientific chart, *IMGT* Repertoire, *IMGT* Education, *IMGT* Index, etc.) and external links (*IMGT* Bloc-notes and *IMGT* other accesses) ***(4)***. Despite the heterogeneity of these different components, all data in the *IMGT®* information system are expertly annotated. The accuracy, the consistency, and the integration of the *IMGT®* data, as well as the coherence between the different *IMGT®* components (databases, tools, and Web resources), are based on *IMGT-ONTOLOGY* ***(5)***, the first ontology in immunogenetics and immunoinformatics. *IMGT-ONTOLOGY* provides a semantic specification of the terms to be used in the domain and, thus, allows the management of immunogenetics knowledge for all vertebrate species.

2. Standardization: *IMGT-ONTOLOGY* and *IMGT* Scientific Chart

IMGT-ONTOLOGY concepts are available, for the biologists and *IMGT®* users, in the *IMGT* Scientific chart ***(4)*** and have been formalized, for the computing scientists, in *IMGT-ML* ***(6,7)***. The *IMGT* Scientific chart ***(4)*** comprises the controlled vocabulary and the annotation rules necessary for the immunogenetics data identification, description, classification, and numbering and for knowledge management in the *IMGT®* information system. All *IMGT®* data are expertly annotated according to the *IMGT* Scientific chart rules. Standardized keywords, labels and annotation rules, standardized IG and TR gene nomenclature, the *IMGT* unique numbering, and standardized origin/methodology were defined, respectively, based on the six main concepts of *IMGT-ONTOLOGY* ***(5)*** (Table 1). The *IMGT* Scientific chart is available as a section of the *IMGT®* Web resources. Examples of *IMGT®* expertised data concepts derived from the *IMGT* Scientific chart rules are summarized in Table 1.

2.1. IDENTIFICATION concept: standardized keywords

IMGT® standardized keywords for IG and TR include the following: (i) general keywords—indispensable for the sequence assignments, they are described in an exhaustive and non-redundant list, and are organized in a tree structure and (ii) specific keywords—they are more specifically associated to particularities of the sequences (orphon, transgene, etc.). The list is not definitive and new specific keywords can easily be added if needed. *IMGT/LIGM-DB* standardized keywords have been assigned to all entries.

2.2. DESCRIPTION concept: standardized sequence annotation

Two hundred and fifteen feature labels are necessary to describe all structural and functional subregions that compose IG and TR sequences, whereas only seven of them are available in *EMBL*, *GenBank* or *DDBJ* ***(14–16)***. Levels of annotation have been defined, which allow the users to query sequences in *IMGT/LIGM-DB* even though they are not fully annotated. Prototypes represent the organizational relationship between labels and give information on the order and expected length (in number of nucleotides) of the labels. This provides rules to verify the manual annotation and to design automatic annotation tool. One hundred and seventy-two additional feature labels have been defined for the 3D structures. Annotation of sequences and 3D structures with these labels

Table 1
***IMGT-ONTOLOGY* main concepts, *IMGT* Scientific chart rules, and examples of *IMGT*® expertised data concepts**

IMGT-ONTOLOGY main concepts ***(5)***	*IMGT* Scientific chart rules ***(4)***	Examples of *IMGT*® expertised data concepts
IDENTIFICATION	Standardized keywords	Species, molecule type, receptor type, chain type, gene type, structure, functionality, specificity
DESCRIPTION	Standardized labels and annotations	Core (V-, D-, J-, C-REGION) Prototypes Labels for sequences Labels for 2D and 3D structures
CLASSIFICATION	Reference sequences Standardized IG and TR gene nomenclature (group, subgroup, gene, and allele)	Nomenclature of the human IG and TR genes ***(1,2)*** [entry in 1999 in GDB ***(8)***, HGNC ***(9)***, and LocusLink and Entrez Gene at NCBI] Alignment of alleles Nomenclature of the IG and TR genes of all vertebrate species
NUMEROTATION	IMGT unique numbering for: V- and V-LIKE-DOMAINs ***(10)*** C- and C-LIKE-DOMAINs ***(11)*** G- and G-LIKE-DOMAINs ***(12)***	Protein displays Colliers de Perles ***(13)*** FR-IMGT and CDR-IMGT delimitations Structural loops and beta strands delimitations
ORIENTATION	Orientation of genomic instances relative to each other	Chromosome orientation Locus orientation Gene orientation DNA strand orientation
OBTENTION	Standardized origin and methodology	

(in capital letters) constitutes the main part of the expertise. Interestingly, 65 *IMGT®*-specific labels have been entered in the newly created Sequence Ontology ***(17)***.

2.3. CLASSIFICATION concept: standardized IG and TR gene nomenclature

The objective is to provide immunologists and geneticists with a standardized nomenclature per locus and per species which allows extraction and comparison of data for the complex B-cell and T-cell antigen receptor molecules. The concepts of classification have been used to set up a unique nomenclature of human IG and TR genes, which was approved by the Human Genome Organization (HUGO) Nomenclature Committee *HGNC* in 1999 ***(9)***. All the human IG and TR genes ***(1,2,18,19)*** have been entered by the *IMGT* Nomenclature Committee in *Genome Database GDB* ***(8)***, *LocusLink* and *Entrez Gene* at NCBI, USA, and in *IMGT/GENE-DB* ***(20)***. *IMGT* reference sequences have been defined for each allele of each gene based on one or, whenever possible, several of the following criteria: germline sequence, first sequence published, longest sequence, and mapped sequence. They are listed in the germline gene tables of the *IMGT* Repertoire. The *IMGT* Protein displays show the translated sequences of the alleles *01 of the functional or ORF genes ***(1,2)***.

2.4. NUMEROTATION concept: the IMGT unique numbering

A uniform numbering system for IG and TR sequences of all species has been established to facilitate sequence comparison and cross-referencing between experiments from different laboratories whatever the antigen receptor (IG or TR), the chain type or the species ***(21,22)***.

This numbering results from the analysis of more than 5,000 IG and TR variable region sequences of vertebrate species from fish to human. It takes into account and combines the definition of the framework (FR) and complementarity determining region (CDR) ***(23)***, structural data from X-ray diffraction studies ***(24)***, and the characterization of the hypervariable loops ***(25)***. In the *IMGT* unique numbering, conserved amino acids from FR always have the same number whatever the IG or TR variable sequence and whatever the species they come from, for example cysteine 23 (in FR1-IMGT), tryptophan 41 (in FR2-IMGT), leucine (or other hydrophobic amino acid) 89, and cysteine 104 (in FR3-IMGT). Tables and two-dimensional (2D) graphical representations designated as *IMGT* Colliers de Perles are available on the *IMGT®* Web site at http://imgt.cines.fr and in the works of M.-P. Lefranc and G. Lefranc ***(1,2)***. The *IMGT* Collier de Perles of a variable domain or V-DOMAIN of an IG light chain is shown, as an example, in Fig. 1.

This *IMGT* unique numbering has several advantages:

1. It has allowed the redefinition of the limits of the FR and CDR of the IG and TR variable domains. The FR-IMGT and CDR-IMGT lengths become in themselves crucial information, which characterize variable regions belonging to a group, a subgroup, and/or a gene.
2. FR amino acids (and codons) located at the same position in different sequences can be compared without requiring sequence alignments. This also holds for amino acids belonging to CDR-IMGT of the same length.
3. The unique numbering is used as the output of the *IMGT/V-QUEST* alignment tool. The aligned sequences are displayed according to the *IMGT* unique numbering and with the FR-IMGT and CDR-IMGT delimitations.
4. The unique numbering has allowed a standardization of the description of mutations and the description of IG and TR allele polymorphisms ***(1,2)***. The mutations and allelic polymorphisms of each gene are described by comparison to the *IMGT* reference sequences of the allele *01 ***(1,2)***.
5. The unique numbering allows the description and comparison of somatic hypermutations of the IG variable domains.

By facilitating the comparison between sequences and by allowing the description of alleles and mutations, the *IMGT* unique numbering represents a big step forward in the analysis of the IG and TR sequences of all vertebrate species. Moreover, it gives insight into the structural configuration of the domains and opens interesting views on the evolution of these sequences, as this numbering can be used for all sequences belonging to the V-set and C-set of the IgSF. Structural and functional domains of the IG and TR chains comprise the V-DOMAIN (9-strand beta-sandwich) (Fig. 2), which corresponds to the V-J-REGION or V-D-J-REGION and is encoded by two or three genes ***(1,2)***, and the constant domain or C-DOMAIN (7-strand beta-sandwich) (Fig. 2). The *IMGT* unique numbering has been initially defined for the V-DOMAINs of the IG and TR and for the V-LIKE-DOMAINs of IgSF proteins other than IG and TR, for example in vertebrates human CD4 and *Xenopus* CTXg1 and in invertebrates *Drosophila* amalgam and *Drosophila* fasciclin II. ***(10,26)***. It has been extended to the C-DOMAINs of the IG and TR and to the C-LIKE-DOMAINs of IgSF proteins other than IG and TR ***(11,26,27)***. More recently, the IMGT unique numbering has also been defined for the groove domain or G-DOMAIN (four beta-strand and one alpha-helix) (Fig. 2) of the MHC class I and class II chains and for the G-LIKE-DOMAINs of MhcSF proteins other than MHC, for example MICA ***(12,28)***.

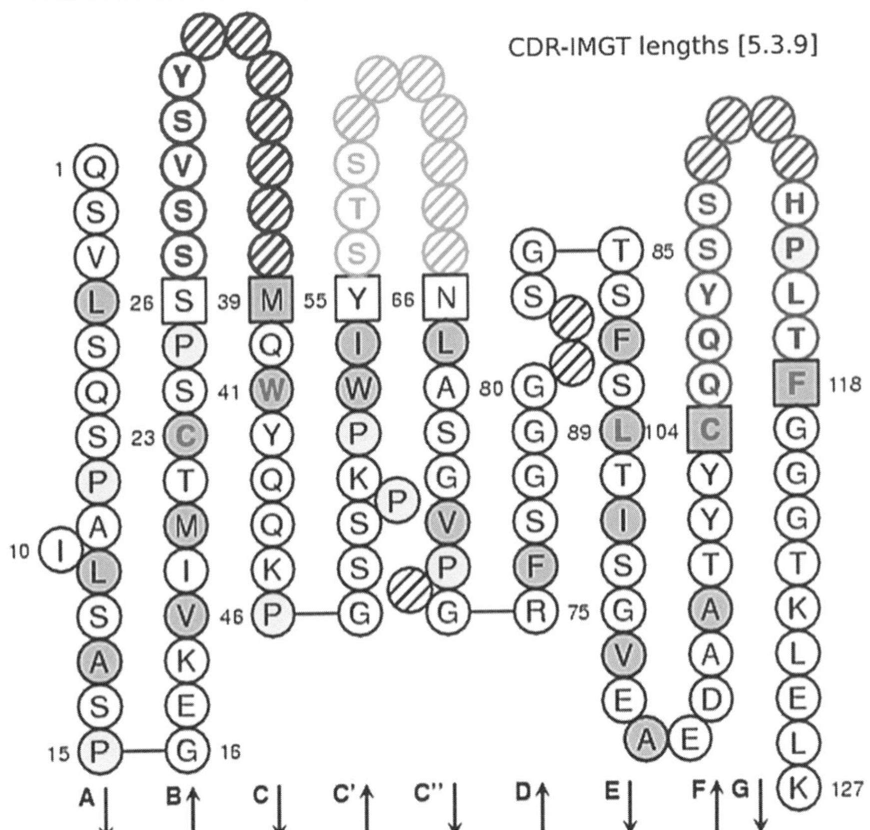

Fig. 1. *IMGT* Collier de Perles of a V-DOMAIN. The *IMGT* Collier de Perles of V-DOMAIN is based on the *IMGT* unique numbering for V-DOMAIN and V-LIKE-DOMAIN ***(10)***. Amino acids are shown in the one-letter abbreviation. The CDR-IMGT are limited by amino acids shown in squares, which belong to the neighbouring FR-IMGT. The CDR3-IMGT extend from position 105 to position 117. Hatched circles correspond to missing positions according to the IMGT unique numbering for V-DOMAIN and V-LIKE-DOMAIN ***(10)***. Arrows indicate the direction of the nine beta strands that form the two beta sheets of the immunoglobulin (IG) fold ***(1,2)***. Positions at which hydrophobic amino acids (hydropathy index with positive value: I, V, L, F, C, M and A) and tryptophan (W) are found in more than 50% of analysed IG and TR sequences are shown in blue. All proline (P) are shown in yellow. The V-DOMAIN chosen as an example is a murine IG light kappa domain or V-KAPPA (*IMGT/3Dstructure-DB*: 1a6t_C). CDR-IMGT regions (for a IG light kappa or lambda, or a TR alpha or gamma V-DOMAIN) are coloured as follows: CDR1-IMGT (blue), CDR2-IMGT (bright green) and CDR3-IMGT (dark green). (*See* Color Plate 1 following p. 32.)

(A) V-DOMAIN (IG, TR)

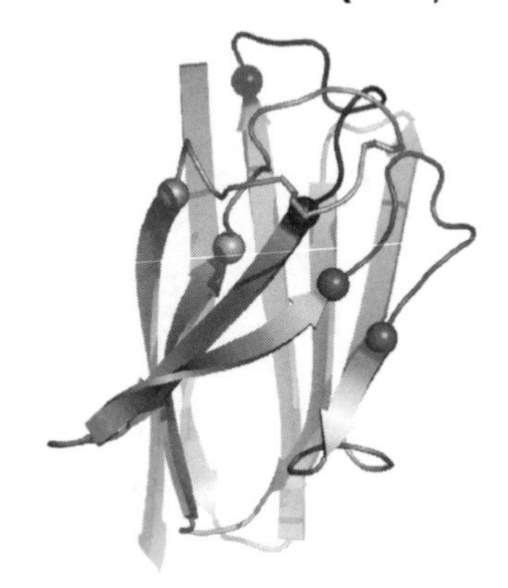

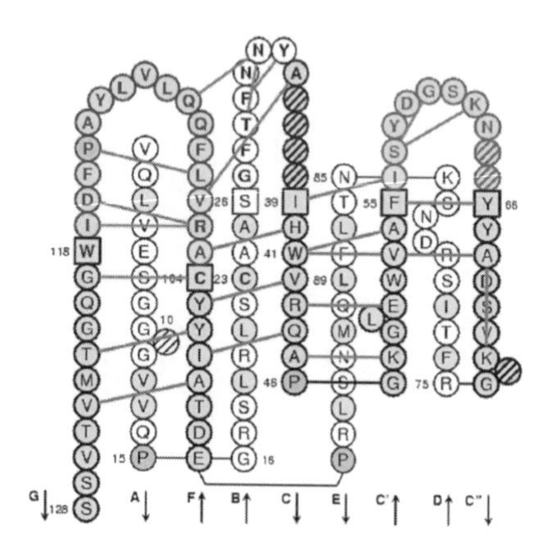

(B) C-DOMAIN (IG, TR)

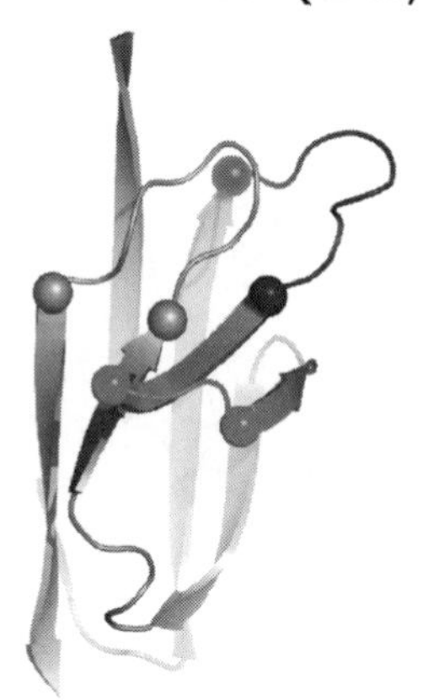

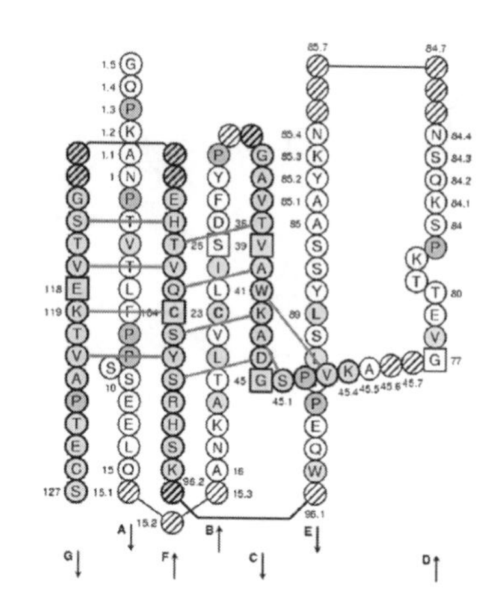

(C) G-DOMAIN (MHC)

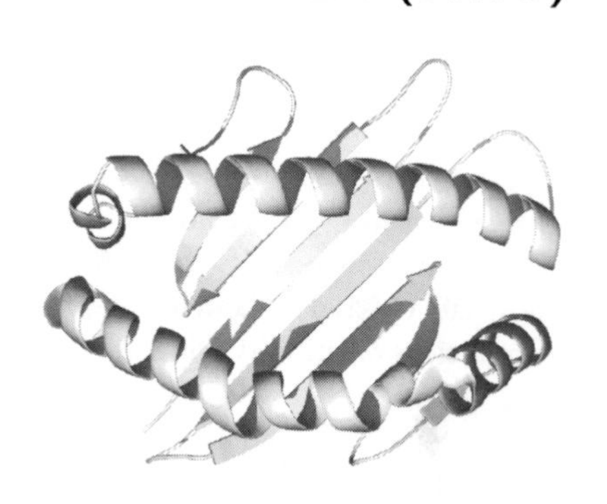

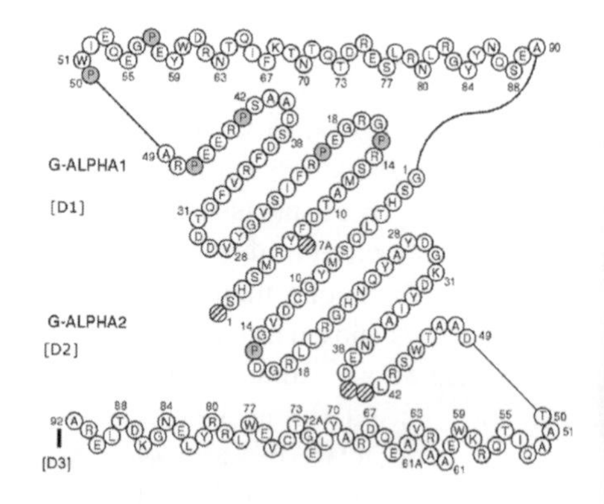

2.5. ORIENTATION concept: orientation of genomic instances relative to each other

The ORIENTATION concept allows to set up genomic orientation (for chromosome, locus, and gene) and DNA strand orientation. It is particularly useful in large genomic projects to localize a gene in a locus and/or a sequence (or a clone) in a contig or on a chromosome.

2.6. OBTENTION concept: controlled vocabulary for biological origin and experimental methodology

The OBTENTION concept, that is still in development, will be particularly useful for clinical data integration. This will help us to compare the repertoires of the IG antibody recognition sites and of the TR recognition sites in normal and pathological situations (autoimmune diseases, infectious diseases, leukemias, lymphomas, and myelomas).

3. *IMGT®* Genomics, Genetics, and Structural Approaches

In order to extract knowledge from *IMGT®* standardized immunogenetics data, three main *IMGT®* biological approaches have been developed: genomics, genetics, and structural approaches (Table 2). The *IMGT®* genomics approach

◄

Fig. 2. Three-dimensional structures and *IMGT* Collier de Perles of a V-DOMAIN, a C-DOMAIN and G-DOMAINs. (**A**) V-DOMAIN. The *IMGT* Collier de Perles is based on the *IMGT* unique numbering for V-DOMAIN and V-LIKE-DOMAIN ***(10)***. The V-DOMAIN chosen as an example is a human immunoglobulin (IG) variable heavy domain or VH (*IMGT/3Dstructure-DB*: 1aqk_H). CDR-IMGT regions (for a IG heavy, or a TR beta or delta V-DOMAIN) are colored as follows: CDR1-IMGT (red), CDR2-IMGT (orange) and CDR3-IMGT (purple). Arrows indicate the direction of the nine beta strands of the V-DOMAIN that form the two beta sheets of the IG fold ***(1,2)***. Hydrogen bonds of the [GFCC′C″] sheet are shown with green lines. (**B**) C-DOMAIN. The *IMGT* Collier de Perles is based on the *IMGT* unique numbering for C-DOMAIN and C-LIKE-DOMAIN ***(11)***. The C-DOMAIN chosen as an example is a human IG constant light lambda domain or C-LAMBDA (*IMGT/3Dstructure-DB*: 1mcd_B). Arrows indicate the direction of the seven beta strands of the C-DOMAIN that form the two beta sheets of the IG fold ***(1,2)***. Hydrogen bonds of the [GFC] sheet are shown with green lines. (**C**) G-DOMAINs. The *IMGT* Colliers de Perles are based on the *IMGT* unique numbering for G-DOMAIN and G-LIKE-DOMAIN ***(12)***. The G-DOMAINs chosen as examples are human major histocompatibility complex (MHC) class I alpha groove domains or G-ALPHA1 and G-ALPHA2 (*IMGT/3Dstructure-DB*: 1agb_A). Amino acids are shown in the one-letter abbreviation. Hatched circles correspond to missing positions according to the IMGT unique numbering ***(10–12)***. (*See* Color Plate 2 following p. 32.)

Table 2
***IMGT*® databases, tools and Web resources for genomics, genetics and structural approaches**

Approaches	Databases	Tools	Web resources[a]
Genomics	*IMGT/GENE-DB* ***(16)***	*IMGT/GeneView* *IMGT/LocusView* *IMGT/CloneSearch* *IMGT/GeneSearch* *IMGT/GeneInfo* ***(29)***	*IMGT* Repertoire "Locus and genes" section: • Chromosomal localizations ***(1,2)*** • Locus representations ***(1,2)*** • Locus description • Gene tables, etc. • Potential germline repertoires • Lists of genes • Correspondence between nomenclatures ***(1,2)***
Genetics	*IMGT/LIGM-DB* ***(30)*** *IMGT/PRIMER-DB* ***(31)*** *IMGT/MHC-DB* ***(32)***	*IMGT/V-QUEST* ***(33)*** *IMGT/JunctionAnalysis* ***(34)*** *IMGT/Allele-Align* *IMGT/PhyloGene* ***(35)***	*IMGT* Repertoire "Proteins and alleles' section: • Alignments of alleles • Protein displays • Tables of alleles, etc.
Structural	*IMGT/3Dstructure-DB* ***(36)***	*IMGT/StructuralQuery* ***(36)***	*IMGT* Repertoire "2D and 3D structures" section: • *IMGT* Colliers de Perles (2D representations on one layer or two layers) • *IMGT*® classes for amino acid characteristics ***(37)*** • *IMGT* Colliers de Perles reference profiles ***(37)*** • 3D representations

[a] Only Web resources examples from the *IMGT* Repertoire section are shown.

is gene-centered and mainly orientated towards the study of the genes within their loci and on the chromosomes. The *IMGT*® genetics approach refers to the study of the genes in relation with their sequence polymorphisms and mutations, their expression, their specificity, and their evolution. The genetics approach heavily relies on the DESCRIPTION concept (and particularly on the V-, D-, J- and C-REGION core concepts for the IG and TR), on the CLASSIFICATION concept (*IMGT*® gene and allele names) and on the NUMEROTATION concept [*IMGT* unique numbering ***(10–12)***]. The *IMGT*® structural approach refers to the study of the 2D and 3D structures of the IG, TR, MHC, and RPI and to the antigen- or ligand-binding characteristics in relationship with the protein functions, polymorphisms and evolution. The structural approach relies on the CLASSIFICATION concept (*IMGT*® gene and allele names), DESCRIPTION concept (receptor and chain description and domain delimitations), and NUMEROTATION concept [amino acid positions according to the *IMGT* unique numbering ***(10–12)***].

For each approach, *IMGT*® provides databases [one genome database (*IMGT/GENE-DB*), three sequence databases (*IMGT/LIGM-DB*, *IMGT/MHC-DB*, and *IMGT/PRIMER-DB*), one 3D structure database (*IMGT/3Dstructure-DB*)], interactive tools (ten on-line tools for genome, sequence and 3D structure analysis), and *IMGT* Repertoire Web resources (providing an easy-to-use interface to carefully and expertly annotated data on the genome, proteome, and polymorphism and structural data of the IG and TR, MHC and RPI) (Table 2). These databases, tools, and Web resources are detailed in the following Sections 4–6. Other *IMGT*® Web resources include:

1. *IMGT* Bloc-notes (Interesting links, etc.) provides numerous hyperlinks towards the Web servers specializing in immunology, genetics, molecular biology, and bioinformatics (associations, collections, companies, databases, immunology themes, journals, molecular biology servers, resources, societies, tools, etc.) ***(38)***.
2. *IMGT* Lexique.
3. The *IMGT* Immunoinformatics page.
4. The *IMGT* Medical page.
5. The *IMGT* Veterinary page.
6. The *IMGT* Biotechnology page.
7. *IMGT* Education (Aide-mémoire, Tutorials, Questions and answers, etc.) provides useful biological resources for students and includes figures and tutorials (in English and/or in French) in immunogenetics.
8. *IMGT* Aide-mémoire provides an easy access to information such as genetic code, splicing sites, amino acid structures, and restriction enzyme sites.
9. *IMGT* Index is a fast way to access data when information has to be retrieved from different parts of the *IMGT* site. For example, "allele" provides links to the

IMGT Scientific chart rules for the allele description and to the *IMGT* Repertoire "Alignments of alleles" and "Tables of alleles" (http://imgt.cines.fr).

4. *IMGT*® Databases, Tools, and Web Resources for Genomics

Genomic data are managed in *IMGT/GENE-DB*, which is the comprehensive IMGT® genome database ***(20)***. In February 2007, *IMGT/GENE-DB* contained 1,512 IG and TR genes and 2,461 alleles from human and mouse IG and TR genes. Based on the *IMGT*® CLASSIFICATION concept, all the human *IMGT*® gene names ***(1,2)***, approved by the HUGO Nomenclature Committee HGNC in 1999, are available in *IMGT/GENE-DB* ***(20)*** and in Entrez Gene at NCBI (USA). All the mouse *IMGT*® gene and allele names and the corresponding IMGT reference sequences were provided to Mouse Genome Informatics MGI Mouse Genome Database MGD in July 2002 and were presented by *IMGT*® at the 19th International Mouse Genome Conference IMGC 2005, in Strasbourg, France. *IMGT-GENE-DB* allows a query per gene and allele name. *IMGT/GENE-DB* interacts dynamically with *IMGT/LIGM-DB* ***(30)*** to download and display human and mouse gene-related sequence data. This is the first example of an interaction between *IMGT*® databases using the CLASSIFICATION concept.

The *IMGT*® genome analysis tools manage the locus organization and gene location and provide the display of physical maps for the human and mouse IG, TR, and MHC loci. They allow to view genes in a locus (*IMGT/GeneView* and *IMGT/LocusView*) to search for clones (*IMGT/CloneSearch*), to search for genes in a locus (*IMGT/GeneSearch* and *IMGT/GeneInfo*) based on *IMGT*® gene names, functionality or localization on the chromosome, to provide information on the clones that were used to build the locus contigs (accession numbers are from *IMGT/LIGM-DB* and gene names from *IMGT/GENE-DB*) or to display information on the human and mouse IG and TR potential rearrangements.

The *IMGT* Repertoire genome data include chromosomal localizations, locus representations, locus description, germline gene tables, potential germline repertoires, lists of IG and TR genes and links between *IMGT*, *HGNC*, *GDB*, *Entrez Gene*, and *OMIM*, and correspondence between nomenclatures ***(1,2)***.

5. *IMGT*® Databases, Tools, and Web Resources for Genetics

IMGT/LIGM-DB ***(30)*** is the comprehensive *IMGT*® database of IG and TR nucleotide sequences from human and other vertebrate species, with translation for fully annotated sequences, created in 1989 by LIGM, Montpellier, France, on the Web since July 1995. IMGT/LIGM-DB is the first and the largest *IMGT*® database. In February 2007, *IMGT/LIGM-DB* contained 105,188 nucleotide sequences of IG and TR from 150 species. The unique source of data for *IMGT/LIGM-DB* is *EMBL* that shares data with the other two generalist databases *GenBank* and *DDBJ*. *IMGT/LIGM-DB* sequence data are identified by the *EMBL/GenBank/DDBJ* accession number. Based on expert analysis, specific detailed annotations are added to *IMGT* flat files.

Since August 1996, the *IMGT/LIGM-DB* content closely follows the *EMBL* one for the IG and TR, with the following advantages: *IMGT/LIGM-DB* does not contain sequences that have previously been wrongly assigned to IG and TR; conversely, *IMGT/LIGM-DB* contains IG and TR entries that have disappeared from the generalist databases [for example, the L36092 accession number that encompasses the complete human TRB locus is still present in *IMGT/LIGM-DB*, whereas it has been deleted from *EMBL/GenBank/DDBJ* due to its too large size (684,973 bp); in 1999, *IMGT/LIGM-DB* detected the disappearance of 20 IG and TR sequences that inadvertently had been lost by *GenBank*, and allowed the recuperation of these sequences in the generalist databases].

The *IMGT/LIGM-DB* annotations (gene and allele name assignment, labels) allow data retrieval not only from *IMGT/LIGM-DB* but also from other *IMGT®* databases. For example, the *IMGT/GENE-DB* entries provide the *IMGT/LIGM-DB* accession numbers of the IG and TR cDNA sequences that contain a given V, D, J or C gene. The automatic annotation of rearranged human and mouse cDNA sequences in *IMGT/LIGM-DB* is performed by *IMGT/Automat* ***(39)***, an internal Java tool that implements *IMGT/V-QUEST* and *IMGT/ JunctionAnalysis*.

Standardized information on oligonucleotides (or primers) and combinations of primers (Sets and Couples) for IG and TR are managed in *IMGT/PRIMER-DB* ***(31)***, the *IMGT®* oligonucleotide database on the Web since February 2002. *IMGT/MHC-DB* ***(32)*** hosted at EBI comprises *IMGT/HLA* for human MHC (or HLA) and *IMGT/MHC-NHP* for MHC of non-human primates.

The *IMGT®* tools for the genetics approach comprise *IMGT/V-QUEST* ***(33, 40)*** for the identification of the V, D, and J genes and of their mutations, *IMGT/JunctionAnalysis* ***(34,40)*** for the analysis of the V-J and V-D-J junctions that confer the antigen receptor specificity, *IMGT/Allele-Align* for the detection of polymorphisms, and *IMGT/Phylogene* ***(35)*** for gene evolution analyses. *IMGT/V-QUEST* (V-QUEry and STandardization) (http://imgt.cines.fr) is an integrated software for IG and TR ***(33,40)***. This tool, easy to use, analyses an input IG or TR germline or rearranged variable nucleotide sequence. *IMGT/V-QUEST* results comprise the identification of the V, D, and J genes and alleles and the nucleotide alignment by comparison with sequences from the *IMGT* reference directory, the delimitations of the FR-IMGT and CDR-IMGT based on the *IMGT* unique numbering, the protein translation of the input sequence, the identification of the JUNCTION, the description of the mutations and amino acid changes of the V-REGION, and the 2D *IMGT* Collier de Perles representation of the V-REGION or V-DOMAIN. The set of sequences from the *IMGT* reference directory, used for *IMGT/V-QUEST*, can be downloaded in *FASTA* format from the *IMGT*® site.

IMGT/JunctionAnalysis ***(34,40)*** is a tool developed by LIGM, complementary to *IMGT/V-QUEST*, which provides a thorough analysis of the V-J and V-D-J junction of IG and TR rearranged genes. The JUNCTION extends from 2nd-CYS 104 to J-PHE or J-TRP 118 inclusive. J-PHE or J-TRP are easily identified for in-frame rearranged sequences when the conserved Phe/Trp-Gly-X-Gly motif of the J-REGION is present. The length of the CDR3-IMGT of rearranged V-J-GENEs or V-D-J-GENEs is a crucial piece of information. It is the number of amino acids or codons from position 105 to 117 (J-PHE or J-TRP non-inclusive). CDR3-IMGT amino acid and codon numbers are according to the IMGT unique numbering for V-DOMAIN ***(10)***. *IMGT/JunctionAnalysis* identifies the D-GENE and allele involved in the IGH, TRB, and TRD V-D-J rearrangements by comparison with the *IMGT* reference directory and delimits precisely the P, N, and D regions ***(1,2)***. Results from *IMGT/JunctionAnalysis* are more accurate than those given by *IMGT/V-QUEST* regarding the D-GENE identification. Indeed, *IMGT/JunctionAnalysis* works on shorter sequences (JUNCTION) and with a higher constraint because the identification of the V-GENE and J-GENE and alleles is a prerequisite to perform the analysis. Several hundreds of junction sequences can be analysed simultaneously.

Other *IMGT*® Tools for sequence analysis comprise *IMGT/Allele-Align* that allows the comparison of two alleles highlighting the nucleotide and amino acid differences and *IMGT/PhyloGene* ***(35)***, an easy-to-use tool for phylogenetic analysis of IMGT standardized reference sequences.

The *IMGT* Repertoire polymorphism data are represented by "Alignments of alleles," "Tables of alleles,"' "Allotypes,", "Protein displays," particularities in protein designations, *IMGT* reference directory in *FASTA* format, correspondence between IG and TR chain and receptor *IMGT* designations ***(1,2)***.

6. *IMGT*® Databases, Tools, and Web Resources for Structural Analysis

Structural data are compiled and annotated in *IMGT/3Dstructure-DB* ***(36)***, the *IMGT*® 3D structure database, created by LIGM, on the Web since November 2001. *IMGT/3Dstructure-DB* comprises IG, TR, MHC, and RPI with known 3D structures. In February 2007, *IMGT/3Dstructure-DB* contained 1,221

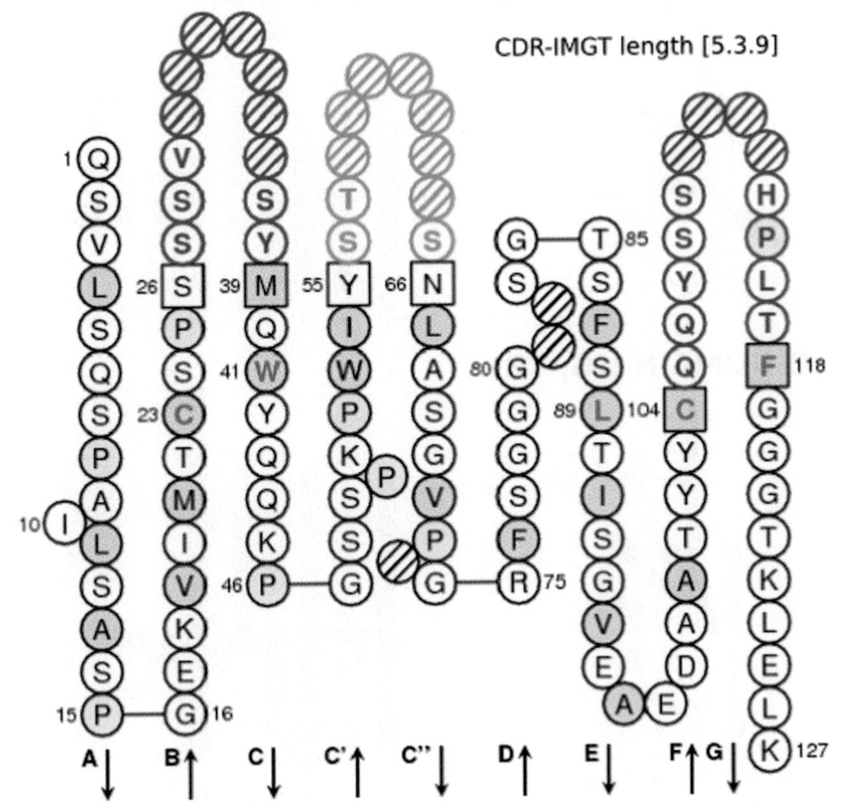

Color Plate 1, *IMGT* Collier de Perles of a V-DOMAIN. The *IMGT* Collier de Perles of V-DOMAIN is based on the *IMGT* unique numbering for V-DOMAIN and V-LIKE-DOMAIN (***10***) (Chapter 2, Fig. 1; *see* full caption on p. 25 and discussion on p. 24.)

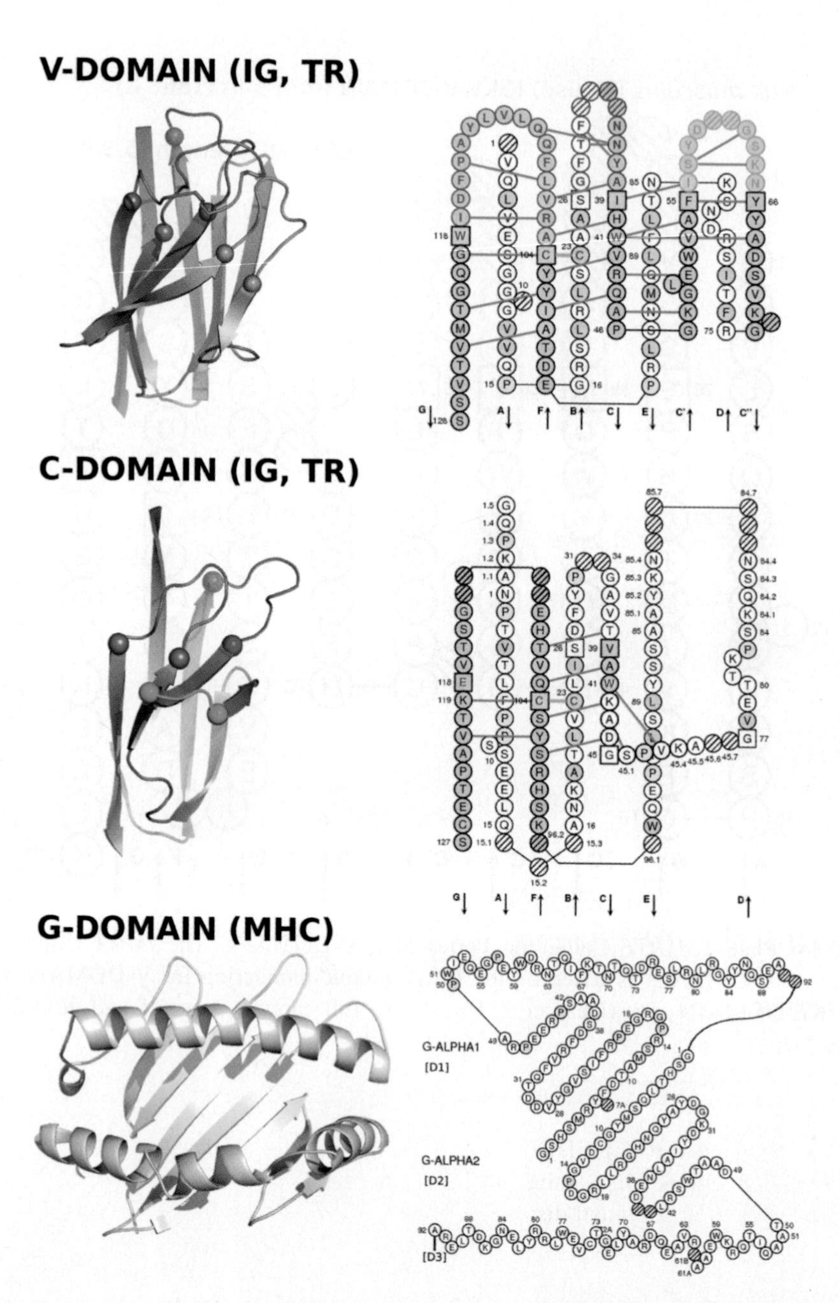

Color Plate 2, Three-dimensional structures and *IMGT* Collier de Perles of a V-DOMAIN, a C-DOMAIN and G-DOMAINs. (Chapter 2, Fig. 2; *see* full caption on p. 27 and discussion on p. 24.)

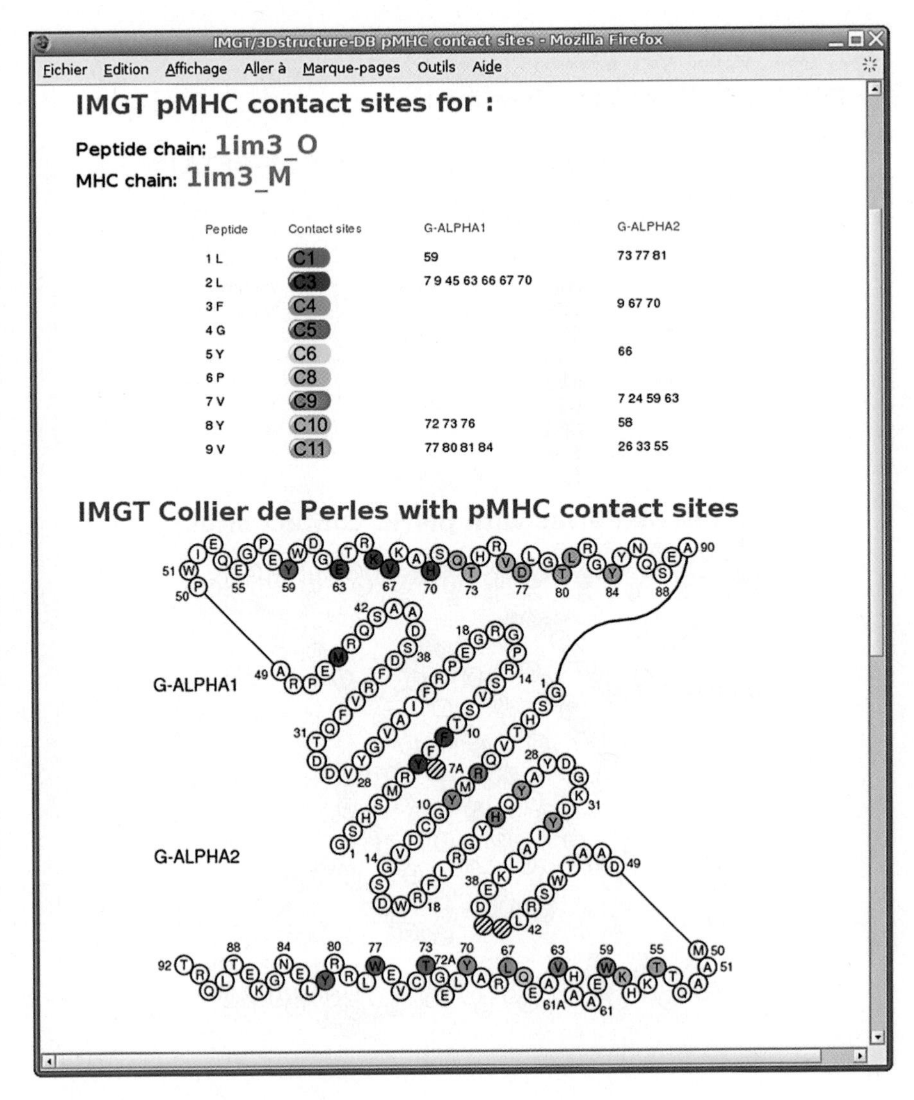
IMGT/3Dstructure-DB pMHC contact sites - Mozilla Firefox

Fichier Edition Affichage Aller à Marque-pages Outils Aide

IMGT pMHC contact sites for :

Peptide chain: 1im3_O
MHC chain: 1im3_M

Peptide	Contact sites	G-ALPHA1	G-ALPHA2
1 L	C1	59	73 77 81
2 L	C3	7 9 45 63 66 67 70	
3 F	C4		9 67 70
4 G	C5		
5 Y	C6		66
6 P	C8		
7 V	C9		7 24 59 63
8 Y	C10	72 73 76	58
9 V	C11	77 80 81 84	26 33 55

IMGT Collier de Perles with pMHC contact sites

Color Plate 3, *IMGT* pMHC contact sites of human HLA-A*0201 MHC-I and a 9-amino acid peptide side chains (*IMGT/3Dstructure-DB*: 1im3). (Chapter 2, Fig. 4; *see* full caption on p. 36 and discussion on p. 35.)

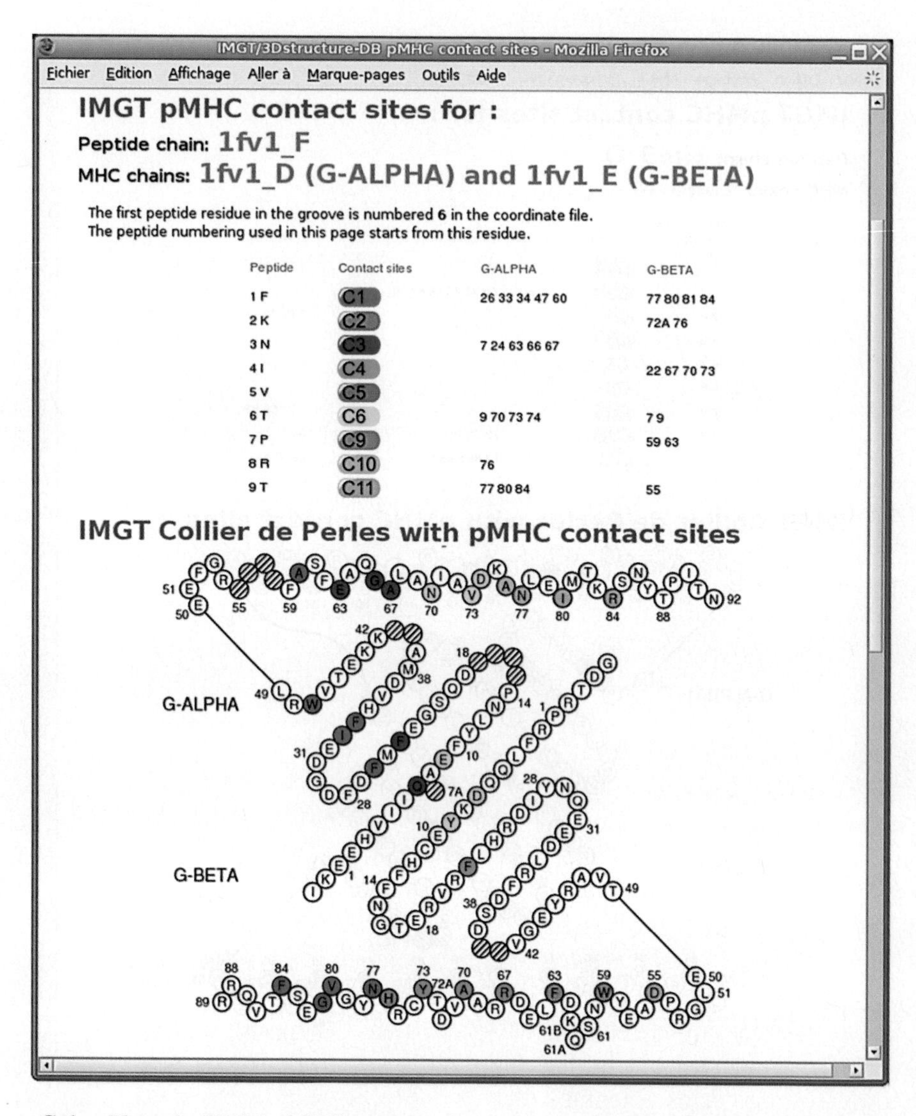

IMGT/3Dstructure-DB pMHC contact sites - Mozilla Firefox

Eichier Edition Affichage Aller à Marque-pages Outils Aide

IMGT pMHC contact sites for :

Peptide chain: 1fv1_F

MHC chains: 1fv1_D (G-ALPHA) and 1fv1_E (G-BETA)

The first peptide residue in the groove is numbered **6** in the coordinate file.
The peptide numbering used in this page starts from this residue.

Peptide	Contact sites	G-ALPHA	G-BETA
1 F	C1	26 33 34 47 60	77 80 81 84
2 K	C2		72A 76
3 N	C3	7 24 63 66 67	
4 I	C4		22 67 70 73
5 V	C5		
6 T	C6	9 70 73 74	7 9
7 P	C9		59 63
8 R	C10	76	
9 T	C11	77 80 84	55

IMGT Collier de Perles with pMHC contact sites

Color Plate 4, *IMGT* pMHC contact sites of human HLA-DRA*0101 and HLA-DRB5*0101 MHC-II and the peptide side chains (9 amino acids located in the groove) (*IMGT/3Dstructure-DB*: 1fv1). (Chapter 2, Fig. 5; *see* full caption on p. 37 and discussion on p. 35.)

atomic coordinate files. These coordinate files, extracted from the *Protein Data Bank* (*PDB*) ***(41)***, are renumbered according to the standardized *IMGT* unique numbering ***(10–12)***. The *IMGT/3Dstructure-DB* cards provide *IMGT*® annotations (assignment of *IMGT*® genes and alleles, *IMGT*® chain and domain labels, and *IMGT* Colliers de Perles on one layer and two layers), downloadable renumbered *IMGT/3Dstructure-DB* flat files, visualization tools and external links. *IMGT/3Dstructure-DB* residue cards provide detailed information on the inter- and intra-domain contacts of each residue position (Fig. 3).

The *IMGT/StructuralQuery* tool ***(36)*** analyses the intramolecular interactions for the V-DOMAINs. The contacts are described per domain (intra- and inter-domain contacts) and annotated in term of *IMGT*® labels (chains and domain), positions (*IMGT* unique numbering), backbone or side-chain implication. *IMGT/StructuralQuery* allows to retrieve the *IMGT/3Dstructure-DB* entries, based on specific structural characteristics: phi and psi angles, accessible surface area (ASA), amino acid type, distance in angstrom between amino acids, and CDR-IMGT lengths ***(36)***.

In order to appropriately analyse the amino acid resemblances and differences between IG, TR, MHC, and RPI chains, 11 *IMGT*® classes were defined for the amino acid "chemical characteristics" properties and used to set up *IMGT* Colliers de Perles reference profiles ***(37)***. The *IMGT* Colliers de Perles reference profiles allow to easily compare amino acid properties at each position whatever the domain, the chain, the receptor or the species ***(37)***. The IG and TR variable and constant domains and the MHC groove domains represent a privileged situation for the analysis of amino acid properties in relation with 3D structures, by the conservation of their 3D structure despite divergent amino acid sequences and by the considerable amount of genomic (*IMGT* Repertoire), structural (*IMGT/3Dstructure-DB*) and functional data available. These data are not only useful to study mutations and allele polymorphisms but are also needed to establish correlations between amino acids in the protein sequences and 3D structures, to analyse the IgSF and MhcSF domain interactions ***(42)*** and to determine amino acids potentially involved in the immunogenicity. One of the key elements in the adaptive immune response is the presentation of peptides by the MHC to the TR at the surface of T cells. The characterization of the TR/peptide/MHC trimolecular complexes (TR/pMHC) is crucial to the fields of immunology, vaccination, and immunotherapy. In *IMGT/3Dstructure-DB*, TR/pMHC molecular characterization, and pMHC contact analysis have been standardized, based on the *IMGT* unique numbering for G-DOMAIN, and 11 *IMGT* pMHC contact sites (C1–C11) have been defined ***(43)***. The *IMGT* pMHC contact sites represent the MHC amino acid positions that have contacts with the peptide side chains. They are particularly useful to compare pMHC interactions whatever the MHC classes or chains, whatever the species and whatever the peptide sequence or length *(43)*. There are no C2, C7, and C8 contact sites for MHC-I with 8-amino acid peptides and no C2 and C7 for MHC-I 3D structures with 9-amino acid peptides. In contrast, for MHC-II, C2 is present but there are no C7 and C8 *(43)*. The *IMGT* pMHC contact sites are provided dynamically for the pMHC and the TR/pMHC 3D structures

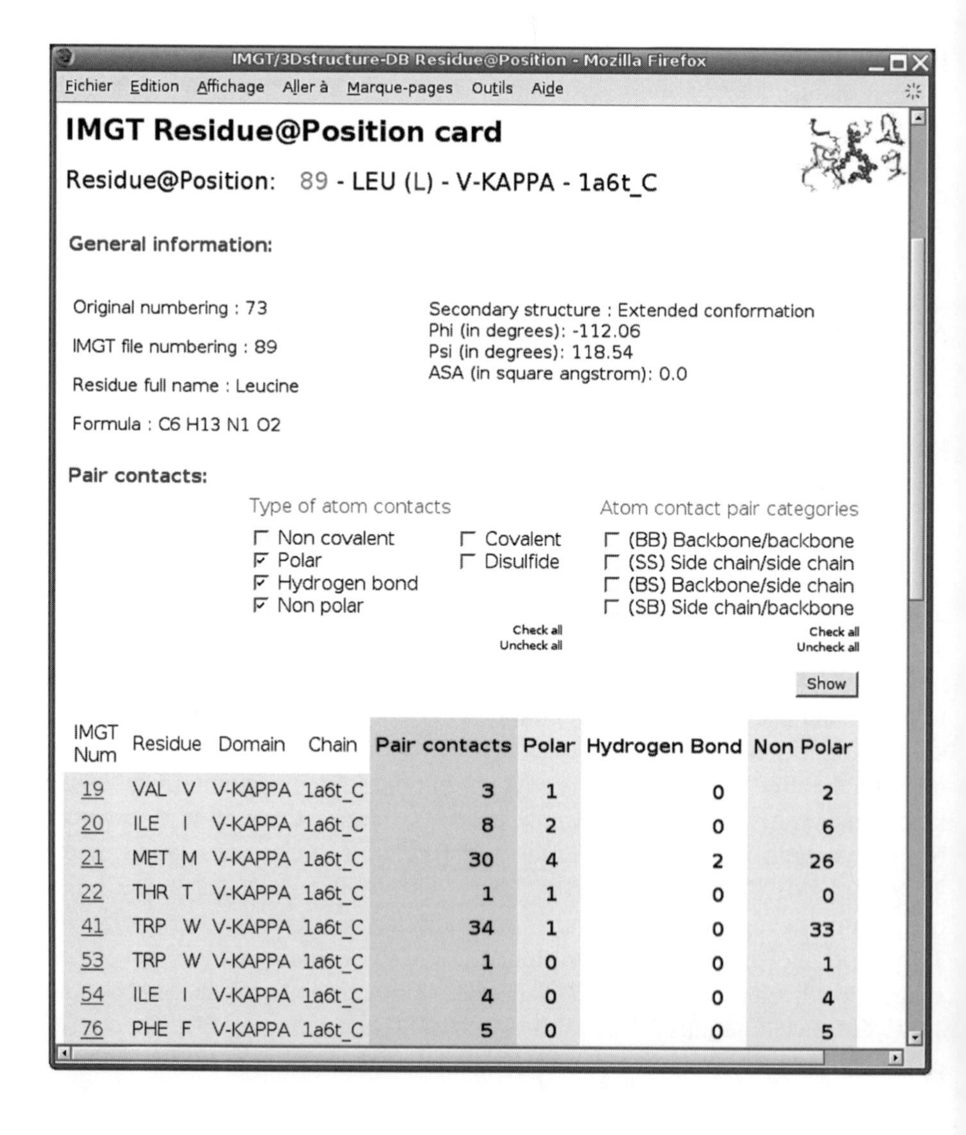

available in *IMGT/3Dstructure-DB*. For example, the *IMGT* pMHC contact sites of a MHC-I (human HLA-A*0201) and a 9-amino acid peptide side chains are shown in Fig. 4 (*IMGT/3Dstructure-DB*: 1im3), and the *IMGT* pMHC contact sites of a MHC-II (human HLA-DRA*0101 and HLA-DRB5*0101) binding 9 amino acids of the peptide in the groove are shown in Fig. 5 (*IMGT/3Dstructure-DB*: 1fv1).

The *IMGT* Repertoire Structural data comprise *IMGT* Colliers de Perles ***(1,2,10–12)***, FR-IMGT and CDR-IMGT lengths, and 3D representations of IG and TR variable domains. This visualization permits rapid correlation between protein sequences and 3D data retrieved from the *PDB*.

7. Conclusion

Since July 1995, *IMGT*® has been available on the Web at http://imgt.cines.fr. *IMGT*® has an exceptional response with more than 140,000 requests a month. The information is of much value to clinicians and biological scientists in general. *IMGT*® databases, tools, and Web resources are extensively queried and used by scientists from both academic and industrial laboratories, who are equally distributed between the United States, Europe, and the remaining world. *IMGT*® is used in very diverse domains: (i) fundamental and medical research (repertoire analysis of the IG antibody recognition sites and of the TR recognition sites in normal and pathological situations such as autoimmune diseases, infectious diseases, AIDS, leukemias, lymphomas, and

Fig. 3. *IMGT* Residue@Position card. The identification of a "*IMGT* Residue@Position" comprises the position number according to the *IMGT* unique numbering ***(10–12)***, the residue name (with three letters and eventually one letter abbreviation), the domain description and the *IMGT/3Dstructure-DB* chain ID. The example shows the contacts of position 89, occupied by a leucine LEU (L), in the V-KAPPA domain of the 1a6t_C chain. The original number in the *PDB* file is indicated. The secondary structure, the phi and psi angles (in degrees) and accessible surface area (ASA) (in square angstroms) are provided. The user can select, for the result display, the types of contacts (non covalent, polar, hydrogen bond, non polar, covalent bond or disulfide bond) and the atom contact pair categories (backbone/backbone, side chain/side chain, backbone/side chain and side chain/backbone atoms). The results are shown as a table with a list of the *IMGT* Residue@Position which are in contact with the *IMGT* Residue@Position at the top of the card, and for each of them, the total number of atom pair contacts and the detailed description of the contacts as selected by the user are also indicated.

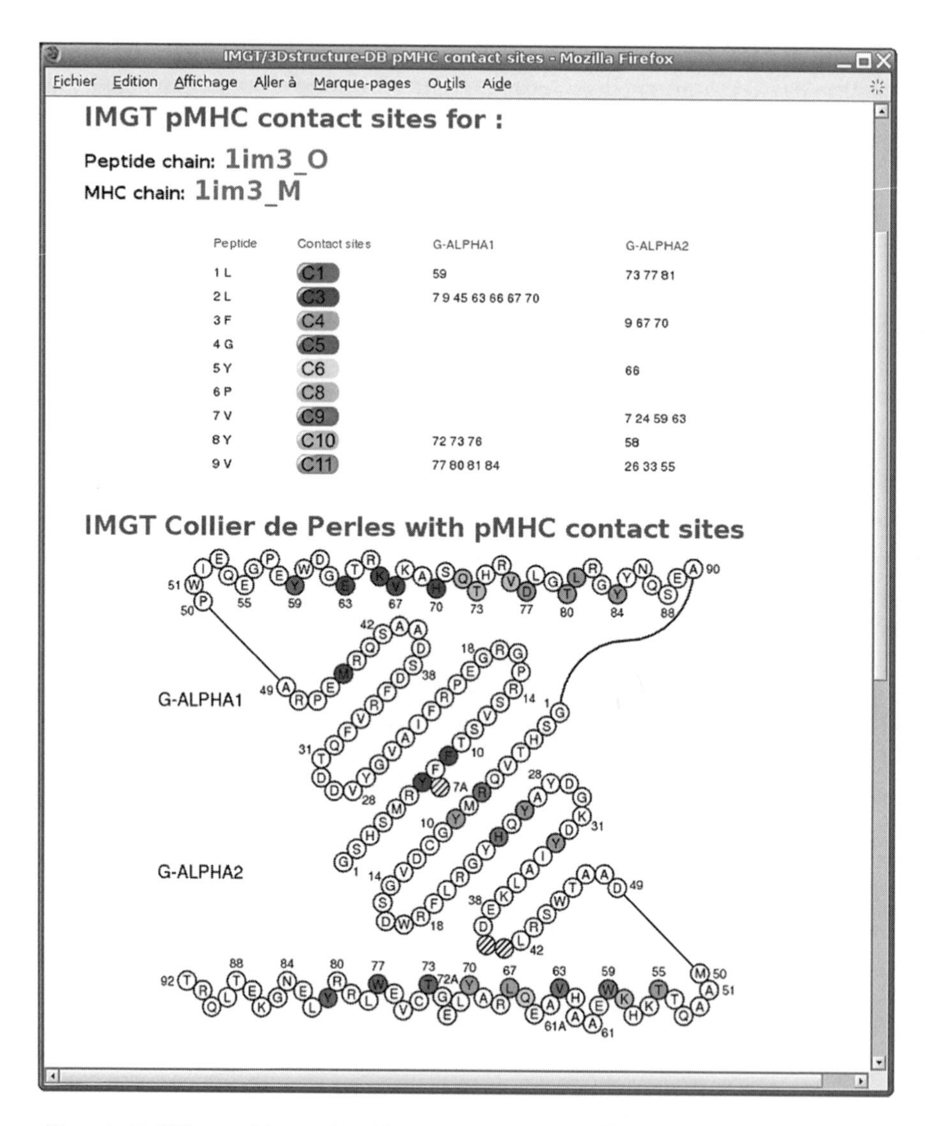

Peptide	Contact sites	G-ALPHA1	G-ALPHA2
1 L	C1	59	73 77 81
2 L	C3	7 9 45 63 66 67 70	
3 F	C4		9 67 70
4 G	C5		
5 Y	C6		66
6 P	C8		
7 V	C9		7 24 59 63
8 Y	C10	72 73 76	58
9 V	C11	77 80 81 84	26 33 55

Fig. 4. IMGT peptide major histocompatibility complex (pMHC) contact sites of human HLA-A*0201 MHC-I and a 9-amino acid peptide side chains (IMGT/3Dstructure-DB: 1im3). The numbers 1–9 refer to the numbering of the peptide amino acids P1–P9. C1–C11 refer to the 11 pMHC contact sites defined by IMGT® ***(43)***. There are no C2 and C7 in MHC-I 3D structures with 9-amino acid peptides. There are no C5 and C8 in this 3D structure as P4 and P6 do not contact MHC amino acids. The view of the IMGT Collier de Perles is from above the cleft, with G-ALPHA1 on top and G-ALPHA2 on bottom of the figure. (*See* Color Plate 3 following p. 32.)

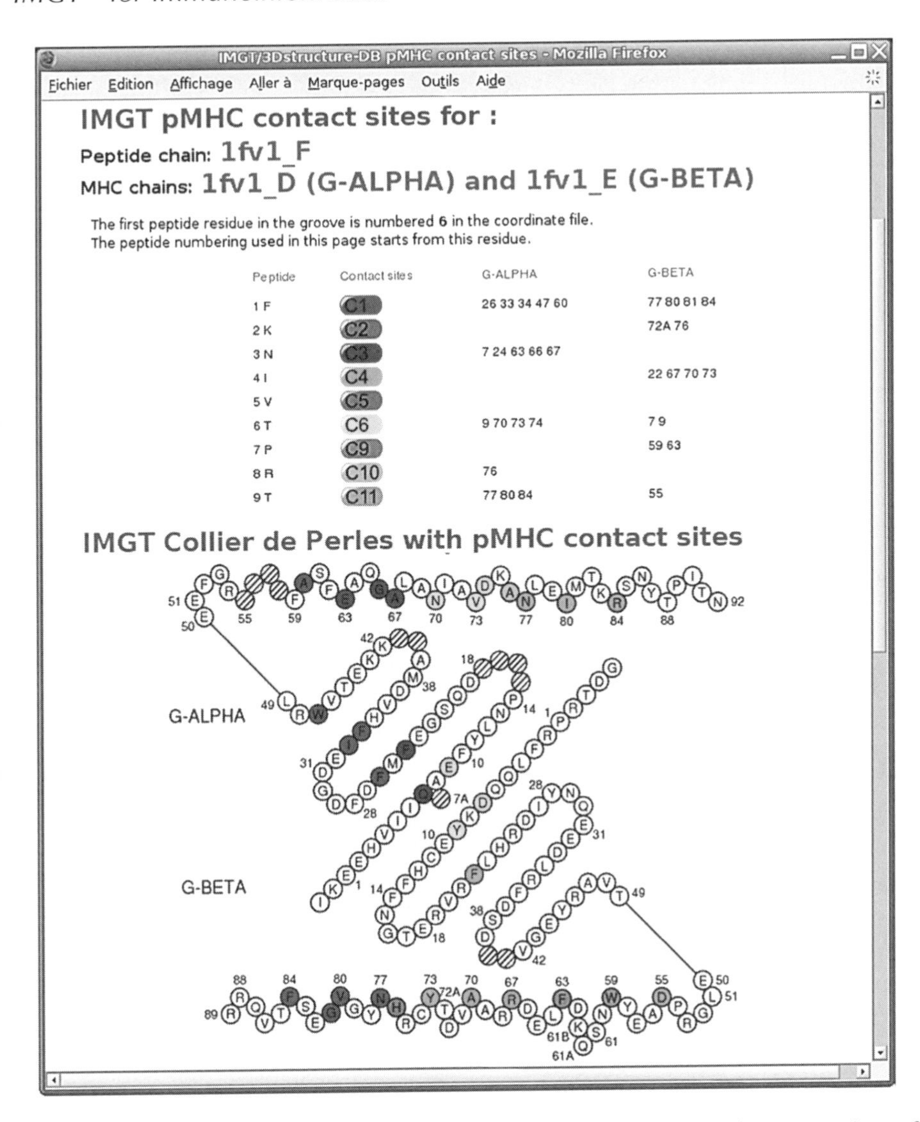

IMGT/3Dstructure-DB pMHC contact sites - Mozilla Firefox

Fichier Edition Affichage Aller à Marque-pages Outils Aide

IMGT pMHC contact sites for :

Peptide chain: 1fv1_F

MHC chains: 1fv1_D (G-ALPHA) and 1fv1_E (G-BETA)

The first peptide residue in the groove is numbered **6** in the coordinate file.
The peptide numbering used in this page starts from this residue.

Peptide	Contact sites	G-ALPHA	G-BETA
1 F	C1	26 33 34 47 60	77 80 81 84
2 K	C2		72A 76
3 N	C3	7 24 63 66 67	
4 I	C4		22 67 70 73
5 V	C5		
6 T	C6	9 70 73 74	7 9
7 P	C9		59 63
8 R	C10	76	
9 T	C11	77 80 84	55

IMGT Collier de Perles with pMHC contact sites

Fig. 5. IMGT peptide major histocompatibility complex (pMHC) contact sites of human HLA-DRA*0101 and HLA-DRB5*0101 MHC-II and the peptide side chains (9 amino acids located in the groove) (IMGT/3Dstructure-DB: 1fv1). The numbers 1–9 refer to the numbering of the peptide amino acids 1–9 located in the groove. C1–C11 refer to the 11 pMHC contact sites defined by IMGT® ***(43)***. There are no C7 and C8 in MHC-II 3D structures with peptide of 9 amino acids located in the groove. There is no C5 in this 3D structure as 5 does not contact MHC amino acids. The view of the IMGT Collier de Perles is from above the cleft, with G-ALPHA on top and G-BETA on bottom of the figure. (*See* Color Plate 4 following p. 32.)

myelomas), (ii) veterinary research (IG and TR repertoires in farm and wildlife species), (iii) genome diversity and genome evolution studies of the adaptive immune responses, (iv) structural evolution of the IgSF and MhcSF proteins, (v) biotechnology related to antibody engineering [single chain Fragment variable (scFv), phage displays, combinatorial libraries, chimeric, humanized, and human antibodies], (vi) diagnostics (clonalities, detection, and follow-up of residual diseases) and (vii) therapeutical approaches (grafts, immunotherapy, and vaccinology). Owing to its high quality and data distribution based on *IMGT-ONTOLOGY*, *IMGT®* has an important role to play in the development of immunogenetics Web services. The creation of dynamic interactions between the *IMGT®* databases and tools, using Web services and *IMGT-ML*, and the design of *IMGT-Choreography* ***(4)***, represents novel and major developments of *IMGT®*, the international reference in immunogenetics and immunoinformatics.

8. Citing *IMGT®*

Authors who make use of the information provided by *IMGT®* should cite **ref.** ***3*** as a general reference for the access to and content of *IMGT®* and quote the *IMGT®* home page URL, http://imgt.cines.fr.

Acknowledgments

I thank Véronique Giudicelli, Patrice Duroux, Quentin Kaas, Joumana Jabado-Michaloud, Géraldine Folch, Chantal Ginestoux, Denys Chaume, and Gérard Lefranc for helpful discussion. I am deeply grateful to the *IMGT®* team for its expertise and constant motivation. *IMGT®* is a registered mark of the Centre National de la Recherche Scientifique (CNRS). *IMGT®* has received the National Bioinformatics Platform RIO label since 2001 (CNRS, INSERM, CEA, and INRA). *IMGT®* was funded in part by the BIOMED1 (BIOCT930038), Biotechnology BIOTECH2 (BIO4CT960037) and 5th PCRDT Quality of Life and Management of Living Resources (QLG2-2000-01287) programmes of the European Union (EU). *IMGT-ML* was developed in collaboration with the EU Online Research Information Environment for the Life Sciences, ORIEL project (IST-2001-32688). IMGT® is currently supported by the CNRS, the Ministère de l'Education Nationale, de l'Enseignement Supérieur et de la Recherche (MENESR) (Université Montpellier II Plan Pluri-Formation, Institut Universitaire de France, ACI-IMPBIO IMP82-2004), the EU ImmunoGrid (IST-028069) programme, GIS AGENAE, Réseau National des Génopoles and the Région Languedoc-Roussillon BIOSTIC-LR2004.

References

1. Lefranc, M.-P. and Lefranc, G. (2001) *The Immunoglobulin FactsBook*. Academic Press, London, UK, ISBN:012441351X, 458 pages.
2. Lefranc, M.-P. and Lefranc, G. (2001) *The T cell Receptor FactsBook*. Academic Press, London, UK, ISBN:0124413528, 398 pages.
3. Lefranc, M.-P., Giudicelli, V., Kaas, Q., Duprat, E., Jabado-Michaloud, J., Scaviner, D., Ginestoux, C., Clément, O., Chaume, D., and Lefranc G. (2005) IMGT, the International ImMunoGeneTics information system. *Nucleic Acids Res.* **33**, D593–D597.
4. Lefranc, M.-P., Clément, O., Kaas, Q., Duprat, E., Chastellan, P., Coelho, I., Combres, K., Ginestoux, C., Giudicelli, V., Chaume, D., and Lefranc, G. (2005) IMGT-Choreography for immunogenetics and immunoinformatics. Epub *In Silico Biol.* (Reference for Epub 5 0006) http://www.bioinfo.de/isb/2004/05/0006/24 December 2004 *In Silico Biol.* **5**, pp. 45–60.
5. Giudicelli, V. and Lefranc, M.-P. (1999) Ontology for immunogenetics: the IMGT-ONTOLOGY. *Bioinformatics* **12**, 1047–1054.
6. Chaume, D., Giudicelli, V., and Lefranc, M.-P. (2001) IMGT-ML a language for IMGT-ONTOLOGY and IMGT/LIGM-DB data. In: *CORBA and XML: Towards a Bioinformatics Integrated Network Environment, Proceedings of NETTAB 2001, Network tools and Applications in Biology,* May 17–18, Gchoa, Italy, pp. 71–75.
7. Chaume, D., Giudicelli, V., Combres, K., and Lefranc, M.-P. (2003) IMGT-ONTOLOGY and IMGT-ML for Immunogenetics and immunoinformatics. In: *Abstract book of the Sequence Databases and Ontologies Satellite Event,* European Congress in Computational Biology ECCB'2003, September 27–30, Paris, France, pp. 22–23.
8. Letovsky, S.I., Cottingham, R.W., Porter, C.J., and Li, P.W. (1998) GDB: the human genome database. *Nucleic Acids Res.* **26**, 94–99.
9. Wain, H.M., Bruford, E.A., Lovering, R.C., Lush, M.J., Wright, M.W., and Povey, S. (2002) Guidelines for human gene nomenclature. *Genomics* **79**, 464–470.
10. Lefranc, M.-P., Pommié, C., Ruiz, M., Giudicelli, V., Foulquier, E., Truong, L., Thouvenin-Contet, V., and Lefranc, G. (2003) IMGT unique numbering for immunoglobulin and T cell receptor variable domains and Ig superfamily V-like domains. *Dev. Comp. Immunol.* **27**, 55–77.
11. Lefranc, M.-P., Pommié, C., Kaas, Q., Duprat, E., Bosc, N., Guiraudou, D., Jean C., Ruiz M., Da Piedade, I., Rouard, M., Foulquier, E., Thouvenin, V., and Lefranc, G. (2005) IMGT unique numbering for immunoglobulin and T cell receptor constant domains and Ig superfamily C-like domains. *Dev. Comp. Immunol.* **29**, 185–203.
12. Lefranc, M.-P., Duprat, E., Kaas, Q., Tranne, M., Thiriot, A., and Lefranc, G. (2005) IMGT unique numbering for MHC groove G-DOMAIN and MHC superfamily (MhcSF) G-LIKE-DOMAIN. *Dev. Comp. Immunol.* **29**, 917–938.

13. Ruiz, M. and Lefranc, M.-P. (2002) IMGT gene identification and Colliers de Perles of human immunoglobulins with known 3D structures. *Immunogenetics* **53**, 857–883.
14. Cochrane, G., Aldebert, P., Althorpe, N., Andersson, M., Baker, W., Baldwin, A., Bates, K., Bhattacharyya, S., Browne, P., van den Broek, A., Castro, M., Duggan, K., Eberhardt, R., Faruque, N., Gamble, J., Kanz, C., Kulikova, T., Lee, C., Leinonen, R., Lin, Q., Lombard, V., Lopez, R., McHale, M., McWilliam, H., Mukherjee, G., Nardone, F., Garcia Pastor, M.P., Sobhany, S., Stoehr, P., Tzouvara, K., Vaughan, R., Wu, D., Zhu, W., and Apweiler, R.(2006) EMBL nucleotide sequence database: developments in 2005. *Nucleic Acids Res.* **34**, D10–D15.
15. Benson, D.A., Karsch-Mizrachi, I., Lipman, D.J., Ostell, J., and Wheeler, D.L. (2006) GenBank. *Nucleic Acids Res.* **34**, D16–D20.
16. Okubo, K., Sugawara, H., Gojobori, T., and Tateno, Y. (2006) DDBJ in preparation for overview of research activities behind data submissions. *Nucleic Acids Res.* **34**, D6–D9.
17. Eilbeck, K., Lewis, S.E., Mungall, C.J., Yandell, M., Stein, L., Durbin, R., and Ashburner, M. (2005) The sequence ontology: a tool for the unification of genome annotations. *Genome Biol.* **6** (5), R44. Epub 29 Apr 2005.
18. Lefranc, M.-P. (2000) Nomenclature of the human immunoglobulin genes. In: *Current Protocols in Immunology* (Coligan, J.E., Bierer, B.E., Margulies, D.E., Shevach, E.M. and Strober W., eds.), John Wiley and Sons, Hoboken, N.J., A.1P.1–A.1P.37.
19. Lefranc, M.-P. (2000) Nomenclature of the human T cell receptor genes. In: *Current Protocols in Immunology* (Coligan, J.E., Bierer, B.E., Margulies, D.E., Shevach, E.M. and Strober, W., eds.), John Wiley and Sons, Hoboken, N.J., A.1O.1–A.1O.23.
20. Giudicelli, V., Chaume, D., and Lefranc, M.-P. (2005) IMGT/GENE-DB: a comprehensive database for human and mouse immunoglobulin and T cell receptor genes. *Nucleic Acids Res.* **33**, D256–D261.
21. Lefranc, M.-P. (1997) Unique database numbering system for immunogenetic analysis. *Immunol. Today* **18,** 509.
22. Lefranc, M.-P. (1999) The IMGT unique numbering for immunoglobulins, T cell receptors and Ig-like domains. *The Immunologist* **7,** 132–136.
23. Kabat, E.A., Wu, T.T., Perry, H.M., Gottesman, K.S., and Foeller, C. (1991) Sequences of proteins of immunological interest. National Institute of Health Publications, Washington D.C., USA, Publication no. 91-3242.
24. Satow, Y., Cohen, G.H., Padlan, E.A., and Davies, D.R. (1986) Phosphocholine binding immunoglobulin Fab McPC603. *J. Mol. Biol.* **190**, 593–604.
25. Chothia, C. and Lesk, A.M. (1987) Canonical structures for the hypervariable regions of immunoglobulins. *J. Mol. Biol.* **196,** 901–917.

26. Duprat, E., Kaas, Q., Garelle, V., Lefranc, G., and Lefranc, M.-P. (2004) IMGT standardization for alleles and mutations of the V-LIKE-DOMAINs and C-LIKE-DOMAINs of the immunoglobulin superfamily. *Recent Research Developments in Human Genetics* (Pandalai, S.G., ed.) Research Signpost, Trivandrum, Kerala, India, **2**, 111–136.
27. Bertrand, G., Duprat, E., Lefranc, M.-P., Marti, J., and Coste, J. (2004) Characterization of human FCGR3B*02 (HNA-1b, NA2) cDNAs and IMGT standardized description of FCGR3B alleles. *Tissue Antigens* **64**, 119–131.
28. Frigoul, A., and Lefranc M.-P. (2005) MICA: standardized IMGT allele nomenclature, polymorphisms and diseases. *Recent Research Developments in Human Genetics* (Pandalai, S.G., ed.) Research Signpost, Trivandrum, Kerala, India, **3**, 95–145.
29. Baum, T.P., Pasqual, N., Thuderoz, F., Hierle, V., Chaume, D., Lefranc, M.-P., Jouvin-Marche, E., Marche, P.N., and Demongeot, J. (2004) IMGT/GeneInfo: enhancing V(D)J recombination database accessibility. *Nucleic Acids Res.* **32**, D51–D54.
30. Giudicelli, V., Duroux, P., Ginestoux, C., Folch, G., Jabado-Michaloud, J., Chaume, D., and Lefranc, M.-P. (2006) IMGT/LIGM-DB, the IMGT® comprehensive database of immunoglobulin and T cell receptor nucleotide sequences. *Nucleic Acids Res.* **34**, D781–D784.
31. Folch G., Bertrand J., Lemaitre M., and Lefranc M.-P. (2004) IMGT/PRIMER-DB. In: *Database Listing* (Galperin, M.Y., ed.), The Molecular Biology Database Collection: 2004 update. *Nucleic Acids Res.* **32**, D3–D22.
32. Robinson, J., Waller, M.J., Parham, P., de Groot, N., Bontrop, R., Kennedy, L. J., Stoehr, P., and Marsh, S.G. (2003) IMGT/HLA and IMGT/MHC sequence databases for the study of the major histocompatibility complex. *Nucleic Acids Res.* **31**, 311–314.
33. Giudicelli, V., Chaume, D., and Lefranc, M.-P. (2004) IMGT/V-QUEST, an integrated software program for immunoglobulin and T cell receptor V-J and V-D-J rearrangement analysis. *Nucleic Acids Res.* **32**, W435–W440.
34. Yousfi Monod, M., Giudicelli, V., Chaume, D., and Lefranc, M.-P. (2004) IMGT/JunctionAnalysis: the first tool for the analysis of the immunoglobulin and T cell receptor complex V-J and V-D-J JUNCTIONs. *Bioinformatics* **20**, 379–385.
35. Elemento, O., and Lefranc, M.-P. (2003) IMGT/PhyloGene: an on-line tool for comparative analysis of immunoglobulin and T cell receptor genes. *Dev. Comp. Immunol.* **27**, 763–779.
36. Kaas, Q., Ruiz, M., and Lefranc, M.-P. (2004) IMGT/3Dstructure-DB and IMGT/StructuralQuery, a database and a tool for immunoglobulin, T cell receptor and MHC structural data. *Nucleic Acids Res.* **32**, D208–D210.
37. Pommié, C., Sabatier, S., Lefranc, G., and Lefranc, M.-P. (2004) IMGT standardized criteria for statistical analysis of immunoglobulin V-REGION amino acid properties. *J. Mol. Recognit.* **17**, 17–32.

38. Lefranc, M.-P. (2006) Web sites of interest to immunologists. *In*: *Current Protocols in Immunology* (Coligan, J.E., Bierer, B.E., Margulies, D.E., Shevach, E.M., and Strober, W., eds.), John Wiley and Sons, Hoboken N.J. pp. A.1J.1–A.1J.74.
39. Giudicelli V., Chaume D., Jabado-Michaloud J., and Lefranc M.-P. (2005) Immunogenetics sequence annotation: the strategy of IMGT based on IMGT-ONTOLOGY. *Stud. Health Technol. Inform.* **116**, 3–8.
40. Lefranc, M.-P. (2004) IMGT, The International ImMunoGeneTics Information System®, http://imgt.cines.fr. In: *Antibody engineering: Methods and Protocols* (Lo, B.K.C., ed.), Humana, Totowa, N.J., *Methods Mol. Biol.* **248**, 27–49.
41. Berman, H.M., Westbrook, J., Feng, Z., Gilliland, G., Bhat, T.N., Weissig, H., Shindyalov, I.N., and Bourne, P.E. (2000) The Protein Data Bank. *Nucleic Acids Res.* **28**, 235–242.
42. Duprat, E., Lefranc, M.-P., and Gascuel, O. (2006) A simple method to predict protein binding from aligned sequences – application to MHC superfamily and beta2-microglobulin. *Bioinformatics*, **22**, 453–459.
43. Kaas, Q., and Lefranc, M.-P. (2005) T cell receptor/peptide/MHC molecular characterization and standardized pMHC contact sites in IMGT/3Dstructure-DB. Epub *In Silico Biol.*, (5 0046 refers to Epub) 0046 20 Oct 2005 In Silico Biol. **5**, 505–528.

3

The IMGT/HLA Database

James Robinson and Steven G. E. Marsh

Summary

The human leukocyte antigen (HLA) complex is located within the 6p21.3 region on the short arm of human chromosome 6 and contains more than 220 genes of diverse function. Many of the genes encode proteins of the immune system and include many highly polymorphic HLA genes. The naming of new HLA genes and allele sequences and their quality control is the responsibility of the WHO Nomenclature Committee for Factors of the HLA System. The IMGT/HLA Database acts as the repository for these sequences and is recognized as the primary source of up-to-date and accurate HLA sequences. The IMGT/HLA website provides a number of tools for accessing the database: these include allele reports, sequence alignments, and sequence similarity searches. The website is updated every 3 months with all the new and confirmatory sequences submitted to the WHO Nomenclature Committee. Submission of HLA sequences to the committee is possible through the tools provided by the IMGT/HLA Database.

Key Words: HLA; MHC; nomenclature; sequences; alleles; database

1. Introduction

1.1. Background

The IMGT/HLA Database is a specialist database for the allelic sequences of the genes in the human leukocyte antigen (HLA) system, also known as the human major histocompatibility complex (MHC). This complex of over 4 Mb is located within the 6p21.3 region of the short arm of human chromosome 6 and contains in excess of 220 genes *(1)*. Genes included in the database are those HLA genes involved in antigen presentation to T cells or are non-functional genes related to them. The core of the HLA system consists of 21 highly polymorphic HLA genes. These influence the outcome of cell and organ

From: *Methods in Molecular Biology, vol. 409: Immunoinformatics: Predicting Immunogenicity In Silico*
Edited by: D. R. Flower © Humana Press Inc., Totowa, NJ

transplants and mediate the host response to infectious disease and are associated with susceptibility to a wide range of chronic, non-infectious diseases *(2,3)*. Nucleotide sequences for more than 2,100 different alleles of these genes have been determined. HLA genes are divided into class I (HLA-A, HLA-B, and HLA-C) or class II (HLA-DR, HLA-DQ, and HLA-DP) genes depending on the structure and function of their protein products. The naming of new HLA genes and allele sequences and their quality control is the responsibility of the WHO Nomenclature Committee for Factors of the HLA System *(4,5)*. The IMGT/HLA Database acts as the repository for these sequences and is recognized as the primary source of up-to-date and accurate HLA sequences.

1.2. A Historical Perspective

The sequencing of HLA alleles first began in the late 1970s using protein sequencing techniques predominately to determine HLA class I alleles. The first complete HLA class I allele sequence, B7.2, now known as B*070201, was published in 1979 *(6)*. It was a few years later in 1982 that the first HLA class II allele, a DRA allele, was determined by more conventional cDNA sequencing *(7)*. In 1987, the first HLA DNA sequences or alleles were named by the WHO Nomenclature Committee for Factors of the HLA System *(8)*. Previous to this only the serologically defined antigens had been given official designations. At this time, some 12 class I alleles were named A*0201-0204, B*0701-0702, and B*2701-2706, together with 9 class II alleles, DRB1*0401-0405, DRB3*0101, 0201, 0301, and DRB4*0101. Although many other alleles had already been defined, they were not considered by the committee at that time. Two years later, in 1989, the Nomenclature Committee met for the first time outside the auspices of an International Histocompatibility Workshop to assign official allele names to the large number of HLA allele sequences that were by that time being published regularly. A total of 56 novel class I alleles and 78 class II alleles were named *(9)*. It soon became apparent that the analysis and assigning of official names to alleles could not wait for either periodic histocompatibility workshops of even annual Nomenclature Committee meetings and so began the process of assessing newly defined HLA allele sequences. This work was carried out by Julia Bodmer and Steven Marsh at the Imperial Cancer Research Fund (ICRF) in collaboration with Peter Parham at Stanford University. Out of the need to record and manage the HLA sequence data being submitted to the Nomenclature Committee came the first incarnation of an HLA Sequence Databank (HLA-DB) *(10)*. Periodically, HLA class I *(11–14)* and class II *(15–20)* sequence alignments were published in various journals, and by 1995, the numbers of new alleles being reported warranted the publication of monthly nomenclature updates *(21)*, something which continues to this day (Fig. 1). By 1995, the expansion of the Internet and the introduction of the World Wide Web (WWW) saw the first distribution of the HLA sequence alignments from the web pages of the Tissue Antigen Laboratory at the ICRF. This work was transferred to the Anthony Nolan Research Institute (ANRI) in 1996 where it continues. The IMGT/HLA Database *(22–25)* began in 1997 as part of a European collaboration involving the ICRF, ANRI, and the European Bioinformatics Institute (EBI) who maintain the European Molecular Biology Laboratory's nucleotide sequence database (EMBL). The work was initially funded by grants from the European Union, BIOMED1 (BIOCT930038), and

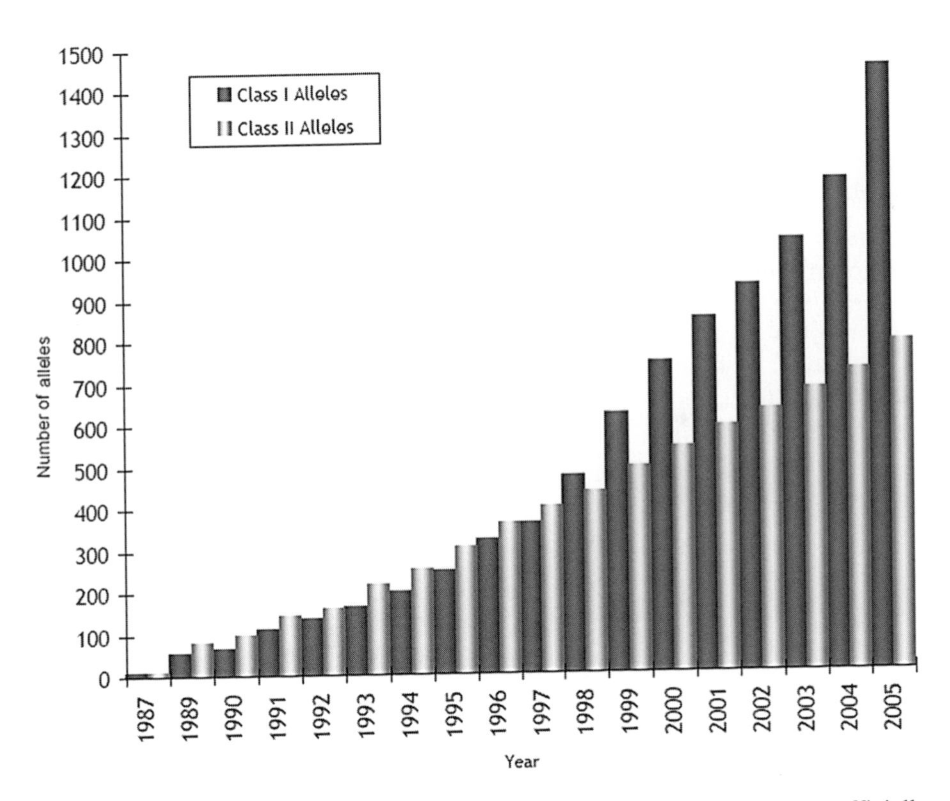

Fig. 1. Numbers of human leukocyte antigen (HLA) class I and II alleles officially recognized by the WHO Nomenclature Committee for Factors of the HLA System 1987–2005. The numbers of HLA class I and II alleles that have been officially named by the WHO Nomenclature Committee for Factors of the HLA System over the period November 1987 to December 2005.

BIOTECH2 (BIO4CT960037), awarded to the ICRF as part of the IMGT database project *(26)*. The IMGT/HLA Database was first released in 1998; the database combines the sequence data and information previously provided to the WHO Nomenclature Committee for Factors of the HLA System and the additional data found in the original EMBL/GenBank/DNA Database of Japan (DDBJ) entries.

1.3. Generalist Databanks

It should be noted that all sequences within the IMGT/HLA Database should also be available from the more general nucleotide sequence databases, EMBL *(27)*, GenBank *(28)*, and DDBJ *(29)*. The generalist databanks are not HLA specific, rather large international data repositories of gene sequences for all organisms. These three databases form an international collaboration and exchange sequences daily so that each contains identical data. Most published sequences can be found in these databases. The advantage of using the large generalist sequence databases is the large number of sequences available, covering a wide range of data relating to HLA. The main problem when accessing HLA sequences from these databases lies in the definition of the sequence. Upon submission, a number of steps are taken to ensure that the sequence is accurately described and up-to-date. The author then assigns keywords and description, and these can vary. Despite the work of the WHO Nomenclature Committee for Factors of the HLA System in monitoring HLA allele designations and maintaining the sequences, they have no control of how sequences are defined in these generalist databases. Readers should therefore be aware that entries may be incorrectly named, contain unofficial designations, or contain sequencing errors.

2. The IMGT/HLA Database

2.1. IMGT/HLA Organization and Content

The database contains entries for all HLA alleles officially named by the WHO Nomenclature Committee for Factors of the HLA System. These entries are derived from expertly annotated copies of the original EMBL/GenBank/DDBJ entries. In order to store all the component entries as well as the nomenclature information and other data, the IMGT/HLA Database utilizes a relational database model system. The database has evolved from an original flat file structure, through an intermediate version held in a Filemaker Pro® database (Filemaker, Inc., USA), to a much more sophisticated structure as required to store the increasingly complex information. The relational database uses the ORACLE® database management system to provide a database server. Access to the relational database is restricted to the HLA Informatics Group at the ANRI only. All user queries access a smaller publicly available copy of the database via a web-based interface. This allows the database to store confidential information, restrict data access, and improve data security. All information is released into the public copy on regular 3 monthly updates. Users of the database who submit new information are encouraged to make their data publicly available as soon as possible and not to maintain long holding times on the data.

2.2. The Virtual Sequence

The IMGT/HLA Database can contain a number of different entries for any single allele. In order to store all the information for each allele in a single entry available to the user, the virtual sequence concept was developed (Fig. 2). In the database, a virtual sequence represents the combination of all the EMBL/GenBank/DDBJ entries for a single allele combining to form a single expertly annotated entry. These component entries are submitted to the database in the form of IMGT/HLA submissions either by the original author or by our curators when sequences of interest have been identified by data mining. The submissions are annotated by our curators to remove known sequence anomalies and incorrect annotations. The nucleotide sequences are then aligned using the ClustalX program ***(30)***. This provides a multiple sequence alignment of the component sequence using a recognized alignment program. The alignment

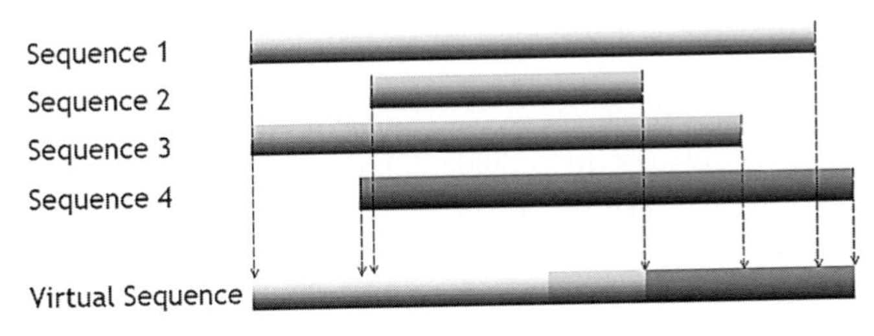

Fig. 2. Construction of a virtual sequence. A virtual sequence is constructed for each allele, representing the longest available sequence for each allele. This is generated by combining the component entries from European Molecular Biology Laboratory's nucleotide sequence database (EMBL)/GenBank/DNA Database of Japan (DDBJ) to create a single entry. The diagram shows four component entries aligned and merged to form a virtual sequence.

produced is then reduced to the single longest contiguous sequence. This nucleotide sequence can be used for the allele sequence; this virtual sequence is then aligned against the reference sequence for that locus. Previous alignments sometimes used a consensus sequence for alignments; however, following guidelines issued by the Human Gene Nomenclature Committee ***(31)***, the IMGT/HLA Database instead uses a reference allele sequence at each locus. Insertions, indicated by periods (.), are added to the virtual sequence to ensure alignment to the reference sequence.

To distinguish the IMGT/HLA entries from the component EMBL entries, all alleles have a unique accession number. The EMBL accession numbers are not used as primary identifiers in IMGT/HLA because many alleles are derived from multiple EMBL sequence entries. The accession numbers follow the format HLA00000, where the "00000" represents a numerical code. Both their unique name and accession number can, therefore, identify all alleles.

2.3. Database Distribution

The first public release of the IMGT/HLA Database was made on the December 16, 1998 and was included on the EBI web server as part of the IMGT project. The main access point for the user is the World Wide Web, which allows the users to employ a number of search tools and other facilities to retrieve, manipulate, and analyze HLA data. This is all done through the custom-written Common Gateway Interface (CGI) scripts available at the IMGT/HLA website. Other access points for the user include the EBI Sequence Retrieval System (SRS) search engine, EBI Basic Local Alignment Search Tool (BLAST) and Fast-All (FASTA) search tools, the EMBL CD-ROM releases, and the EBI public File Transfer Protocol (FTP) directory. A copy of some data stored in the IMGT/HLA Database is also provided in a text-based format at the HLA Informatics Group web pages of the ANRI website. The database is updated every 3 months to include all the publicly available sequences officially named by the WHO Nomenclature Committee since the last release of the database. With each release, all the tools are updated to include the new sequences, and information on all the new and modified sequences is reported. The previous release is archived for reference.

2.4. The IMGT/HLA Website

The website can be split into three main areas. First, information and help pages that provide background on the database, provide in-depth help on the tools and data available and documentation of the IMGT/HLA file formats. Secondly, the website includes the tools designed specifically for the IMGT/HLA Database. These allow the user to perform sequence analysis and retrieval. Third, the final pages are links to sequence-analysis tools at the EBI, including SRS, BLAST, and FASTA. The tools available from the website will be discussed in detail later. The core tools allow the users to perform sequence alignments, allele queries, and sequence searches. This is done by combining the custom-built tools, with existing tools already available from the EBI such as BLAST, FASTA, and SRS. Access to all of the pages and tools is via the IMGT/HLA Database homepage (see Fig. 3).

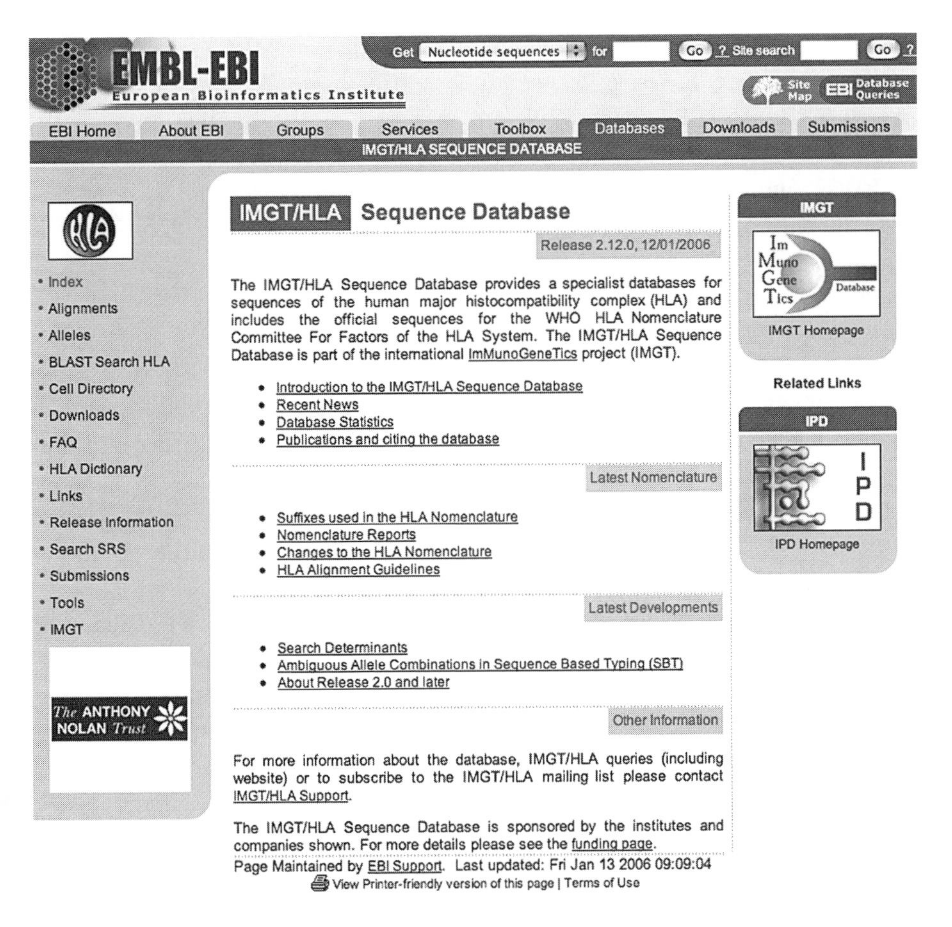

Fig. 3. The IMGT/HLA Database Homepage. The database homepage is the main access point to the online tools provided by the database.

2.4.1. IMGT/HLA tools

2.4.1.1. Allele Search Tools

The most used tool on the IMGT/HLA website is the "Allele Query Tool." This is designed to allow the user to retrieve the full sequence and information pertaining to any officially named allele. The tool is available directly from the homepage or from the tools section of the website. The tool provides a simple-to-use interface for retrieving allele information. Similar searches can be performed using the SRS interface (discussed later), but this requires more knowledge of the tools, file structure, and data. The "Allele Query Tool" requires only a search term in order to retrieve a report on any allele in the database. This can be either the allele name or a single EMBL/GenBank/DDBJ accession number. The search tool allows different resolutions of the alleles numeric code to allow flexibility in searches. For example, the A*0101 designation could refer to one of four alleles, A*01010101, A*01010102N, A*010102, and A*010103, and entering "A*0101" will retrieve all these alleles. Working back from this, entering just "A*01" would retrieve the 19 alleles with the "A*01" designation. This facility can be exploited to gain a list of all alleles at any locus by entering a "locus*," for example, "B*" for HLA-B, in the box to retrieve the full list. Other shortcuts include the retrieval of null alleles for a specific locus; by entering "locus*N" into the box you can retrieve a list of all non-expressed or null alleles. The output of the search lists all the allele names that correspond to the search term provided; these then provide a hypertext link to the full entry for each allele. Searching by accession number is exactly the same as for allele designations; by entering a recognized accession number (or part of) into the tool, the allele information can be retrieved.

The output provided for each allele includes the official allele designation, previously used designations, and the unique IMGT/HLA accession number that is a link to the IMGT/HLA flat file (Fig. 4). Other information provided includes the date that the allele was named and its current status (as some allele designations have been deleted) and information on the individual or cell line from which the sequence was derived. Links to all component EMBL/GenBank/DDBJ entries are also included. Any published references are also included with, wherever possible, a link to the PubMed entry for that citation. PubMed provides an online version of the abstract as well as links to other citations by the author and to similar papers. The final section of the output details the official nucleotide and protein sequence.

The IMGT/HLA Database is also available in a flat file format, which is standard format for text files. These flat files follow the EMBL flat file format and provide a standard release format. The flat files utilize the unique accession number and assign a standardized description and keywords to all entries. This accession number is used in the EBI tools to link back to the flat file entry. The flat files also contain the first release of the IMGT/HLA features. The sequence features currently used are a small subset of the standard set used by EMBL, but as the database continues to develop, further features may be added. The initial feature qualifiers cover the source (cell of origin), the coding sequence (cds), exon boundaries, and the protein translation. Other information provided by the flat file replicates the allele query tool output. The only additional information included is a list of all the component sequence entries and other

```
ID   HLA00001   standard; DNA; HUM; 1098 BP.
XX
AC   HLA00001;
XX
DT   01-AUG-1989 (Rel. 2.12.0, Created)
DT   04-FEB-2003 (Rel. 2.12.0, Last Sequence Update, Version 1)
DT   12-FEB-2008 (Rel. 2.12.0, Last Annotation Update, Version 8)
XX
DE   HLA-A*01010101 Human MHC Class I sequence
XX
KW   Human MHC; HLA; Class I; HLA-A; Allele; A*01010101;
XX
OS   Homo Sapiens (human)
OC   Eukaryota; Metazoa; Chordata; Vertebrata; Mammalia; Eutheria; Primates;
OC   Catarrhini; Hominidae; Homo.
XX
RN   [1]
RP   1-1098
RX   PUBMED; 2251127.
RA   Girdlestone J;
RT   "Nucleotide sequence of an HLA-A1 gene";
RL   Nucleic Acids Res 18:6701-6701(1990).
XX
RN   [2]
RP   1-1098
RX   PUBMED; 3375250.
RA   Parham P, Lomen CE, Lawlor DA, Ways JP, Holmes N, Coppin HL, Salter RD,
RA   Wan AM, Ennis PD;
RT   "Nature of polymorphism in HLA-A, -B, and -C molecules";
RL   Proc Natl Acad Sci U S A 85:4005-9(1988).
XX
RN   [3]
RP   1-1098
RX   PUBMED; 9249617.
RA   Laforet M, Froelich N, Parissiadis A, Pfeiffer B, Schell A, Faller B,
RA   Woehl-Jaegle ML, Cazenave JP, Tongio MM;
RT   "A nucleotide insertion in exon 4 is responsible for the absence of
RT   expression of an HLA-A*01 allele";
RL   Tissue Antigens 50:347-50(1997).
XX
DR   EMBL; AJ278305; AJ278305.1.
DR   EMBL; X55710; X55710.1.
DR   EMBL; Z93949; Z93949.1.
DR   EMBL; X55710; X55710.1.
DR   EMBL; M24043; M24043.1.
DR   EMBL; Z93949; Z93949.1.
DR   EMBL; AJ278305; AJ278305.1.
DR   GDB; 119310;
DR   GOA; P30443;
DR   SWISS-PROT; P30443;
XX
CC   The nucleotide sequence provided is a CDS sequence, constructed from a
CC   number of different sequence entries. As a result the cross references to
CC   the original sequence may differ. The sequence below is the official
CC   sequence for HLA-A*01010101 approved by the WHO Nomenclature Committee
CC   for Factors of the HLA system.
XX
FH   Key             Location/Qualifiers
FH
FT   source          1..1098
FT                   /organism="Homo sapiens"
FT                   /ethnic="Caucasoid"
FT                   /cell_line="LCL721"
FT                   /cell_line="MOLT-4"
FT                   /cell_line="PP"
FT   CDS             1..1098
FT                   /codon_start=1
FT                   /product="MHC Class I HLA A*01010101 sequence"
FT                   /translation=MAVMAPRTLLLLLSGALALTQTWAGSHSMRYFFTSVSRPGRGEPRF
FT                   IAVGYVDDTQFVRFDSDAASQKMEPRAPWIEQEGPEYWDQETRNMKAHSQTDRANLGTL
FT                   RGYYNQSEDGSHTIQIMYGCDVGPDGRFLRGYRQDAYDGKDYIALNEDLRSWTAADMAA
FT                   QITKRKWEAVHAAEQRRVYLEGRCVDGLRRYLENGKETLQRTDPPKTHMTHHPISDHEA
FT                   TLRCWALGFYPAEITLTWQRDGEDQTQDTELVETRPAGDGTFQKWAAVVVPSGEEQRYT
FT                   CHVQHEGLPKPLTLRWELSSQPTIPIVGIIAGLVLLGAVITGAVVAAVMWRRKSSDRKG
FT                   GSYTQAASSDSAQGSDVSLTACKV
FT   exon            1..73
FT                   /number="1"
FT   exon            74..343
FT                   /number="2"
FT   exon            344..619
FT                   /number="3"
FT   exon            620..895
FT                   /number="4"
FT   exon            896..1012
FT                   /number="5"
FT   exon            1013..1045
FT                   /number="6"
FT   exon            1046..1093
FT                   /number="7"
FT   exon            1094..1098
FT                   /number="8"
SQ   Sequence 1098 BP; 221 A; 332 C; 368 G; 177 T; 0 other;
     atggccgtca tggcgccccg aaccctcctc ctgctactct cgggggccct ggccctgacc        60
     cagacctggg cgggctccca ctccatgagg tatttcttca catccgtgtc ccggcccggc       120
     cgcggggagc cccgcttcat cgccgtgggc tacgtggacg acacgcagtt cgtgcggttc       180
     gacagcgacg ccgcgagcca gaagatggag ccgcgggcgc cgtggataga gcaggagggg       240
     ccggagtatt gggaccagga gacacggaat atgaaggccc actcacagac tgaccgagcg       300
     aacctgggga ccctgcgcgg ctactacaac cagagcgagg acggttctca cactatccag       360
     ataatgtatg gctgcgacgt ggggccggac gggcgcttcc tccgcgggta ccggcaggac       420
     gcctacgacg gcaaggatta catcgccctg aacgaggacc tgcgctcttg gaccgcggcg       480
     gacatggcag ctcagatcac caagcgcaag tgggaggcgg tccatgcggc ggagcagcgg       540
     agagtctacc tggagggccg gtgcgtggac gggctccgca gatacctgga gaacgggaag       600
     gagacgctgc agcgcacgga cccccccaag acacatatga cccaccaccc catctctgac       660
     catgaggcca ccctgaggtg ctgggccctg ggcttctacc ctgcggagat cacactgacc       720
     tggcagcggg atggggagga ccagacccag gacacggagc tcgtggagac caggcctgca       780
     ggggatggaa ccttccagaa gtgggcggct gtggtggtgc cttctggaga ggagcagaga       840
     tacacctgcc atgtgcagca tgagggtctg cccaagcccc tcaccctgag atgggagctg       900
     tcttcccagc ccaccatccc catcgtgggc atcattgctg gcctggttct ccttggagct       960
     gtgatcactg gagctgtggt cgctgccgtg atgtggagga ggaagagctc agatagaaaa      1020
     ggagggagtt acactcaggc tgcaagcagt gacagtgccc agggctctga tgtgtctctc      1080
     acagcttgta aagtgtga                                                    1098
//
```

Fig. 4. An IMGT/HLA flat file entry. An IMGT/HLA flat file is shown that can be accessed through the EBI's SRS browser. The flat file is a text-based file that identifies different sections by the line headings, for example, AC = accession number and lines beginning with "R" relate to references. The format is consistent with that used for the European Molecular Biology Laboratory's nucleotide sequence database (EMBL). Lines beginning with "FT" denote the sequence features; in these examples, they include the source cell and exon information.

cross-references to sequence databases such as SWISSPROT, TREMBL, and PDB. These links are to the original entries, and so, the files retrieved may differ from the IMGT/HLA entry due to the annotation procedure. The information in the IMGT/HLA entry should therefore be taken as the definitive source in cases of disagreement.

The SRS tool ***(32)*** is an advanced search tool for interrogating flat files, and the IMGT/HLA flat files are included in the EBI SRS libraries. These interrogations can range from very simple queries, such as searching for an accession number or keyword, through to more complex searches such as for authors of papers describing an allele. In order to use SRS, some familiarity with the flat file format is required. This tool allows the user to search on any of the sequence features, the accession numbers, keywords assigned to the sequences, or the description, and is probably the best method of retrieving HLA sequences from a general nucleotide database. Once users are accustomed to the way data are presented, they can quickly build up very complex queries. Another advantage of the SRS tool is that it can also be used to launch other applications, for example, BLAST, Clustal, on any query results. Therefore, it is possible to retrieve all the flat files for a certain subset of data and automatically load these into the Clustal alignment tool for example. The SRS tool also allows the users to customize the output of searches, so that they can quickly see how relevant entries are to your search criteria. Tutorials for the SRS search engine are available from the EBI and SRS website. SRS can be found at the EBI website and can be used to search a number of different databases.

2.4.1.2. Sequence Search Tools

The first type of search that many people do is to look for a particular allele by name or by a certain characteristic such as the cell or author. The main alternative to this is to search on the actual sequence and not on the name or keywords. Sequence similarity searches look for sequence similarities in a query sequence against a reference database of known sequences. The accuracy of these matches is based on a number of similarity measures that in general retrieve identical or highly similar sequences. The IMGT/HLA Database is included as a library for searching within the EBI's Similarity & Homology toolset. These include well-known tools such as BLAST and FASTA. BLAST is used to compare a sequence with those contained in nucleotide and protein databases by aligning the novel sequence with previously characterized genes. The emphasis of this tool is to find regions of sequence similarity, which will yield functional and evolutionary clues about the structure and function of this novel sequence. FASTA is used for a fast protein comparison or a fast nucleotide comparison. This program achieves a high level of sensitivity for similarity searching at high speed. Both these tools can be used for searching libraries of HLA nucleotide and protein sequences.

2.4.1.3. IMGT/HLA SEQUENCE ALIGNMENTS

The other main use of the IMGT/HLA Database is viewing multiple sequence alignments. HLA allele sequences may differ from each other by as little as a single nucleotide, over a genomic sequence of 3,300 bases. These alignments allow a visual interpretation of sequence similarity, so that polymorphic positions can easily be identified, and motifs found in multiple alleles are easily identified (Fig. 5). The representation of HLA sequences in this manner can be useful when designing reagents for HLA typing, such as primers or oligonucleotide probes. The sequence alignments are available via a link from the main IMGT/HLA homepage; they can also be found using the tools page of the website. The interface provided lets the user define a number of key variables for the alignments, before producing an online output, which can be printed or downloaded. The first step in any alignment is to select the locus of interest. The tool provides a drop-down list of all loci available and also includes some additional options such as class I (all HLA-A, HLA-B, and HLA-C alleles) and different HLA-DRB gene combinations. The selection of a locus automatically updates the list of features that can be aligned, as well as the default reference sequence used for the alignment. The types of feature available for alignment are the nucleotide cds and individual exons, the genomic sequences and individual introns (where available); and the signal peptide, mature protein and full length protein sequence. In addition, there are some commonly requested regions like a single alignment of both exons 2 and 3 or exons 2, 3, and 4, which have been included to aid in the analysis of sequence-based typing (SBT) results. Genomic sequence is currently available only for the HLA class I loci; however, work is underway on class II genomic sequences for inclusion into the alignment tool. The alignment tool options also allow the user to display a subset of alleles of a particular locus, omit alleles unsequenced for a particular region, and align against a particular reference or consensus sequence. The alignment tool uses standard formatting conventions for the display of sequence alignments. The alignment tool does not perform a sequence alignment each time it is used, but it extracts pre-aligned sequences, allowing for faster access.

The alignments adhere to a number of conventions for displaying evolutionary events and numbering. The numbering of the alignments is based on the sequence of the reference allele. For a nucleotide sequence, the A of the

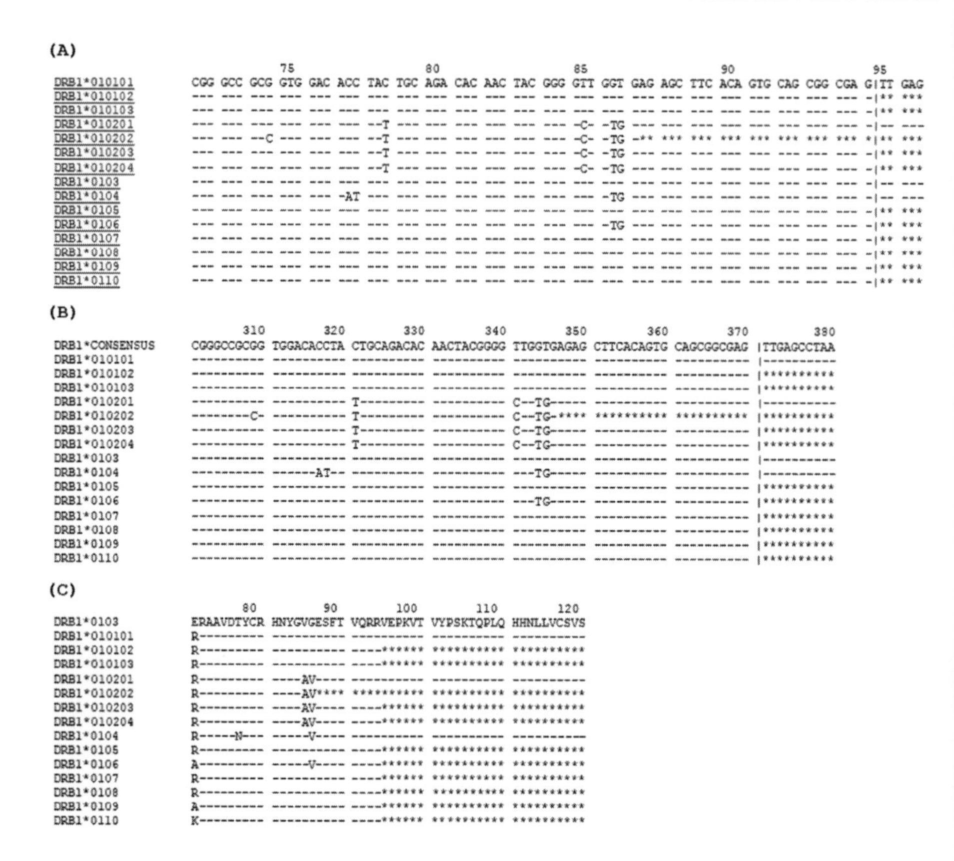

Fig. 5. Alignment formats available from the IMGT/HLA Database. The examples shown are based on alignment **A** that shows 15 DRB1*01 alleles. In these alignments, a dash (-) indicates identity to the reference sequence, and an asterisk (*) denotes an unsequenced base. In alignment **A**, the allele names are underlined, as they are hyperlinks to the allele's entry, and the nucleotide sequence is displayed in codons. Alignment **B** shows how an alternative reference sequence can be used; here for example, we have used a DRB1 consensus sequence. The sequence is displayed in blocks of ten nucleotides. Alignment **C** represents a translation of the nucleotide sequence to produce a protein sequence alignment. In this final example, the DRB1*0103 allele has been used as the reference sequence.

initiation Methionine codon is denoted nucleotide +1 and the nucleotide 5′ to +1 is numbered −1. There is no nucleotide zero (0). All numbering is based on the ATG of the reference sequence. If a nucleotide sequence is displayed in codons, then the protein numbering is applied.

For amino acid-based alignments, the first codon of the mature protein, after cleavage of the signal sequence, is labeled codon 1, and the codon 5′ to this is numbered −1. In all sequences, the following conventions are used. Where identity to the reference sequence is present, the base will be displayed as a hyphen (-). Nonidentity to the reference sequence is shown by displaying the appropriate base at that position. Where an insertion or deletion has occurred, this is represented by a period (.). If the sequence is unknown at any point in the alignment, this is represented by an asterisk (*). In protein alignments for null alleles, the "Stop" codons are represented by an X. In protein alignments, sequence following the termination codon is not marked and appears blank.

The flexibility of the new alignment tool means that unlike previous alignments you can now display a small subset of sequences against an allele of your choice, using a number of display options. The previous text alignments are still requested and as a result are available from the ANRI website and in a zipped file in the FTP directory.

2.4.1.4. Cell Search Tools

The IMGT/HLA Database contains an entry for the source material of each allele submitted to the database. This source material is normally in the form of a cell line or DNA from which each HLA allele in the database was isolated and characterized. The information contained within this data set can be used independently from the HLA allele data, as well as through more obvious links from the allele entries. In order to promote use of this data set and to provide a resource listing the HLA types of known cell lines, a specialized source tool has been developed. The cell query tool is used to interrogate this accompanying cell database. The interface can be used without prior knowledge of which alleles are linked to the cells or from the allele reports. The interface allows the user to search on known cell fields, in a similar manner to SRS and flat files. The cells all have a primary name and accession number that are unique within the database. Searching these fields can easily retrieve cells. As some cells are sequenced by different groups, or certain names are repeated, the database also contains a list of aliases for each cell. These aliases are automatically searched whenever the cell name field is queried. Other cell fields that are available for searching include the HLA typing, serology, ethnic origin, geographical location, sex, consanguineous status, workshop numbers, and availability from a number of cell repositories. The search tool provides an interface that can be used constructing both simple queries, for example, a cell name, and complex queries that is, "A*0101" positive "Caucasoid" "males."

The output page for the cell search tool lists all the relevant hits. These are displayed with their primary name, aliases, and accession number, which links to the individual entry. The individual entry for each cell details a large amount of information on the cell including any typing and serological information. The entry also provides a list of all alleles for which a cell has been sequenced. This may vary from the typing profile originally submitted by the author due to a number of reasons: incomplete typing, unknown allele designations, or additional information from other groups. There are no "virtual" cells as with sequence entries, but if a number of groups have sequenced the same cell then the information can be pooled into a single entry. The cell database currently contains around 3,540 cells.

2.4.1.5. SBT Ambiguities

The IMGT/HLA Database contains the data to provide a number of tools to support the work of tissue typing laboratories. The use of SBT as a method for defining the HLA type is well documented *(33)*. Most SBT strategies currently employed use the exon 2 and exon 3 sequences for HLA class I analysis and exon 2 alone for HLA class II analysis. Due to the heterozygous nature of the SBT analysis, the combinations of many pairs of alleles may give an ambiguous typing result. In order to aid in analysis, the database makes available a list of all alleles that are identical over exons 2 and 3 for HLA class I and exon 2 for HLA class II. In addition, all ambiguous results obtained when using the alleles included in this release are also included. This information is presented on the website in a portable document format (PDF) or Excel spreadsheets. This document can then be easily downloaded or printed out for reference. These data are updated with each release of the database to ensure that the combinations are up-to-date.

2.4.1.6. HLA Dictionary and Search Determinants

The IMGT/HLA Database allows investigators to retrieve information from the HLA Dictionary *(34)*. This dictionary presents the serological equivalents of HLA-A, HLA-B, HLA-C, HLA-DRB1, HLA-DRB3, HLA-DRB4, HLA-DRB5, and HLA-DQB1 alleles. The data summarizes equivalents obtained by the WHO Nomenclature Committee for Factors of the HLA System, the International Cell Exchange (UCLA), the National Marrow Donor Program (NMDP), the 13th International Histocompatibilty Workshop, recent publications, and individual laboratories.

To complement the HLA Dictionary, details of the search determinants used by different bone marrow registries are also stored in the database. Registries and cord blood banks around the world collect and store the HLA types of volunteers in order to identify matched unrelated donors for patients requiring hematopoietic stem cell transplantation ***(35)***. The HLA assignments of volunteers are received in many formats depending on whether the types were obtained by serology or by DNA-based methods and by the panel of reagents used in the assay, which determines the resolution of the test results (high vs. intermediate vs. low resolution). Registries must take these diverse assignments and identify which of them is most likely to actually match the HLA assignments of a searching patient. This comparison is usually facilitated by the conversion of assignments to "search determinants" before matching algorithms are performed.

2.4.1.7. Sequence Submissions

In addition to providing HLA sequences for retrieval, the IMGT/HLA website also provides the tools for submitting both new and confirmatory sequences to the WHO HLA Nomenclature Committee. This is now the only accepted method for submitting new sequences. Submissions are processed, which incorporates automated analysis and annotation of the sequence, and then given an official name, before being loaded into the IMGT/HLA Database and included in the monthly nomenclature reports ***(21)***. The submission tool can be used for both new and confirmatory sequences and is capable of holding confidential entries until a set time, thus allowing alleles to be named before publication. The submission of new HLA sequences to the IMGT/HLA Database does not replace the submission of these sequences to EMBL/GenBank/DDBJ, as the submission criteria state that the sequences must also have been submitted to these databases.

2.4.2. Other tools

The IMGT/HLA Database provides access to a number of other tools and facilities. These include a public FTP server. The FTP server contains the HLA sequences in a number of common formats that can then be downloaded directly. The FTP directory includes all the IMGT/HLA flat files and full documentation to accompany these. All nucleotide and protein sequences are provided in the three common sequence file formats: FASTA, PIR, and MSF. These file formats are accepted by most sequence analysis programs, for example, Clustal, GeneDoc, and GCG.

2.4.3. Documentation

The IMGT/HLA Database also includes documentation to aid and support users with the tools and concepts behind the database. Each tool is accompanied by a help page that details all the options available and breaks down the output into individual fields to aid interpretation. Documentation is also provided for some of the external tools such as BLAST to aid new users.

The second section of the documentation covers the quarterly releases. Each release is accompanied by a number of text documents. These update the user with the contents of each release and provide explanations of file formats. Also included is a version report. This file details all new alleles included in each release. Details of any corrections or extensions to sequences since the last release are also included. This is particularly important as it ensures that users are notified of any changes in the alignments. Changes are made when additional sequence for partially sequenced alleles is submitted, or if an error in a published sequence is reported. The version report also includes the names of any deleted alleles. The deletion of an allele is a rare event but can happen when a sequence is shown to be in error.

3. Conclusions

IMGT/HLA Database provides a centralized resource for everybody interested, clinically or scientifically, in the HLA system. The database and accompanying tools allow the study of all HLA alleles from a single site on the World Wide Web. It should aid in the management and continual expansion of HLA nomenclature, providing an ongoing resource for the WHO Nomenclature Committee.

Acknowledgments

We would like to thank Matthew Waller and Sylvie Fail for their curation of the database. We would like to acknowledge the support of the following organizations: Abbott Laboratories Inc., the American Society for Histocompatibility and Immunogenetics (ASHI), the Anthony Nolan Bone Marrow Trust (ANBMT), Biotest, Dynal, the European Federation for Immunogenetics (EFI), Genovision, LabCorp, Histogenetics, Innogenetics, the NMDP, the Marrow Foundation, and One Lambda Inc. In particular, we thank Ram Ray at the NMDP for his support and effort in co-ordinating funding for this project. Initial support for the IMGT/HLA Database project was from the ICRF (now Cancer Research UK) and an EU Biotech grant (BIO4CT960037).

Appendix: Access and Contact

IMGT/HLA Homepage: http://www.ebi.ac.uk/imgt/hla/
IMGT/HLA FTP Site: ftp://ftp.ebi.ac.uk/pub/databases/imgt/mhc/hla/
Contact: hladb@ebi.ac.uk

References

1. Horton, R., Wilming, L., Rand, V., Lovering, R. C., Bruford, E. A., Khodiyar, V. K., Lush, M. J., Povey, S., Talbot, C. C., Jr., Wright, M. W., Wain, H. M., Trowsdale, J., Ziegler, A., and Beck, S. (2004) *Nat Rev Genet* **5,** 889–99.
2. Charron, D. (Ed.) (1997) Genetic Diversity of HLA: Functional and Medical Implication, EDK, Paris.
3. Marsh, S. G. E., Parham, P., and Barber, L. D. (2000) HLA FactsBook, Academic Press, London.
4. Bodmer, J. G., Marsh, S. G. E., Albert, E. D., Bodmer, W. F., Bontrop, R. E., Dupont, B., Erlich, H. A., Hansen, J. A., Mach, B., Mayr, W. R., Parham, P., Petersdorf, E. W., Sasazuki, T., Schreuder, G. M. Th., Strominger, J. L., Svejgaard, A., and Terasaki, P. I. (1999) *Tissue Antigens* **53,** 407–46.
5. Marsh, S. G. E., Bodmer, J. G., Albert, E. D., Bodmer, W. F., Bontrop, R. E., Dupont, B., Erlich, H. A., Hansen, J. A., Mach, B., Mayr, W. R., Parham, P., Petersdorf, E. W., Sasazuki, T., Schreuder, G. M. Th., Strominger, J. L., Svejgaard, A., and Terasaki, P. I. (2001) *Tissue Antigens* **57,** 236–83.
6. Orr, H. T., Lopez de Castro, J. A., Lancet, D., and Strominger, J. L. (1979) *Biochemistry* **18,** 5711–20.
7. Lee, J. S., Trowsdale, J., Travers, P. J., Carey, J., Grosveld, F., Jenkins, J., and Bodmer, W. F. (1982) *Nature* **299,** 750–52.
8. Bodmer, W. F., Albert, E., Bodmer, J. G., Dupont, B., Mach, B., Mayr, W. R., Sasazuki, T., Schreuder, G. M. T., Svejgaard, A., and Terasaki, P. I. (1989) Immunobiology of HLA (Dupont, B., Ed.), Vol. 1, pp. 72–9, Springer-Verlag, New York.
9. Bodmer, J. G., Marsh, S. G. E., Parham, P., Erlich, H. A., Albert, E., Bodmer, W. F., Dupont, B., Mach, B., Mayr, W. R., Sasasuki, T., Schreuder, G. M. Th., Strominger, J. L., Svejgaard, A., and Terasaki, P. I. (1990) *Tissue Antigens* **35,** 1–8.
10. Marsh, S. G. E., and Bodmer, J. G. (1993) *Hum Immunol* **36,** 44.
11. Zemmour, J., and Parham, P. (1991) *Tissue Antigens* **37,** 174–80.
12. Zemmour, J., and Parham, P. (1992) *Tissue Antigens* **40,** 221–8.
13. Arnett, K. L., and Parham, P. (1995) *Tissue Antigens* **46,** 217–57.
14. Mason, P. M., and Parham, P. (1998) *Tissue Antigens* **51,** 417–66.
15. Marsh, S. G. E., and Bodmer, J. G. (1990) *Immunogenetics* **31,** 141–4.
16. Marsh, S. G. E., and Bodmer, J. G. (1991) *Tissue Antigens* **37,** 181–9.

17. Marsh, S. G. E., and Bodmer, J. G. (1992) *Tissue Antigens* **40,** 229–43.
18. Marsh, S. G. E., and Bodmer, J. G. (1994) *Eur J Immunogenet* **21,** 519–51.
19. Marsh, S. G. E., and Bodmer, J. G. (1995) *Tissue Antigens* **46,** 258–80.
20. Marsh, S. G. E. (1998) *Tissue Antigens* **51,** 467–507.
21. Marsh, S. G. E. (2005) *Tissue Antigens* **67,** 94–5.
22. Robinson, J., Malik, A., Parham, P., Bodmer, J. G., and Marsh, S. G. E. (2000) *Tissue Antigens* **55,** 280–7.
23. Robinson, J., Waller, M. J., Parham, P., Bodmer, J. G., and Marsh, S. G. E. (2001) *Nucleic Acids Res* **29,** 210–3.
24. Robinson, J., Waller, M. J., Parham, P., de Groot, N., Bontrop, R., Kennedy, L. J., Stoehr, P., and Marsh, S. G. E. (2003) *Nucleic Acids Res* **31,** 311–4.
25. Robinson, J., and Marsh, S. G. E. (2000) *Rev Immunogenet* **2,** 518–31.
26. Giudicelli, V., Chaume, D., Bodmer, J., Muller, W., Busin, C., Marsh, S. G. E., Bontrop, R., Marc, L., Malik, A., and Lefranc, M. P. (1997) *Nucleic Acids Res* **25,** 206–11.
27. Kanz, C., Aldebert, P., Althorpe, N., Baker, W., Baldwin, A., Bates, K., Browne, P., van den Broek, A., Castro, M., Cochrane, G., Duggan, K., Eberhardt, R., Faruque, N., Gamble, J., Diez, F. G., Harte, N., Kulikova, T., Lin, Q., Lombard, V., Lopez, R., Mancuso, R., McHale, M., Nardone, F., Silventoinen, V., Sobhany, S., Stoehr, P., Tuli, M. A., Tzouvara, K., Vaughan, R., Wu, D., Zhu, W., and Apweiler, R. (2005) *Nucleic Acids Res* **33,** D29–33.
28. Benson, D. A., Karsch-Mizrachi, I., Lipman, D. J., Ostell, J., and Wheeler, D. L. (2005) *Nucleic Acids Res* **33,** D34–8.
29. Tateno, Y., Saitou, N., Okubo, K., Sugawara, H., and Gojobori, T. (2005) *Nucleic Acids Res* **33,** D25–8.
30. Thompson, J. D., Higgins, D. G., and Gibson, T. J. (1994) *Comput Appl Biosci* **10,** 19–29.
31. Antonarakis, S. E. (1998) *Hum Mutat* **11,** 1–3.
32. Etzold, T., Ulyanov, A., and Argos, P. (1996) *Methods Enzymol* **266,** 114–28.
33. Rozemuller, E. H., and Tilanus, M. G. (2000) *Rev Immunogenet* **2,** 492–517.
34. Schreuder, G. M. Th., Hurley, C. K., Marsh, S. G. E., Lau, M., Fernandez-Vina, M., Noreen, H. J., Setterholm, M., and Maiers, M. (2005) *Tissue Antigens* **65,** 1–55.
35. Hurley, C. K., Setterholm, M., Lau, M., Pollack, M. S., Noreen, H., Howard, A., Fernandez-Vina, M., Kukuruga, D., Muller, C. R., Venance, M., Wade, J. A., Oudshoorn, M., Raffoux, C., Enczmann, J., Wernet, P., and Maiers, M. (2004) *Bone Marrow Transplant* **33,** 443–50.

4

IPD

The Immuno Polymorphism Database

James Robinson and Steven G. E. Marsh

Summary

The Immuno Polymorphism Database (IPD) (http://www.ebi.ac.uk/ipd/) is a set of specialist databases related to the study of polymorphic genes in the immune system. IPD currently consists of four databases: IPD-KIR, contains the allelic sequences of killer cell immunoglobulin-like receptors (KIRs); IPD-MHC, a database of sequences of the major histocompatibility complex (MHC) of different species; IPD-HPA, alloantigens expressed only on platelets; and IPD-ESTAB, which provides access to the European Searchable Tumour Cell Line Database, a cell bank of immunologically characterized melanoma cell lines. The IPD project works with specialist groups or nomenclature committees who provide and curate individual sections before they are submitted to IPD for online publication. The IPD project stores all the data in a set of related databases. Those sections with similar data, such as IPD-KIR and IPD-MHC, share the same database structure.

Key Words: KIR; MHC; nomenclature; sequences; alleles; database

1. Introduction

The Immuno Polymorphism Database (IPD) is a set of specialist databases related to the study of polymorphic genes in the immune system. The IPD project works with specialist groups or nomenclature committees, with each curating a different section of the project. IPD currently consists of four databases: IPD-KIR, contains the allelic sequences of killer cell immunoglobulin-like receptors (KIRs); IPD-MHC, a database of sequences of the major histocompatibility complex (MHC) of different species; IPD-HPA, alloantigens expressed only on platelets; and IPD-ESTAB, which provides

From: *Methods in Molecular Biology, vol. 409: Immunoinformatics: Predicting Immunogenicity In Silico*
Edited by: D. R. Flower © Humana Press Inc., Totowa, NJ

access to the European Searchable Tumour Cell Line Database (ESTDAB), a cell bank of immunologically characterized melanoma cell lines. Access to all sections of the IPD database is through the homepage (see Fig. 1).

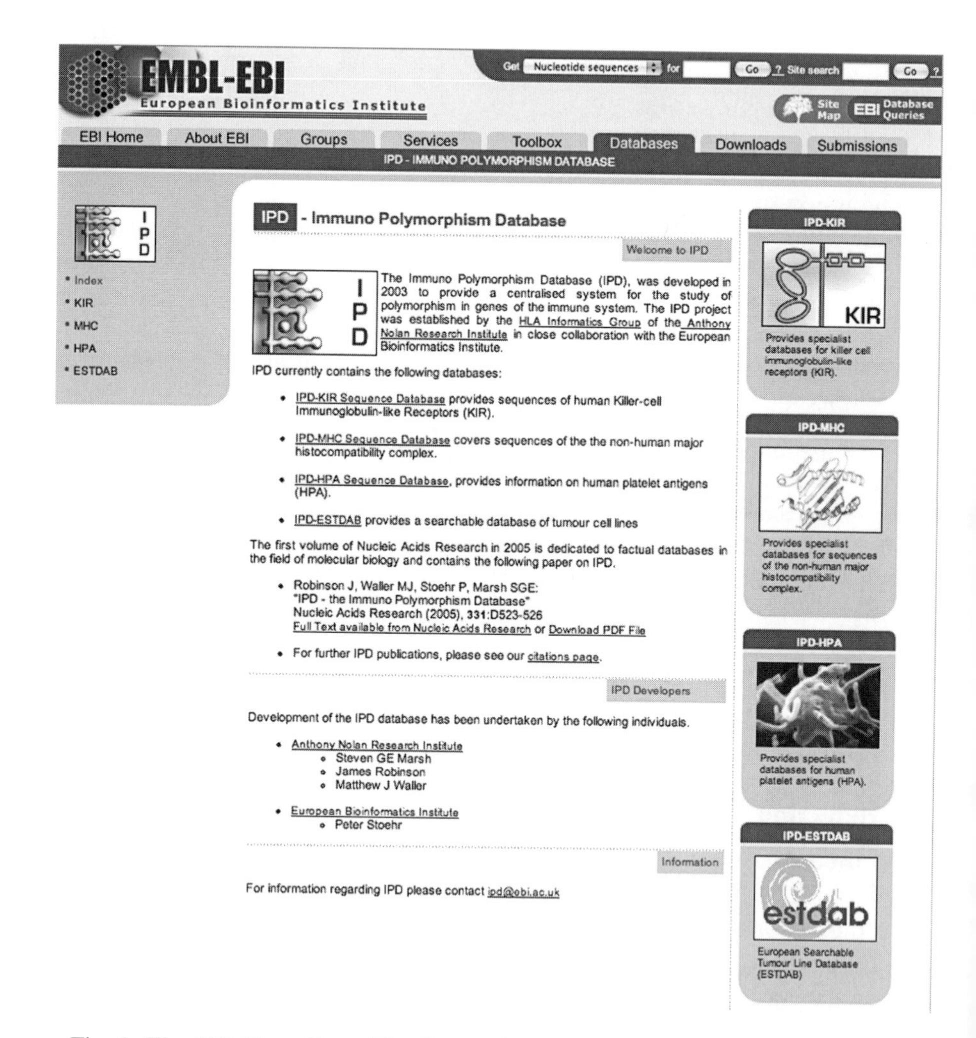

Fig. 1. The IPD HomePage. The IPD HomePage provides access to the component databases: IPD-KIR, contains the allelic sequences of killer cell immunoglobulin-like receptors (KIRs); IPD-MHC, a database of sequences of the major histocompatibility complex of different species; IPD-HPA, alloantigens expressed only on platelets; and IPD-ESTAB, which provides access to the European Searchable Tumour Cell Line Database (ESTDAB).

The study of the immune system constitutes many different complex areas of research. The aim of the IPD Database project is to provide a centralized resource for information pertaining to polymorphic genes of the immune system, by coupling the expertise of various research groups or nomenclature committees with the informatics experience of the HLA Informatics Group at the Anthony Nolan Research Institute. The individual experts or nomenclature committees are established within their own fields; each has a role in assessing the quality and validity of new data submitted to their own section of the database. This may be in the identification and naming of new alleles based on the submission of new sequences to generalist databases such as the European Molecular Biology Laboratory's nucleotide Database (EMBL), the National Center for Biotechnology Information's GenBank, and the DNA DataBank of Japan (DDBJ) ***(1–3)***, or in the collation and validation of data from a variety of different cell characterization methods, such as the cases for the IPD-ESTDAB database. The resulting data sets held within this specialist system differ from that available in more generalist databases in its quality and in the further curation by the experts in the relevant field. This has been discussed in more detail in Chapter 3. The result being the specialist database should be treated as the most reliable and accurate source of information, this is particularly important when it comes to the sequences of polymorphic genes.

One advantage of using a centralized system is the ability to share or reuse elements of database structure, when dealing with similar data sets. Much of the database structure of IPD-KIR and IPD–MHC sections are shared with the IMGT/HLA Database. This has also enabled cross database implementation of some of the core tools, particularly those for data analysis, submission, and retrieval.

2. IPD Projects

2.1. IPD-KIR

The KIRs are members of the immunoglobulin super family (IgSF) formerly called killer-cell inhibitory receptors. KIRs have been shown to be highly polymorphic at both the allelic and haplotypic levels ***(4)***. They are composed of two or three Ig domains, a transmembrane region and cytoplasmic tail, which can in turn be short (activatory) or long (inhibitory). The leukocyte receptor complex (LRC), which encodes KIR genes, has been shown to be polymorphic, polygenic, and complex in a manner similar to the MHC. Because of the complexity in the KIR region and KIR sequences, a KIR Nomenclature Committee was established in 2002 to undertake the naming of KIR allele sequences. The first KIR Nomenclature report was published in 2002 *(5)*, which coincided with the first release of the IPD-KIR database. To aid in the analysis and naming of new alleles, an online submission tool is provided on the IPD-KIR website. New alleles are added to the database once they have successfully completed the submission procedure and their quality assured. Periodic new releases of the database contain newly submitted sequences together with conformations and extensions of those already available.

The KIR Nomenclature Committee is also involved in the naming of the complex haplotypes and genotypes currently seen in KIR research. Proposals for such a nomenclature have been published, but as yet, this nomenclature has not been implemented, although it is planned to include this data once available. The online tools available for IPD-KIR include allele queries, sequence alignments, and cell queries. As the database is based on the work of a nomenclature committee, the website includes links to a portable document format (PDF) file of recent nomenclature reports. From the data contained within these reports, the database is also able to provide individual allele reports (Fig. 2). These pages contain the official allele name, any previous designations, the EMBL, GenBank, or DDBJ accession number(s), and a reference linked, wherever possible, to the PubMed abstract. Where possible additional details on the source of sequence are also provided. This source material is normally in the form of a cell line or DNA from which each allele in the database was isolated and characterized. The information contained within this data set can be searched independently from the allele data. The cell query tool is used to interrogate this accompanying cell database. The interface can be used without prior knowledge of which alleles are linked to the cells or from the allele reports. The interface allows the user to search on known cell fields. The cells all have a primary name and accession number that are unique within the database. As some cells are sequenced by different groups, or certain names are repeated, the database also contains a list of aliases for each cell. These aliases are automatically searched whenever the cell name field is queried. Other cell fields that are available for searching include HLA and KIR typing, serology, ethnic origin, and geographical location.

The other main use of the database is to view multiple sequence alignments. Within each IPD section, allele sequences may differ from each other by as little as a single nucleotide. As previously discussed for IMGT/HLA (see Chapter 3), these alignments allow a visual interpretation of sequence similarity, so that polymorphic positions can easily be identified and motifs found in multiple alleles are also easily identified. The sequence alignments are available via a link from the section homepage. The interface provided lets the user define a

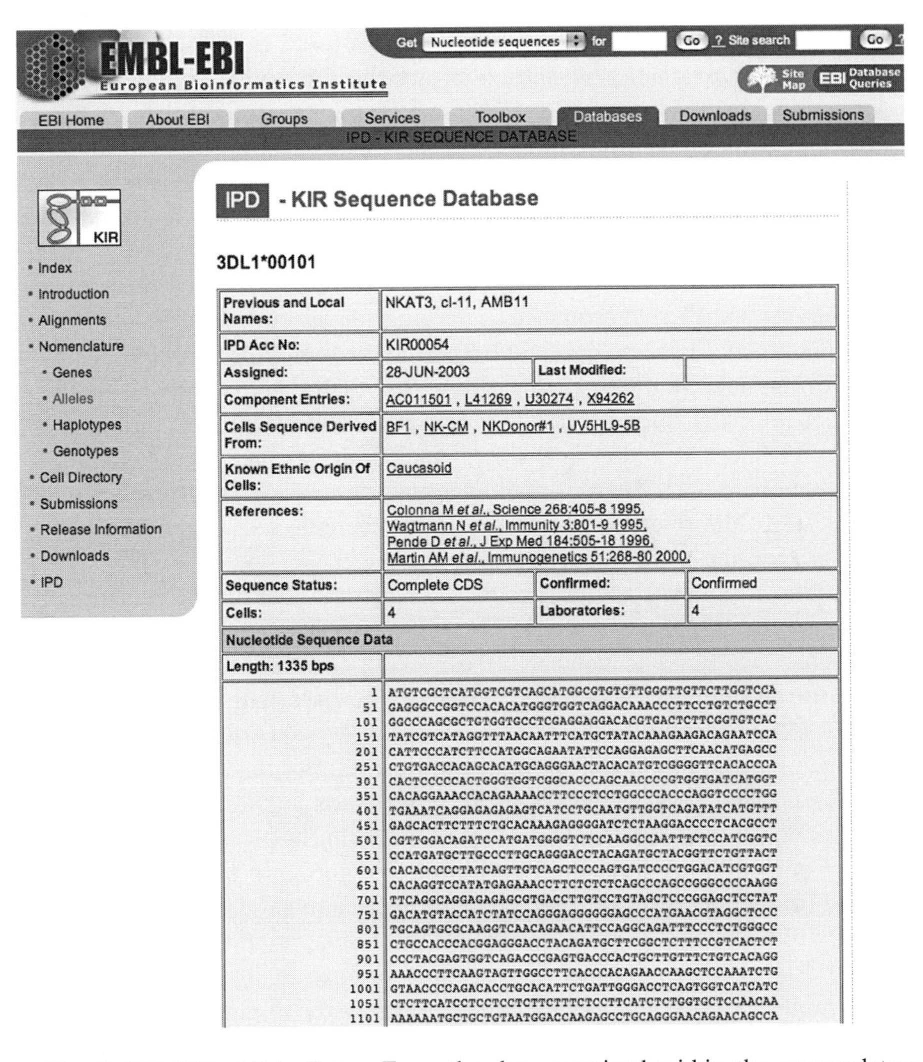

Fig. 2. IPD-KIR Allele Entry. From the data contained within the nomenclature reports, the database is able to provide individual allele reports. The report shows part of the KIR3DL1*00101 entry. The underlined text is a link to further information on the web both within IPD and in external sources such as PubMed.

number of key variables for the alignments, before producing an online output, which can be printed or downloaded. The first step in any alignment is to select the locus of interest. The tool provides a drop-down list of all loci. The selection of a locus automatically updates the list of features that can be aligned, as well as the default reference sequence used for the alignment. The types of feature available for alignment are the nucleotide coding sequence and individual exons, the signal peptide, mature protein, and full-length protein sequence. The alignment tool options also allow the user to display a subset of alleles of a particular locus, omit alleles unsequenced for a particular region, and align against a particular reference or consensus sequence. The alignment tool uses standard formatting conventions for the display of sequence alignments. The alignment tool does not perform a sequence alignment each time it is used, but it extracts pre-aligned sequences, allowing for faster access.

The alignments adhere to a number of conventions for displaying evolutionary events and numbering. The numbering of the alignments is based on the sequence of the reference allele. For a nucleotide sequence, the A of the initiation Methionine codon is denoted nucleotide +1 and the nucleotide 5′ to +1 is numbered −1. There is no nucleotide zero (0). All numbering is based on the ATG of the reference sequence. If a nucleotide sequences is displayed in codons, then the protein numbering is applied.

For amino acid-based alignments, the first codon of the mature protein, after cleavage of the signal sequence, is labeled codon 1 and the codon 5′ to this is numbered −1. In all sequences, the following conventions are used. Where identity to the reference sequence is present, the base will be displayed as a hyphen (-). Non-identity to the reference sequence is shown by displaying the appropriate base at that position. Where an insertion or deletion has occurred, this is represented by a period (.). If the sequence is unknown at any point in the alignment, this is represented by an asterisk (*). In protein alignments for null alleles, the "Stop" codons are represented by an X. In protein alignments, sequence following the termination codon is not marked and appears blank. The flexibility of the new alignment tool means that unlike previous alignments you can now display a small subset of sequences against an allele of your choice, using a number of display options. Figure 3 shows the alignment tool used for the IPD-KIR section.

2.2. IPD-MHC

The MHC sequences of many species have been reported in the literature and are represented in the generalist sequence databases. For some species or related species groups, such as the bovines, non-human primates, and dog, there have been efforts to use a standardized nomenclature system and establish comprehensive data sets ***(6–8)***. The availability of these data sets for use by other groups has often been limited by a lack of informatics resources available to the researchers compiling the data sets. The aim of

IPD - KIR Sequence Database Alignment Tool		
Select Locus :	2DL1	Help
Select the feature to align :	Nucleotide - CDS	Help
Enter any specific sequences required :		Help
Enter the reference sequence :	001	Help
Select how you wish to view any mismatches :	Show mismatches between sequences	Help
Select how the alignment will be numbered :	Nucleotide - nucleotide sequence displayed in blocks of 10 bases	Help
Do you want to omit alleles unsequenced for this region :	Show all alleles	Help
Proceed with the alignment :	Align Sequence Now / Reset Form	

Fig. 3. Alignment Interface. The alignment interface provides a user-friendly method of viewing sequence alignments with output options easily selected.

the IPD-MHC database is to address this issue and provide a centralized database that will facilitate the comparative analysis of these sequences that are highly conserved between different closely related species ***(9)***. In addition, the formation of the International Society of Animal Genetics (ISAG)/International Union of Immunological Society (IUIS)-Veterinary Immunology Symposium (VIC) Comparative MHN Nomenclature Committee, which brings together representatives of many nomenclature committees covering different species, will aid in the establishment of standardized nomenclature practices across species ***(10)***.

For each species, there are differences in the spectrum of data covered, but all sections provide the core nomenclature pages and sequence alignments. The nomenclature and alignments follow a structure similar to that of the IPD-KIR section, and the same basic tools are used in both sections. Some nomenclature committees may provide additional information, but the core components of any nomenclature reported are the allele names, accession numbers, and publications.

Currently, the IPD-MHC sequence alignments are limited to species-specific alignments; however, we are working to allow cross-species alignments and the inclusion of human sequences from the IMGT/HLA Database ***(11)*** for comparative purposes. The alignments for non-human primates can use a human HLA sequence as a reference sequence but this is only a single sequence in each alignment. The sequence alignment tool is similar to that used in the IPD-KIR database, see Fig. 3, and it is possible to define the output parameters to display the sequence alignments in various formats (Fig. 4). The IPD-MHC

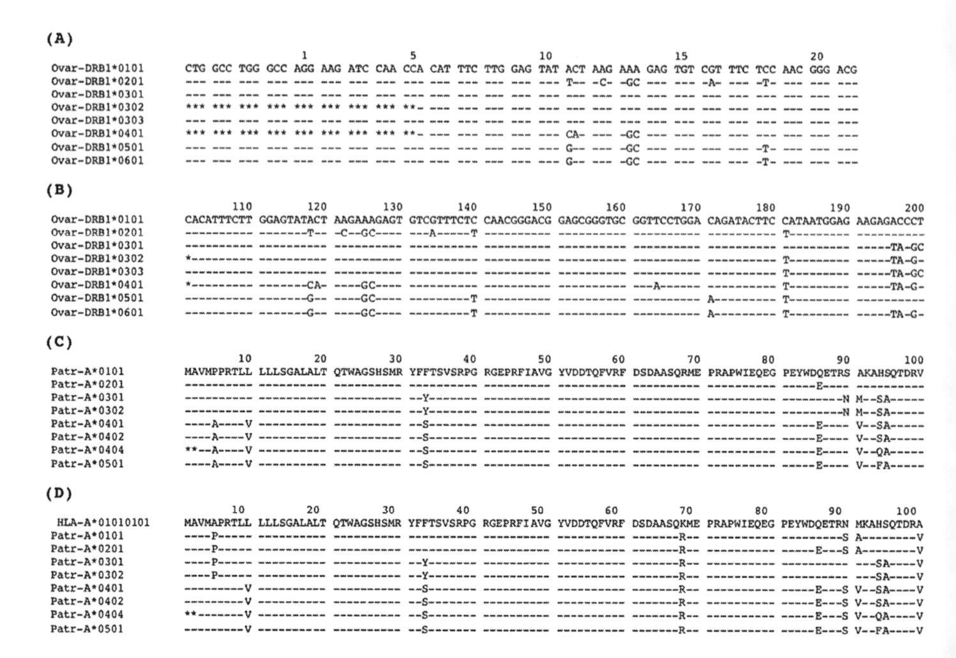

Fig. 4. Alignment formats available from IPD-MHC. The examples shown are all alignments of DRB alleles from different species. In these alignments, a dash (–) indicates identity to the reference sequence and an asterisk (*) denotes an unsequenced base. The first two alignments (**A** and **B**) show the nucleotide sequences of sheep (*Ovis aries*) DRB1 alleles using different display parameters. The second set (**C** and **D**) shows chimpanzee (*Pan troglodytes*) Patr-A protein sequences. The same sets of alleles are used for both C and D, but in D, the chimpanzee sequences are aligned against a human reference sequence (HLA-A*01010101).

Database also contains a submission tool for online submission of new and confirmatory sequences to the appropriate nomenclature committee.

The first release of the IPD-MHC database involved the work of groups specializing in non-human primates, canines, and felines and incorporated all data previously available in the IMGT/MHC Database *(12)*. Since this release, sequences from cattle, fish, rat, sheep, and swine have been added; work is also underway on the inclusion of chicken and horse sequences.

2.3. IPD-HPA

Human platelet antigens (HPAs) are alloantigens expressed on platelets, specifically on platelet membrane glycoproteins. These platelet-specific antigens are immunogenic and can result in pathological reactions to transfusion therapy. The HPA nomenclature system was adopted in 1990 ***(13,14)*** to overcome problems with the previous nomenclature. Since then, more antigens have been described, and the molecular basis of many has been resolved. As a result the nomenclature was revised in 2003 ***(15)*** and included in the IPD project. The IPD-HPA section contains nomenclature information and additional background material. The different genes in the HPA system have not been sequenced to the same level as some of the other projects, and so currently only single-nucleotide polymorphisms (SNPs) are used to determine alleles. This information is presented in a grid of SNP for each gene. The HPA Nomenclature Committee hopes to expand this to provide full sequence alignments when possible. The IPD-HPA section also provides data on the frequency of different HPA alleles in a number of populations. These tables contain allele frequencies as well as the ethnic origins of the samples, typing methodology, and relevant publications. This table is regularly updated and is now considered one of the main resources for HPA frequency data (Fig. 5).

2.4. IPD-ESTDAB

IPD-ESTDAB is a database of immunologically well-characterized melanoma cell lines. The database works in conjunction with the ESTDAB cell bank, which is housed in Tübingen, Germany and provides access to the immunologically characterized tumor cells ***(16)***. The ESTDAB consortium is made up of seven laboratories from countries around the European Union, with each lab responsible for the generation of data on a prearranged set of immunological or genetic markers that reflect the laboratory's expertise and technical specialities. The central facility and physical location of the cell bank was established at the Centre for Medical Research (ZMF) of the University of Tübingen in Germany, where cells lines were gathered from a variety of sources around Europe, Australia, and the United States of America. Since the project began, cells have been acquired as they become available, this has meant in particular that a number of cell lines have been sourced from melanoma samples collected from patients entered into clinical immunotherapy trials associated with ESTDAB's sister project, OISTER (Outcome and Impact of Specific Treatment in European Research on Melanoma, http://www.dkfz.de/oister/). Consequently, ESTDAB now provides access to many more cell lines of melanoma origin than are available from other non-specialist cell banks.

The IPD-ESTDAB section of the website provides an online search facility for cells stored in this cell bank. This enables investigators to identify cells

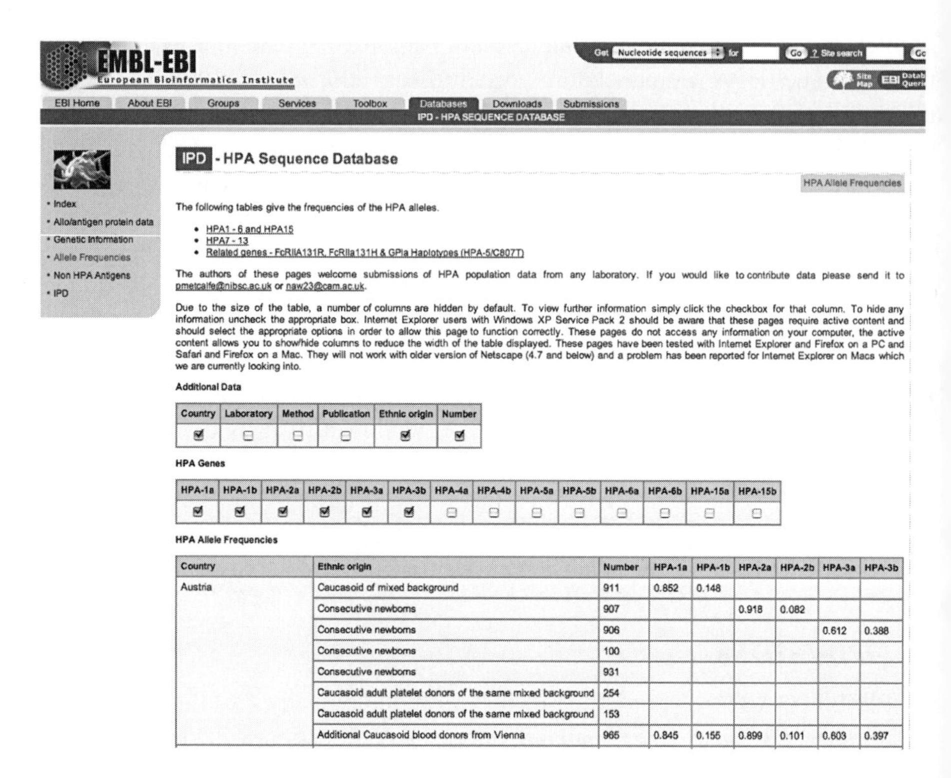

Country	Ethnic origin	Number	HPA-1a	HPA-1b	HPA-2a	HPA-2b	HPA-3a	HPA-3b
Austria	Caucasoid of mixed background	911	0.852	0.148				
	Consecutive newborns	907			0.918	0.082		
	Consecutive newborns	906					0.612	0.388
	Consecutive newborns	100						
	Consecutive newborns	931						
	Caucasoid adult platelet donors of the same mixed background	254						
	Caucasoid adult platelet donors of the same mixed background	153						
	Additional Caucasoid blood donors from Vienna	965	0.845	0.155	0.899	0.101	0.603	0.397

Fig. 5. IPD-HPA frequency data. The IPD-HPA database contains frequency data on different HPA alleles. A selection of data is shown. Further columns can be added to the table by selecting the appropriate column from the menus above the main report. The columns already selected are identified by a tick mark in the relevant box.

possessing specific parameters important for studies of immunity, immunogenetics, gene expression, metastasis, response to chemotherapy, and other tumor biological experimentation. The search tool allows for searches based on a single parameter, or clusters of parameters on over 250 different markers for each cell. The detailed reports produced can then be used to identify cells of interest, which can then be obtained from the cell bank. Some elements of the design of the ESTDAB database are borrowed from that used in the IMGT/HLA Database (see Chapter 3).

There are several ways to search the ESTDAB database, all of which are based on a system of expanding lists and all of which return results data in a format identical to that presented in the search tools. Using the quick-search tool can find data based on an individual cell line's name or unique accession number. Because of the number of markers studied by the ESTDAB consortium, the main search tool is split into two sections. The primary search tool allows users to find cells with a particular HLA genotype or surface expression pattern, or otherwise according to the cell's expression of tumor-associated antigens or secretion of cytokines (Fig. 6). These primary search determinants are more likely to return positive hits than the remaining markers which are included in a second search tool, given the current data set available those factors included in the secondary search tool, given the current data set available. Both search tools allow the use of the Boolean operators "AND" and "OR" when searching for combinations of markers. A complete list of markers studied in ESTDAB is available in the data dictionary along with information on the acceptable values for their related fields. Protocols and/or information sufficient to reproduce the experimental conditions applied in the characterization of the ESTDAB cells is available in the help section for each methodology. Contributing laboratories can be contacted directly for more information.

Fig. 6. IPD-ESTDAB cell entry. Part of a cell report available from IPD-ESTDAB. The figure shows only the general information on a cell and some of the high resolution HLA typing. Each cell entry can contain data on over 250 individual markers.

3. Discussion

The IPD project provides a new resource for those interested in the study of polymorphic sequences in the immune system. By accommodating related systems in a single database, data can be made available in common formats aiding use and interpretation. As the projects grow and more sections are added, the benefit of having expertly curated sequences from related areas stored in a single location will become more apparent. This is particularly true of the IPD-MHC project, which already contains 2,270 sequences from over 50 different species, where cross-species studies will be able to utilize the high-quality sequences provided by the different nomenclature committees in a common format, ready for use. The initial release of the IPD Database contained only four sections and a small number of tools; however, as the database grows and more sections and species are added, more tools will be added to the website. We plan to use the existing database structures to house data for new sections of the IPD project as they become available. Data will also be made available in different formats to download from the website and ftp server and included into Sequence Retrieval System (SRS), Basic Local Alignment Search Tool (BLAST), and FAST-All (FASTA) search engines at the European Bioinformatics Institute *(17)*.

Acknowledgments

We would like to thank Peter Parham and Libby Guethlein for their work on IPD-KIR, our IPD-MHC collaborators Shirley Ellis, Lorna Kennedy, Ronald Bontrop, Natasja de Groot, Lutz Walter, Douglas Smith, Keith Ballingall, Mike Stear and René Stet. Also Nick Watkins and Paul Metcalfe for IPD-HPA and Graham Pawelec for ESTDAB. Our thanks also go to Matthew Waller and Sylvie Fail for their work on the tools and processing of submissions for the various databases and finally Peter Stoehr at the EBI for assisting in the continued collaboration between our group and the EBI.

Appendix: access and contact

IPD Homepage: http://www.ebi.ac.uk/ipd/
IPD-KIR Homepage: http://www.ebi.ac.uk/ipd/kir/
IPD-MHC Homepage: http://www.ebi.ac.uk/ipd/mhc/
IPD-HPA Homepage: http://www.ebi.ac.uk/ipd/hpa/
IPD-ESTDAB Homepage: http://www.ebi.ac.uk/ipd/estdab/
Contact: ipd@ebi.ac.uk

If you are interested in contributing to the project, there are specific guidelines for the inclusion of new sections, and interested parties should contact Dr. S. G. E. Marsh, (E-mail: marsh@ebi.ac.uk) for further information.

References

1. Benson, D. A., Karsch-Mizrachi, I., Lipman, D. J., Ostell, J., and Wheeler, D. L. (2005) *Nucleic Acids Res* **33**, D34–8.
2. Kanz, C., Aldebert, P., Althorpe, N., Baker, W., Baldwin, A., Bates, K., Browne, P., van den Broek, A., Castro, M., Cochrane, G., Duggan, K., Eberhardt, R., Faruque, N., Gamble, J., Diez, F. G., Harte, N., Kulikova, T., Lin, Q., Lombard, V., Lopez, R., Mancuso, R., McHale, M., Nardone, F., Silventoinen, V., Sobhany, S., Stoehr, P., Tuli, M. A., Tzouvara, K., Vaughan, R., Wu, D., Zhu, W., and Apweiler, R. (2005) *Nucleic Acids Res* **33**, D29–33.
3. Miyazaki, S., Sugawara, H., Ikeo, K., Gojobori, T., and Tateno, Y. (2004) *Nucleic Acids Res* **32**, D31–4.
4. Garcia, C. A., Robinson, J., Guethlein, L. A., Parham, P., Madrigal, J. A., and Marsh, S. G. E. (2003) *Immunogenetics* **55**, 227–39.
5. Marsh, S. G. E., Parham, P., Dupont, B., Geraghty, D. E., Trowsdale, J., Middleton, D., Vilches, C., Carrington, M., Witt, C., Guethlein, L. A., Shilling, H., Garcia, C. A., Hsu, K. C., and Wain, H. (2003) *Immunogenetics* **55**, 220–6.
6. Kennedy, L. J., Altet, L., Angles, J. M., Barnes, A., Carter, S. D., Francino, O., Gerlach, J. A., Happ, G. M., Ollier, W. E., Polvi, A., Thomson, W., and Wagner, J. L. (1999) *Tissue Antigens* **54**, 312–21.
7. Klein, J., Bontrop, R. E., Dawkins, R. L., Erlich, H. A., Gyllensten, U. B., Heise, E. R., Jones, P. P., Parham, P., Wakeland, E. K., and Watkins, D. I. (1990) *Immunogenetics* **31**, 217–9.
8. Stear, M. J., Pokorny, T. S., Fryda-Bradley, S., Lie, O., and Bull, R. W. (1990) *J Immunogenet* **17**, 21–8.
9. Parham, P. (1999) *Immunol Rev* **167**, 5–15.
10. Ellis, S. A., Bontrop, R. E., Antczak, D. F., Ballingall, K., Davies, C. J., Kaufman, J., Kennedy, L. J., Robinson, J., Smith, D. M., Stear, M. J., Stet, R. J., Waller, M. J., Walter, L., and Marsh, S. G. (2006) *Immunogenetics* **57**, 953–8.
11. Robinson, J., Malik, A., Parham, P., Bodmer, J. G., and Marsh, S. G. E. (2000) *Tissue Antigens* **55**, 280–7.
12. Robinson, J., Waller, M. J., Parham, P., de Groot, N., Bontrop, R., Kennedy, L. J., Stoehr, P., and Marsh, S. G. E. (2003) *Nucleic Acids Res* **31**, 311–4.
13. von dem Borne, A. E., and Decary, F. (1990) *Vox Sang* **58**, 176.
14. von dem Borne, A. E., and Decary, F. (1990) *Hum Immunol* **29**, 1–2.

15. Metcalfe, P., Watkins, N. A., Ouwehand, W. H., Kaplan, C., Newman, P., Kekomaki, R., De Haas, M., Aster, R., Shibata, Y., Smith, J., Kiefel, V., and Santoso, S. (2003) *Vox Sang* **85**, 240–5.
16. Pawelec, G., and Marsh, S. G. (2006) *Cancer Immunol Immunother* 1–5.
17. Harte, N., Silventoinen, V., Quevillon, E., Robinson, S., Kallio, K., Fustero, X., Patel, P., Jokinen, P., and Lopez, R. (2004) *Nucleic Acids Res* **32**, W3–9.

5

SYFPEITHI

Database for Searching and T-Cell Epitope Prediction

Mathias M. Schuler, Maria-Dorothea Nastke, and Stefan Stevanović

Summary

Reverse immunology has been used for about 12 years in order to identify T-cell epitopes from pathogens or tumor-associated antigens. In this chapter, we discuss the advantages and pitfalls of T-cell epitope prediction compared to classical experimental procedures such as epitope mapping and cloning experiments. We introduce our three established programs, SYFPEITHI, PAProc, and SNEP, which are freely accessible at no cost in the World Wide Web for the prediction of either HLA–peptide binding or proteasomal processing of antigens. We demonstrate the performance of our epitope prediction programs with several examples and in comparison to other epitope prediction programs available. We also reflect the actual possibilities and limitations of such computer-aided work.

Key Words: MHC; peptide motif; T-cell epitope; SNP; minor histocompatibility antigen; epitope prediction; CMV; HLA; proteasome; reverse immunology

1. Introduction

Right from the start when allele-specific peptide motifs were first identified *(1)*, rules for major histocompatibility complex (MHC)-mediated antigen presentation were exploited and applied to the prediction of viral or bacterial T-cell epitopes *(2,3)*. In the meantime, the peptide motifs of most MHC alleles that are widely expressed in mice and humans have been determined, and prediction strategies and programs have been developed. With the existing possibilities of the World Wide Web, epitope prediction programs can now be found at many different sites, and—most important—many of them are available without any restriction and at no cost for the scientific community.

From: *Methods in Molecular Biology, vol. 409: Immunoinformatics: Predicting Immunogenicity In Silico*
Edited by: D. R. Flower © Humana Press Inc., Totowa, NJ

The first program of this kind has been established by Ken Parker's group *(4)*. "BIMAS"(bimas.dcrt.nih.gov) still represents the classical epitope prediction program which contains most of the classical mouse peptide motifs and also a broad range of human MHC class I motifs. The next program to follow was "SYFPEITHI" (www.syfpeithi.de) *(5)*, which offers prediction of epitopes for some MHC class II restrictions in addition to MHC class I predictions. After the millennium, many more programs evolved and became available via the World Wide Web. At present, there are more than ten programs freely available and others with restricted access. The latest program is "SNEP" (www.elchtools.de/SNEP) *(6)*, a program which allows the prediction of single-nucleotide polymorphism (SNP)-derived T-cell epitopes and has been developed together with Pierre Dönnes from Oliver Kohlbacher's group at the Wilhelm-Schickard-Institute for Computer Science. Pierre Dönnes developed also an MHC class I peptide prediction program called SVMHC *(7)*.

Apart from algorithms that predict the mere presentation of peptide sequences (which most basically means binding of peptides by a given MHC allotype), another checkpoint of antigen processing has caught the eye of biomathematicians: the proteolytic activities of the proteasome. Because the majority of T-cell epitopes presented by MHC class I molecules have to be processed by this multimeric multispecific protease complex, several groups investigated the rules of such processing events. Either by digesting short model substances that release chromophores after being cleaved, short peptide sequences of 20–30 amino acids that also may contain known T-cell epitopes, or complete proteins whose peptide fragments are then separated and analyzed, the preference of proteasomal processing was analyzed and served as a basis for prediction algorithms. Such a prediction program called "PAProC" (www.paproc.de) *(8)* has been developed in our department together with Karl-Peter Hadeler's biomathematic group In this chapter, we will discuss the advantages and disadvantages of epitope prediction compared to experimental methods and then describe some of the most prominent prediction procedures for human cytomegalovirus (HCMV) T-cell epitope identification by MHC binding and/or proteasomal processing predictions. We will introduce some of the more widely used programs and compare the performance of these by "predicting" in a retrospective way a number of well-known T-cell epitopes from the immunodominant pp65 protein of HCMV.

2. Epitope Identification

2.1. Epitope Mapping Using Synthetic Overlapping Peptides

This method for the determination of T-cell epitopes is based on the stimulation of fresh peripheral blood mononuclear cells (PBMCs) with synthetic peptides, usually applied as pools, which represent the complete sequence of a viral antigen and therefore contain all potential CD8 T-cell epitopes of the respective antigen. The peptide pool may, for example, consist of peptides with a length of 15 amino acids whereby at least 9 amino acids overlap between neighboring peptides ***(9)***. T cells recognizing epitopes contained in these pools have the ability to secrete large amounts of interferon γ (IFN-γ), which can be measured and quantified by flow cytometry. Because the peptide pools provide all possible CD8 T-cell epitopes in the protein, this strategy depicts an approach which is not dependent on HLA types. Therefore, this strategy represents a useful tool in clinical terms with the option to record T-cell response to a complete peptide pool over time without knowing the responsible epitope(s) and presenting HLA molecule(s). The enormous effort involved in peptide preparations is partly compensated by the use of 15-mer peptides instead of smaller ones, which reduces the number of synthesized peptides. On the other hand, the use of 15-mer peptides in a search for MHC class I-presented peptides involves a certain risk of losing relevant epitopes because the affinity of 15-amino acid peptides is reduced by a factor of 10–1,000 if compared to optimal MHC class I ligands. For this reason, the peptides have to be applied in rather high concentrations ($> 1\mu M$), which often causes irrelevant signals by unspecific interactions–relevant T-cell epitopes should easily work at concentrations of 10 nM or less. Another critical point is the potential competition between peptides that represent relevant epitopes in the pool and peptides that are nonrelevant binders. However, such competition effects do not seem to play a major role if the peptides are applied in about equimolar concentrations, as has been demonstrated by systematic studies using combinatorial peptide libraries ***(10)***.

After a 15-mer peptide has been defined as a target of T-cell recognition, the optimal epitope and the corresponding HLA restriction still have to be defined; otherwise, this finding remains useless information. At this point, either time-consuming, tedious truncation variant analyses have to be carried out or epitope prediction has to be employed to define the optimal T-cell epitope contained in the 15-mer peptide.

2.2. Epitope Determination by Cloning Experiments

By this strategy, PBMCs of HCMV-seropositive donors are stimulated with virion-infected fibroblasts to increase the precursor frequency of virus-specific T cells ***(11)***. With a subsequent limiting dilution, including a depletion of CD4+ cells and cloning CD8+ cytotoxic T lymphocytes (CTLs), a further characterization of the specific allelic restriction and epitope specificity is possible. The transfection of truncated variants of the original viral gene allows one to narrow down the region of the gene that is recognized by the T cells. As for the overlapping peptides, the final steps of epitope identification have to employ stepwise truncation of minigenes which leads to an enormous amount of plasmids to be generated and tested ***(12)***. For this reason, epitope prediction usually represents the method of choice at the final stage of all strategies.

2.3. Epitope Prediction and Verification

After the discovery that natural MHC ligands usually share an allele-specific motif, the path for an exact prediction of T-cell epitopes, especially for MHC class I molecules, was paved. The determination of MHC class II epitopes proved to be more difficult than with the MHC class I counterpart because of the variable length of the MHC class II ligands and more degenerated anchor positions.

During the past decade, several strategies for the determination of peptide motifs have been published. One possibility is to collect and register information about natural ligands, with individual ligands being analyzed on an individual basis or as pools representing the entity of peptides being presented by a distinct MHC allotype. This characterization can be done by Edman degradation ***(13–15)*** or tandem mass spectrometry (MSMS) ***(16)***. The resulting so-called natural ligand motifs contain information about the natural peptide repertoire presented by a certain MHC molecule. SYFPEITHI depicts such a computational predictive method of MHC–peptide binding and is based on allele-specific peptide motifs ***(5)***. Prediction of MHC–peptide binding can also base on experimental results of binding assays that are transformed into quantitative matrices ***(4,17)***. The most prominent example of this kind of epitope prediction is the Bioinformatics and Molecular Analysis Section (BIMAS) matrix that underlies measurements of half-time dissociation rates of peptide–HLA complexes ***(4)***. Also, structural information may form the basis of the data set used for epitope prediction. The structures of MHC–peptide complexes obtained by X-ray crystallography design a basis for computer models of certain MHC molecules ***(18)***. All these data sets are used for different computational predictions that are based on predictive algorithms. In some cases, the database works with motif matrices deduced from natural ligands. These motif-based algorithms are quite simple programs that regard each single amino acid position in a peptide. Therefore, these programs are called "specific position scoring matrices" (SPSM). Otherwise, artificial neural networks (ANN) ***(19)*** are in use which afford the consideration of amino acid preferences that depend on the properties of amino acids in other positions of the peptides. ANNs were successfully applied to the prediction of MHC class I binding peptides ***(19–21)***. An advantage of ANN is that generalization and capturing relationships within data are possible, while at the same time, a tolerization and a filtration of erroneous or noisy data takes place. Recapitulating each prediction method has its advantages and drawbacks. Binding motifs encode the most important rules of MHC/peptide interaction, but do not generalize well. Quantitative matrices are able to predict large subsets of binding peptides, but they are not adaptive or self-learning, what often requires a redesigning of the matrix for integration of new data. These matrices also cannot handle nonlinearities within data and because of this may miss distinct subsets of binders. ANNs can deal with such nonlinearity and are adoptive and self-learning, but to guarantee this, a large amount of preprocessed data are required. There is one major advantage of epitope prediction if compared to other strategies: it is a quick and simple procedure—once the prediction program is available—requiring only a small set of peptides to be synthesized and tested, and ending up directly with the optimal epitope. The availability of a peptide motif, however, is an important prerequisite. The disadvantage of epitope prediction lies in the fact that epitopes might be missed by the prediction because they do not fit well with the peptide motif under investigation. Next to the prediction of epitopes, which focus mainly on MHC-peptide binding, there is the possibility of performing predictions of antigen processing which involve a certain knowledge of cellular components, such as the proteasome. For the prediction of proteasomal cleavages, different algorithms were created. These algorithms are based either on published peptide cleavage data ***(22)*** or consist of an evolutionary algorithm trained on cleavage data of digests of a whole protein substrate ***(23,24)***. One of these programs using data of processing is PAProC, the first version of which is based on experimental cleavage data of a small set of proteins by human and yeast proteasomes ***(8,25)***. PAProC allows for the prediction of cleavages carried out by the constitutive proteasome; a second version has been announced which will also enable cleavage predictions if immunoproteasomes; a trial version is already available. Another program for proteasomal cleavage prediction is NetChop (www.cbs.dtu.dk/services/NetChop) ***(24)***, based on a similar data set as PAProC. Recently, the network MAPPP (www.mpiib-berlin.mpg.de/MAPPP) has been made available publicly as an additional proteasomal cleavage prediction program. The verification of the predicted epitopes is achieved by the identification of T cells that recognize one of the candidate epitopes. A successful finding of T cells mirrors the correct prediction of a T-cell epitope because the T cells recognize the naturally processed epitope in an HLA-restricted fashion. T cells specific for HCMV epitopes are usually prepared from PBMC of healthy HCMV-seropositive donors, but sometimes also from patients suffering from HCMV reactivation. Such T cells have already been primed in vivo and are readily detected by tetramer staining, lytic activity, proliferation, or cytokine production. As the CTL response against HCMV epitopes is among the most vigorous immune responses known in humans, there is usually no need to restimulate T cells in vitro, but immunodominant responses can be analyzed ex vivo.

2.4. Prediction of T-cell Epitopes From HCMV Antigens

It is not surprising that the majority of MHC class I-restricted T-cell epitopes known today have been determined by means of epitope predictions. The programs are comparatively easy to use, they work very quickly, and the results are reliable to a high degree. For instance, SYFPEITHI guarantees that in more than 80% of the predictions, the relevant epitope can be identified. Two criteria have to be fulfilled for epitope prediction: the sequence of interest and the desired peptide motif must be known.

- The sequence of interest. With the steadily growing number of entries in the protein and nucleotide databases, the availability of sequences poses no serious problem. Many genomes of pathogens have been sequenced completely, and most protein or gene sequences can be accessed without any problems in any of the famous databases such as Genbank, EMBL nucleotide database, SWISS-PROT, or others. Two problems may arise if the prediction of new T-cell epitopes has to be carried out. First, the knowledge of immunodominant proteins is inevitable. HCMV, with its rather large genome coding for more than 200 proteins, represents a special problem for epitope prediction. Only if we know that the immune response focuses on a small set of antigens—mainly pp65, but also some of the glycoproteins, the immediate-early protein, and pp150—epitope prediction is feasible. In fact, it is not the prediction from large genomes itself that will cause problems, but as a result, hundreds of interesting candidate peptides will appear in the listing, and the verification of the predicted peptides by experimental work will hardly be possible. The second problem lies in the occurrence of different viral strains. The most widely used HCMS strain is AD169, from which most epitopes are derived. In order to deal with naturally occurring infections, the variations in protein sequences between different viral strains have to be considered. Usually, researchers and clinicians try to identify T-cell epitopes in conserved regions of immunodominant antigens in order to obtain precious tools for diagnostic therapeutic purposes.
- The peptide motif. Although the rules of peptide presentation have been published for the most prominent HLA allomorphs, there are some problems encountered with the application of prediction programs. First of all, the correct designation of the HLA allele of interest has to be considered. In the prediction programs (as well as in literature), there is some disagreement with respect to HLA nomenclature. While the modern naming of HLA alleles demands four-digit typing,

HLA-A*0201 for example, which precisely defines the protein sequence of this peptide-presenting receptor, very often imprecise and old-fashioned names such as HLA-A2 can still be found. The scientist who seeks epitope prediction in the web may well be puzzled by the wide offer of several variants of HLA-A*02 epitope prediction.

A number of CTL epitopes have been published from sequences of HCMV antigens. Table 1 shows a listing of epitopes derived from the pp65 protein and recognized by cytotoxic T cells in the context of HLA class I molecules. The epitopes compiled here have been defined by one of three strategies discussed above: by using overlapping synthetic peptides, by cloning experiments, or by epitope prediction. Importantly, we are convinced that not all of the epitopes described so far have the optimal length, as we expect T-cell epitopes to have 8–11 amino acids in the overwhelming majority of cases.

In the following section, the various prediction programs will be introduced and examples of their performance will be demonstrated using the peptides from Table 1.

Table 1
Cytotoxic T-lymphocyte (CTL) epitopes from pp65 of HCMV strain AD169

HLA	Sequence	Protein/position	Optimal	Reference
A*0101	YSEHPTFTSQY	pp65 363–373	+	***(12)***
A*0201	NLVPMVATV	pp65 495–503	+	***(26)***
A*1101	GPISGHVLK	pp65 16–24	+	***(27)***
A*2402	QYDPVAALF	pp65 341–349	+	***(28)***
	VYALPLKML	pp65 113–121	+	***(29)***
	FTSQYRIQGKL	pp65 369–379		***(12)***
A*6801	FVFPTKDVALR	pp65 186–196		***(12)***
B*0702	TPRVTGGGAM	pp65 417–426	+	***(27)***
	RPHERNGFTVL	pp65 265–275	+	***(12)***
B*3501	DDVWTSGSDSDEELV	pp65 397–411		***(26)***
	IPSINVHHY	pp65 123–131	+	***(30)***
B*3502	FPTKDVAL	pp65 188–195	+	***(12)***
B*38	PTFTSQYRIQGKL	Pp65 367–379		***(12)***
B*4402	EFFWDANDIY	Pp65 512–521		***(12)***

For a number of HLA restrictions, epitopes have been reported. In column 4, the "+" indicates that the epitope fits the reported peptide motif and probably cannot be elongated or truncated any further for optimal T-cell recognition.

3. Available Programs: An Overview

3.1. SYFPEITHI

SYFPEITHI is a database for MHC ligands and peptide motifs for MHC class I and class II as well as for humans and other species such as mouse, rat, cattle, and chicken. The database contains MHC–peptide motifs, MHC ligands, and T-cell epitopes. The prediction of T-cell epitopes is based on published motifs derived from pool sequencing and analysis of individual natural ligands, and especially considers amino acids in anchor and auxiliary anchor positions, as well as other frequent amino acids *(5)*. The entries are directly linked to respective sequences of the EMBL database and publications in PubMed. The algorithm used for epitope prediction is written in Object Pascal and is based on motif matrices deduced from refined motifs. In a two-dimensional data array, the letters of the amino acid represent the row index and the position numbers the column index. Any entered sequence is divided into octamers, nonamers, or decamers. A calculation of the sum of the scores from the containing amino acids for each oligomer follows. This process is repeated until the end of the sequence is reached. Different values are given to the amino acids according to their occurrence in natural ligands. Value 10 is allocated to amino acids which occur frequently in anchor positions, value 8 is given to amino acids being present in a still significant number of ligands, and value 6 is assigned to amino acids in auxiliary anchor positions. Less frequent residues in auxiliary anchor positions have the coefficient 4, and preferred amino acids have coefficients of 1–4, depending on the strength of signals in pool sequencing or the frequency in individual sequences. Finally, there are coefficients of −1 to −3 which are given to amino acids that usually do not occur in the respective sequence position of natural ligands. Epitope prediction by SYFPEITHI results in a list of peptides that are presented with high probability by MHC molecules, as indicated by the reliability of at least 80% in retrieving the most qualified epitope. This means that the naturally presented epitope should be among the top scoring 2% of all peptides predicted. Because of the more degenerate peptide motifs and the variable pocket usage, prediction of MHC class II-restricted T-cell epitopes turns out to be more complicated. SYFPEITHI as the first website that offers class I and class II predictions estimates a reliability of approximately 50%. For this reason, only MHC class I-restricted T-cell epitopes will be discussed in this chapter.

3.2. SNEP

SNEP is based on the SYFPEITHI prediction algorithms. The program predicts minor histocompatibility antigens, which are T-cell epitopes containing polymorphic residues, from proteins listed in the SWISS-PROT database. SNEP recognizes polymorphisms and predicts potential T-cell epitopes within a chosen distance around the polymorphic residue. The predictions are available for a number of HLA class I and class II allelic products, which allows for a rapid and precise evaluation of potential minor histocompatibility antigens with polymorphic antigens. There are two ways of predicting possible minor histocompatibility (miHAgs): on the one hand, it is possible to search for specific proteins and their polymorphisms, and on the other hand, the program enables the user to enter own sequence and SNP data.

3.3. Other Prediction Methods

The first predictive algorithm that was taken up by the World Wide Web and is still freely available is BIMAS that was developed by Ken Parker and colleagues *(4)*. This program was developed using data from binding studies with synthetic peptide variants and MHC molecules. Interestingly, BIMAS ranks potential MHC binders according to predicted half-time dissociation of MHC–peptide complexes. Another freely available prediction program has been placed in the World Wide Web, RANKPEP (www.mifoundation.org/Tools/rankpep.html), which can be used for the prediction of peptides binding to MHC class I and MHC class II molecules. It ranks all possible peptides from an input protein sequence according to their similarity to a set of peptides known to bind to a given MHC molecule. Using a position-specific scoring matrix (PSSM), which was assembled from a collection of aligned peptides binding to that focused MHC molecule, the similarities are scored. As a special feature, RANKPEP indicates whether the predicted peptide sequences are probably created by proteasomal processing. While SYFPEITHI and RANKPEP combine the prediction of MHC class I and MHC class II epitopes in one program, the PROPRED programs work separately for MHC class I and MHC class II. PROPRED (www.imtech.res.in/raghava/propred) *(31)* is a server for the prediction of MHC class II binding regions in an antigen sequence using quantitative matrices that are based on published data from Sturniolo *(32)*. PROPRED-I (www.imtech.res.in/raghava/propred1) offers the possibility of identifying MHC class I binding regions in antigens by using 47 MHC I alleles *(33)*. Also in PROPRED-I, proteasome filters can be selected. The sixth prediction program mentioned here which is freely available in the World Wide Web is the HLA Ligand/Motif Database (HLALIG) (hlaligand.ouhsc.edu). It offers inter alia HLA-epitope binding prediction and peptide amino acid frequency calculation *(34)*.

3.4. Prediction Methods Using Data on Processing

Three programs for the prediction of proteasomal cleavages are discussed here. PAProC is an evolutionary algorithm based on experimental cleavage data for human and yeast proteasomes. This approach enables the user to predict proteasomal cleavage sites in amino acid sequences with the possibility of getting information about the general cleavability of amino acid sequences (cuts per amino acid) and individual cleavages (positions and estimated strength). In addition, the website of PAProC offers links, for example to SYFPEITHI, to combine PAProC with the prediction of MHC class I ligands. Another neural network for proteasomal cleavage prediction is NetChop, established by Can Kesmir and colleagues *(24)*. Assuming that MHC class I ligands are not only produced by the immunoproteasome, the authors enlarged the training set of MHC class I ligands by including ligands from the MHCPEP and SYFPEITHI databases. This enlarged data set is able to predict the C-termini of MHC I epitopes. The third network reported is MAPPP *(22)*, which is able to predict proteasomal cleavage of proteins into smaller fragments and the binding of peptide sequences to MHC class I molecules.

4. The Performance of Epitope Prediction Programs, as Shown by a Retrospective Analysis

In all programs offered via the World Wide Web, the user is asked to enter or paste sequence of interest and then to select the MHC molecule of choice. Some programs allow for additional parameters, mostly concerning the output; two of them also enable the inclusion of proteasomal processing. This section demonstrates how T-cell epitope prediction from the sequence of pp65 (SWISS-PROT accession number PO6725) can be achieved, which results are obtained, and which pitfalls encountered. In our attempt to predict CTL epitopes from pp65, the published sequences shown in Table 1 were the candidates being sought. The following steps are necessary:

1. Pasting the pp65 sequence in the respective program window.
2. Selection of the appropriate HLA matrix. As discussed above, this may cause confusion. In order to find immunodominant CTL epitope NLVPMVATV ($pp65_{493-503}$), predictions entitled HLA-A2, HLA-A2.1, HLA-A*02, or HLA-A*0201 may be used. If possible, our search was carried out for HLA-A*0201 ligands.
3. Unfortunately, all the programs are not able to simultaneously calculate peptides of different length. Different matrices are used for prediction of nonamers or decamers, respectively. Therefore, in order to obtain comprehensive information, the user has to perform several predictions which will result in a listing of octamers (if possible),

one listing of nonamers, one of decamers, and so on. Afterwards, the scores of peptides contained in each listing have to be compared, which results in a combined table of candidate epitopes. In our example, the predictions of NLVPMVATV were carried out using the matrices HLA-A*0201 (8 mers), HLA-A*0201 (9 mers), HLA-A*0201 (10 mers), and HLA-A*0201 (11 mers), if offered by the respective program, and the score of NLVPMVATV compared to the scores of all predicted peptides.

4. Often, additional parameters can be selected, for example, the number of top scoring peptides to be listed. We recommend using the default setting of the programs.

The results of epitope prediction using the programs BIMAS, SYFPEITHI, RANKPEP, PROPRED-I, and HLALIG are shown in Table 2 . The numbers indicate the rank of the respective peptide among all potential epitopes within the protein sequence. For example, the peptide QYDPVAALF is the peptide with the highest score in the SYFPEITHI prediction. It was placed in position 2 of the ranking by BIMAS and PROPRED-I, in position 4 by the RANKPEP prediction, and it was not the top scoring 10 peptides, as this roughly corresponds to the top scoring 2% of peptides. The predictions were carried out with all possible length matrices offered by the programs. Rather often it was not possible to calculate epitopes because a matrix of the desired length was not available. As an example, although all programs offer HLA-A*0101-restricted epitope prediction, none of them was able to calculate 11 mers. For this reason, the epitope YSEHPTFTSQY was not identified by any of the programs.

From Table 2, we learn that epitope prediction programs work well within the limits of prediction. The six peptides ranked by BIMAS are all among the top scoring 2%, for SYFPEITHI predictions the ratio is 4/5, in RANKPEP 4/8 rankings were successful. With PROPRED-I, 5/5 epitopes would have been predicted correctly, and with HLALIG 3/5. The striking problem of all programs is indicated by the many "-" annotations: pp65 epitopes often have a rather unusual length and therefore escape epitope prediction—even one keeps in mind that some of the published epitopes have no optimal length. Especially the immunodominant epitopes, YSEHPTFTSQY (HLA-A*0101) and RPHERNGFTVL (HLA-B*0702), are unusually long: 11-mer peptides are not covered by any of the prediction programs.

4.1. Proteasomal Processing of pp65 as Predicted by World Wide Web Programs

The sequence of pp65 from the AD169 strain of HCMV was entered into the three processing prediction programs PAProC, NETChop, and MAPPP. In addition, the prediction of cleavage sites as indicated by RANKPEP is

Table 2
Prediction of cytotoxic T-lymphocyte (CTL) epitopes from pp65 of human cytomegalovirus (HCMV), strain AD169

HLA	Sequence	BIMAS	SYFPEITHI	RANKPEP	PROPRED	HLALIG
A*0101	YSEHPTFTSQSY	–	–	–	–	–
A*0201	NLVPMVATV	9	1	2	4	2
A*1101	GPISGHVLK	10		>10	5	3
A*2402	QYDPVAALF	2	1	4	2	>10
	VYALPLKML	1	3	2	1	>10
	FTSQYRIQGKL	–	–	–	–	–
A*6801	FVFPTKDVALR	–		7	–	–
B*0702	TPRVTGGGAM	2	8	>10	–	–
	RPHERNGFTVL	–	–	–	–	–
B*3501	DDVWTSGSDSDEELV	–	–	–	–	–
	IPSINVHHY	2		>10	2	3
B*3502	FPTKDVAL					
B*38	PTFTSQYRIQGKL	–		–	–	
B*4402	EFFWDANDIY		>10	>10		

Numbers indicate the ranking among high-scoring peptides; "–" indicates that the prediction failed because the correct length could not be calculated by the program.

indicated in Table 3 . All programs offer a wide range of parameters that can be altered. However, as the normal user is not expected to be an expert in proteasomal processing, all programs were used with the default settings. It has to be noted that PAProC, NETChop, and MAPPP are easy-to-use programs with comprehensive output. In contrast, from RANKPEP or PROPRED-I where proteasomal processing can be included in epitope prediction, the results of proteasomal processing are hard to extract. Nevertheless, as a comparison to the processing algorithms, the RANKPEP results are also shown in Table 3.

It is generally agreed that the C-terminal residues of T-cell epitopes (which usually serve as anchor residues in allele-specific peptide motifs) have to be released from the antigen by the proteasome. Although there is still much discussion whether many internal cleavage sites may destroy potential epitopes and whether the N-terminal residue of an epitope also has to be cut properly from the protein sequence, these two features of proteasomal processing are not regarded as relevant. From experimental evidence we know that peptides containing internal cleavage sites may nevertheless be presented by HLA

Table 3
Processing prediction using four World Wide Web programs

HLA	Sequence	PAProC	NETChop	MAPPP	RANKPEP
A*0101	YSEHPTFTSQY	+	+	–	+
A*0201	NLVPMVATV	+	+	+	+
A*1101	GPISGHVLK	–	+	–	+
A*2402	QYDPVAALF	–	+	+	–
	VYALPLKML	+	+	+	+
	FTSQYRIQGKL	+	+	–	+
A*6801	FVFPTKDVALR	+	+	+	+
B*0702	TPRVTGGGAM	+	+	+	–
	RPHERNGFTVL	+	+	–	+
B*3501	DDVWTSGSDSDEELV	–	+	–	–
	IPSINVHHY	+	+	–	+
B*3502	FPTKDVAL	+	–	+	+
B*38	PTFTSQYRIQGKL	+	+	–	+
B*4402	EFFWDANDIY	+	+	–	+

In the sequence of pp65 from HVMC, strain AD169, proteasomal cleavage sites were predicted and compared to the sequences of known CTL epitopes. Each "+" indicates that the programs predict a cleave C-terminal to the peptide sequence, which is a prerequisite for epitope generation by the proteasome.

molecules because there is no 100% efficiency in the cleavage of any single cleavage site, and some intact nonamers or decamers may escape destruction and reach the MHC molecule for subsequent presentation at the cell surface. On the other hand, proteasomal activity seems not to be important for the generation of the N-terminal site of T-cell epitopes, which became evident from two observations. First, the TAP transporter is unable to transport some of the relevant T-cell epitopes, especially if they carry a proline in the second position (such as the HLA-B*0702-presented epitopes, TPRVTGGGAM and RPHERNGFTVL). Such peptides are produced in the cytosol as precursors with additional amino acids at the N-terminal site. Second, the long-expected discovery of a trimming protease in the ER proved the hypothesis that precursors of optimal epitopes may exist in the ER and are trimmed from the N terminus until they possess the correct length. As can be seen in Table 3, NETChop performs best with this data set, with only one C-terminal cleavage site missed. PAProC and RANKPEP would have missed three cleavage sites of HCMV epitopes, while MAPP predicts only 6 of 14 cleavage sites correctly. Of course, this small data set cannot be considered for the overall performance of these programs.

4.2. SNEP-Predicted Possible Minor H Antigens

As an example for the use of SNEP *(6)*, we have predicted two known miHAgs. Table 4 shows the results of this prediction.

The first one originates from the KIAA0020 protein *(35)* and has two alternative sequences, PTLDKVLEV and RTLDKVLEV. Screening the sequence stretch around amino acid 149 in KIAA0020 for HLA-A*0201 epitopes, the P variant scores 22 and the R variant 25. This indicates that the R peptide has a greater probability of being presented to T cells on the cell surface. Both scores are above the half of the maximal score of 18 for HLA-A*0201 nonamers. One other peptide has a score above threshold, ADHPTLDKV, whereas the score of the alternative ADHRTLDKV meets the threshold precisely. The possibility that T cells from a donor carrying the R allele recognize the P allele in the recipient is therefore greater than in lower scored peptides. Table 5 shows the prediction of another miHAg derived from the BFL-1 protein and presented by HLA-A*2402 *(36)*. This peptide is, in contrast to PTLDKVLEV, an example where an amino acid exchange does not influence the score. Therefore, one may expect no difference in the HLA binding behavior. But because the exchange takes place at the central sequence position of the DYLQCVLQI peptide, DYLQYVLQI respectively, T cells could detect it due to its exposure to the outside of the HLA binding cleft.

Table 4
Prediction results for *KIAA0020*

P	Score	Above threshold	R	Score	Above threshold	Difference
PTLDKVLEL	2	+	**R**TLDKVLEL	25	+	3
H**P**TLDKVLE	1		H**R**TLDKVLE	1		0
DH**P**TLDKVL	9		DH**R**TLDKVL	9		0
ADH**P**TLDKV	20	+	ADH**R**TLDKV	18	+	2
SADH**P**TLDK	10		SADH**R**TLDK	9		1
KSADH**P**TLD	4		KSADH**R**TLD	4		0
YKSADH**P**TL	15		YKSADH**R**TL	14		1
LYKSADH**P**T	6		LYKSADH**R**T	6		0
QLYKSADH**P**	11		QLYKSADH**R**	11		0

The P149R polymorphism generates the miHAg HA-8 that was readily identified by a prediction of HLA-A*0201 nonameric peptides. Note that both variants are predicted; the R variant has a higher score than the P variant. The letters given in bold mark the relevant polymorphic spot within the peptide.

The polymorphism I26M in the CD44 protein is not yet known as miHAg. The prediction is done for HLA-A*0201 nonamers. There are only two pairs of peptides with a score above the threshold of 18. The first pair is NITCRFAGV and NMTCRFAGV, both of them scoring 20. The amino acid

Table 5
Prediction results for *BFL-1*

C	Score	Above threshold	Y	Score	Above threshold	Difference
CVLQIPQPG	4		**Y**VLQIPQPG	4		0
Q**C**VLQIPQP	1		Q**Y**VLQIPQP	11		10
LQ**C**VLQIPQ	0		LQ**Y**VLQIPQ	0		0
YLQ**C**VLQIP	1		YLQ**Y**VLQIP	1		0
DYLQ**C**VLQI	23	+	DYLQ**Y**VLQI	23	+	0
QDYLQ**C**VLQ	2		QDYLQ**Y**VLQ	2		0
AQDYLQ**C**VL	10		AQDYLQ**Y**VL	10		0
LAQDYLQ**C**V	1		LAQDYLQ**Y**V	1		0
RLAQDYLQ**C**	0		RLAQDYLQ**Y**	0		0

The C19Y polymorphism generates a miHAg pair (BCL2A1) whose variants receive identical scores. According to the prediction, the miHAg character is not caused by peptide–HLA affinity. The letters given in bold mark the relevant polymorphic spot within the peptide.

exchange does not influence the predicted binding score, just like in the BFL-1 example shown above. The scores of the second pair, SLAQIDLNI (score 23) and SLAQIDLNM (score 19), differ by 4, which suggests that there is a higher possibility of SLAQIDLNI being presented on HLA-A*0201 than SLAQIDLNM. Thus, one has to bear in mind that two factors may decide about polymorphic peptides becoming miHAgs. On one hand, HLA affinity—as reflected by the scores—influences presentation at the cell surface. On the other hand, the repertoire of T-cell receptors, which is modeled by thymic or peripheral selection processes, is responsible for the exclusive recognition of only one peptide of a polymorphic pair, even if both peptides share the same affinity.

5. Conclusion

We have listed a number of CTL epitopes from the pp65 protein of HCMV and demonstrated the performance of epitope prediction and processing prediction programs. While processing predictions worked rather efficiently, the unusual length of several HCMV epitopes posed major problems in HCMV epitope prediction. In addition, for some HLA restrictions, no predictions have yet been offered. Nevertheless, if we see such in silico work as a rapid strategy for first screening procedures, a number of the CTL epitopes investigated here would have been identified without intensive experimental efforts. As epitope prediction procedures are steadily expanded and refined, we expect that in the near future most of HCMV epitopes will be identified with the help of such prediction programs.

Acknowledgments

Our programs were established with the kind support of the European Union programs EPI-PEP-VAC and GenomesToVaccines. We also acknowledge the support of Merck KGaA, Darmstadt, Germany.

References

1. Falk K, Rötzschke O, Stevanovic S, Jung G, Rammensee HG: Allele-specific motifs revealed by sequencing of self-peptides eluted from MHC molecules. Nature 351:290–296, 1991.
2. Pamer EG, Harty JT, Bevan MJ: Precise prediction of a dominant class I MHC-restricted epitope of Listeria monocytogenes. Nature 353:852–855, 1991.
3. Rötzschke O, Falk K, Stevanovic S, Jung G, Walden P, Rammensee HG: Exact prediction of a natural T cell epitope. Eur J Immunol 21:2891–2894, 1991.

4. Parker KC, Bednarek MA, Coligan JE: Scheme for ranking potential HLA-A2 binding peptides based on independent binding of individual peptide side-chains. J Immunol 152:163–175, 1994.
5. Rammensee H, Bachmann J, Emmerich NP, Bachor OA, Stevanovic S: SYFPEITHI: database for MHC ligands and peptide motifs. Immunogenetics 50:213–219, 1999.
6. Schuler MM, Dönnes P, Nastke MD, Kohlacher O, Rammensee H, Stevanovic S: SNEP: SNP-derived Epitope Prediction program for minor H Antigens. Immunogenetics 57:816–820, 2005.
7. Dönnes P, Elofsson A: Prediction of MHC class I binding peptides, using SVMHC. BMC Bioinformatics 3:25, 2002.
8. Nussbaum AK, Kuttler C, Hadeler KP, Rammensee HG, Schild H: PAProC: a prediction algorithm for proteasomal cleavages available on the WWW.Immunogenetics 53:87–94, 2001.
9. Kern F, Faulhaber N, Frommel C, Khatamzas E, Prosch S, Schonemann C, Kretzschmar I, Volkmer-Engert R, Volk HD, Reinke P: Analysis of CD8 T cell reactivity to cytomegalovirus using protein-spanning pools of overlapping pentadecapeptides. Eur J Immunol 30:1676–1682, 2000.
10. Udaka K: Decrypting class I MHC-bound peptides with peptide libraries. Trends Biochem Sci 21:7–11, 1996.
11. Diamond DJ, York J, Sun JY, Wright CL, Forman SJ: Development of a candidate HLA A*0201 restricted peptide-based vaccine against human cytomegalovirus infection. Blood 90:1751–1767, 1997.
12. Longmate J, York J, La Rosa C, Krishnan R, Zhang M, Senitzer D, Diamond DJ: Population coverage by HLA class-I restricted cytotoxic T-lymphocyte epitopes. Immunogenetics 52:165–173, 2001.
13. DiBrino M, Parker KC, Shiloach J, Knierman M, Lukszo J, Turner RV, Biddison WE, Coligan JE: Endogenous peptides bound to HLA-A3 possess a specific combination of anchor residues that permit identification of potential antigenic peptides. Proc Natl Acad Sci USA 90:1508–1512, 1993.
14. Corr M, Boyd LF, Padlan EA, Margulies DH: H-2Dd exploits a four residue peptide binding motif. J Exp Med 178:1877–1892, 1993.
15. Falk K, Rötzschke O, Stevanovic S, Gnau V, Sparbier K, Jung G, Rammensee HG, Walden P: Analysis of a naturally occurring HLA class I-restricted viral epitope. Immunology 82:337–342, 1994.
16. Hunt DF, Henderson RA, Shabanowitz J, Sakaguchi K, Michel H, Sevilir N, Cox AL, Appella E, Engelhard VH: Characterization of peptides bound to the class I MHC molecule HLA-A2.1 by mass spectrometry. Science 255:1261–1263, 1992.
17. Schönbach C, Ibe M, Shiga H, Takamiya Y, Miwa K, Nokihara K, Takiguchi M: Fine tuning of peptide binding to HLA-B*3501 molecules by nonanchor residues. J Immunol 154:5951–5958, 1995.

18. Sette A, Grey HM: Chemistry of peptide interactions with MHC proteins. Curr Opin Immunol 4:79–86, 1992.
19. Brusic V, Rudy G., Harrison L.C.: Complex Systems: Mechanisms of Adaption. Amsterdam, IOS Press., 1994.
20. Adams HP, Koziol JA: Prediction of binding to MHC class I molecules. J Immunol Methods 185:181–190, 1995.
21. Gulukota K, Sidney J, Sette A, DeLisi C: Two complementary methods for predicting peptides binding major histocompatibility complex molecules. J Mol Biol 267:1258–1267, 1997.
22. Holzhütter HG, Frommel C, Kloetzel PM: A theoretical approach towards the identification of cleavage-determining amino acid motifs of the 20 S proteasome. J Mol Biol 286:1251–1265, 1999.
23. Nussbaum AK, Dick TP, Keilholz W, Schirle M, Stevanovic S, Dietz K, Heinemeyer W, Groll M, Wolf DH, Huber R, Rammensee HG, Schild H: Cleavage motifs of the yeast 20S proteasome beta subunits deduced from digests of enolase 1. Proc Natl Acad Sci USA 95:12504–12509, 1998.
24. Kesmir C, Nussbaum AK, Schild H, Detours V, Brunak S: Prediction of proteasome cleavage motifs by neural networks. Protein Eng 15:287–296, 2002.
25. Kuttler C, Nussbaum AK, Dick TP, Rammensee HG, Schild H, Hadeler KP: An algorithm for the prediction of proteasomal cleavages. J Mol Biol 298:417–429, 2000.
26. Wills MR, Carmichael AJ, Mynard K, Jin X, Weekes MP, Plachter B, Sissons JG: The human cytotoxic T-lymphocyte (CTL) response to cytomegalovirus is dominated by structural protein pp65: frequency, specificity, and T-cell receptor usage of pp65-specific CTL. J Virol 70:7569–7579, 1996.
27. Hebart H, Daginik S, Stevanovic S, Grigoleit U, Dobler A, Baur M, Rauser G, Sinzger C, Jahn G, Loeffler J, Kanz L, Rammensee HG, Einsele H: Sensitive detection of human cytomegalovirus peptide-specific cytotoxic T-lymphocyte responses by interferon-gamma-enzyme-linked immunospot assay and flow cytometry in healthy individuals and in patients after allogeneic stem cell transplantation. Blood 99:3830–3837, 2002.
28. Akiyama Y, Maruyama K, Mochizuki T, Sasaki K, Takaue Y, Yamaguchi K: Identification of HLA-A24-restricted CTL epitope encoded by the matrix protein pp65 of human cytomegalovirus. Immunol Lett 83:21–30, 2002.
29. Masuoka M, Yoshimuta T, Hamada M, Okamoto M, Fumimori T, Honda J, Oizumi K, Itoh K: Identification of the HLA-A24 peptide epitope within cytomegalovirus protein pp65 recognized by CMV-specific cytotoxic T lymphocytes. Viral Immunol 14:369–377, 2001.
30. Gavin MA, Gilbert MJ, Riddell SR, Greenberg PD, Bevan MJ: Alkali hydrolysis of recombinant proteins allows for the rapid identification of class I MHC-restricted CTL epitopes. J Immunol 151:3971-3980, 1993.

31. Singh H, Raghava GP: ProPred: prediction of HLA-DR binding sites. Bioinformatics 17:1236–1237, 2001.
32. Sturniolo T, Bono E, Ding J, Raddrizzani L, Tuereci O, Sahin U, Braxenthaler M, Gallazzi F, Protti MP, Sinigaglia F, Hammer J: Generation of tissue-specific and promiscuous HLA ligand databases using DNA microarrays and virtual HLA class II matrices. Nat Biotechnol 17:555–561, 1999.
33. Singh H, Raghava GP: ProPred1: prediction of promiscuous MHC Class-I binding sites. Bioinformatics 19:1009–1014, 2003.
34. Sathiamurthy M, Hickman HD, Cavett JW, Zahoor A, Prilliman K, Metcalf S, Fernandez VM, Hildebrand WH: Population of the HLA ligand database. Tissue Antigens 61:12–19, 2003.
35. Brickner AG, Warren EH, Caldwell JA, Akatsuka Y, Golovina TN, Zarling AL, Shabanowitz J, Eisenlohr LC, Hunt DF, Engelhard VH, Riddell SR: The immunogenicity of a new human minor histocompatibility antigen results from differential antigen processing. J Exp Med 193:195–206, 2001.
36. Akatsuka Y, Nishida T, Kondo E, Miyazaki M, Taji H, Iida H, Tsujimura K, Yazaki M, Naoe T, Morishima Y, Kodera Y, Kuzushima K, Takahashi T: Identification of a polymorphic gene, BCL2A1, encoding two novel hematopoietic lineage-specific minor histocompatibility antigens. J Exp Med 197:1489–1500, 2003.

6

Searching and Mapping of T-Cell Epitopes, MHC Binders, and TAP Binders

Manoj Bhasin, Sneh Lata, and Gajendra P. S. Raghava

Summary

This chapter describes searching and mapping tools of MHCBN database, which is a curated database. It comprises over 23,000 peptide sequences, whose binding affinity with major histocompatibility complex (MHC) or transporter associated with antigen processing (TAP) molecules has been assayed experimentally. Each entry of the database provides full information (such as sequence, its MHC- or TAP-binding specificity, and source protein) about peptide whose binding affinity (IC_{50}) and T-cell activity is experimentally determined. MHCBN has number of web-based tools for analyzing and retrieving information. In this chapter, we describe how to use web tools integrated in MHCBN that include (i) mapping of experimentally determined antigenic regions on the query sequence, (ii) creation of allele-specific peptide data set, and (iii) BLAST search against MHC or antigen databases.

Key Words: database; MHC binders; TAP binders; T-cell epitopes

1. Introduction

To elicit immune response (T cells) for eradicating the self-altered or foreign antigens involves a series of the complex steps that include (i) degradation of antigens to peptides through proteolytic activity, (ii) transport of the peptides to endoplasmic reticulum from cytoplasm through transporter associated with antigen processing (TAP), (iii) binding of transported peptides to major histocompatibility complex (MHC), and (iv) recognition of MHC–peptide complexes by T-cell receptors. The information about binding affinity of peptides with MHC or TAP molecules and its ability to activate T-cell response can play a

From: *Methods in Molecular Biology, vol. 409: Immunoinformatics: Predicting Immunogenicity In Silico*
Edited by: D. R. Flower © Humana Press Inc., Totowa, NJ

pivotal role in developing computational methods for subunit vaccine design. One of the important problems in subunit vaccine design is to search antigenic regions in an antigen *(1)* that can stimulate T cells called T-cell epitopes. In literature, fortunately, a large amount of data about such peptides is available. In past, a number of databases have been developed to provide comprehensive information related to T-cell epitopes *(2–5)* that includes MHCBN developed by our group. In this chapter, we describe how to use MHCBN efficiently.

2. Materials

The database contains the comprehensive information about each entry. The information has been collected mainly from literature and other public databases.

2.1. General Information

The database is accessible from the address http://www.imtech.res.in/raghava/mhcbn (Fig. 1), and it provides detailed information about a peptide, which includes (i) sequence, (ii) MHC- or TAP-binding specificity, (iii) binding affinity with MHC/TAP molecules in terms of IC_{50} value, (iv) T-cell activity, and (v) source protein. The MHC-binding affinity of peptides has been divided into four semi-quantitative groups (high, moderate, low, and unknown) using notation of MHCPEP *(6)*. The information about experimental methods used for exploring MHC–peptide-binding properties and T-cell activity has also been included in each entry. The binding of peptides to MHC is mainly studied by competitive binding assay, peptide elution, MHC reconstitution assay, stabilization assay, mass spectroscopy, and so on. The T-cell activity of peptides can be determined by assays such as cytotoxicity assay, proliferation assay, cytokine release assay, and ELISPOT assay. The MHC-binding properties can be indirectly determined by measuring T-cell activity, as all the T-cell epitopes are MHC binders. MHCBN also provides information about anchor positions (positions of peptides crucial for its interaction with TAP or MHC molecule). The miscellaneous information about peptide is provided in comment field. This field mostly contains the IC_{50} and reference IC_{50} values mentioned in the research paper to discriminate high, moderate, low, and nonbinders. The information about the published literature from where the data relevant to an entry can be obtained is also maintained in the publication reference field of the entry *(7)*. The information includes title of the paper, authors and year of publication, and name of the journal. The published literature has been linked to PubMed database at NCBI for more detailed information. The overall structure of each entry is shown in Fig. 1.

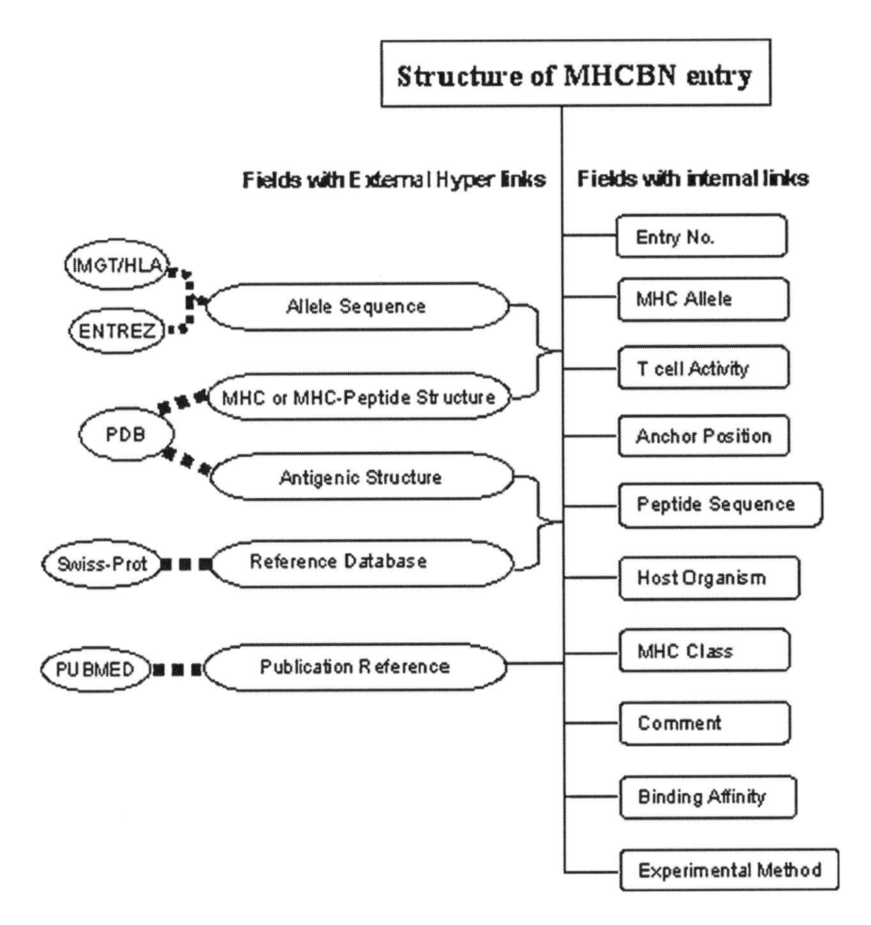

Fig. 1. Structure of MHCBN.

2.2. Sequence and Structure of Antigenic Proteins

The database also contains sequence and structural information of proteins containing regions identical to antigenic peptides. All peptides of the database were searched in SWISS-PROT Version 40, and the proteins having matching peptides have been extracted. These sequences have been stored in FASTA format in database with hyperlinks to GenBank and SWISS-PROT *(8,9)*. To provide structural insight into antigenic peptides, all the antigenic peptides were searched against the primary sequence of proteins whose three-dimensional (3D) structure has been solved by either X-ray crystallography, NMR, or molecular modeling. The summary of the 3D structure of proteins having matching peptides is available in database with hyperlink to PDB through OCA

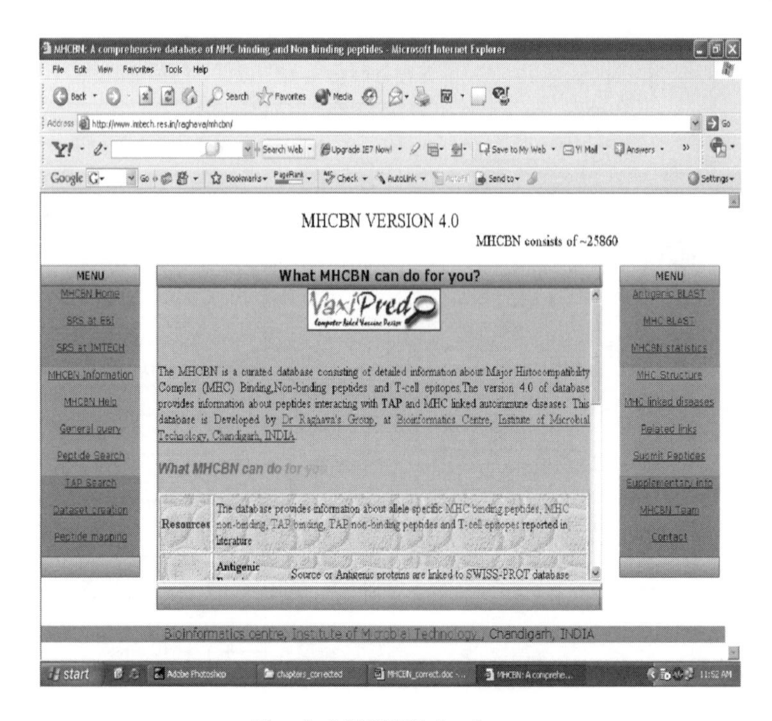

Fig. 2. MHCBN database.

browser *(10)*. This information can provide an insight into structural features of antigenic and nonantigenic regions.

All these data have been maintained in the tables of relation database created using the Postgres SQL. The structured query language is crucial in creating a relation database; it permits query on any field of tables and extracts the information about the linked fields.

3. Methods

The MHCBN has a set of web tools for making interactive and complex queries to retrieve specific information. Following is a brief description of menu options and tools available at MHCBN.

1. MHCBN Home: Clicking at this option leads to the home page of the MHCBN database (Fig. 2).
2. SRS at EBI: This option, which is linked to the EBI site, allows web-based searching and retrieval of nucleotide and protein sequence at EBI site.
3. SRS at IMTECH: The web-based searching is done at Institute Of Microbial Technology's (IMTECH) site through clicking this option.

4. MHCBN Information: This link leads to a section of the server that contains the descriptive account of the information about the architecture of MHCBN database, recent developments at MHCBN, aims and objectives, data management at MHCBN, system requirement to access, data submission and updates, acknowledgements, disclaimer and limitation of liability, and copyright.
5. MHCBN Help: This option is linked to the web page that provides complete information about the database and stepwise guidance to use it. A table of contents is displayed at the top of the page, the contents of which are internally linked to the relevant, detailed descriptions for the corresponding titles.
6. General query: The general query search options allow users to perform keyword search on any field of the database by selecting appropriate options in the general query search form (Fig. 3). The search can be performed on the basis of source protein, published references, source organism, etc. The general search allows search and customization of results on the following options:

   - Search by keyword: Any keyword, for example, a five-digit entry number, MHC allele, and name of protein, entered in this textbox, is searched in all fields of entry to extract the entire information related to that keyword from the database. The search can also be limited to specific field of entry to obtain more precise information as given below. In order to use the database, a user is first required to get registered (*see* **Note 1**).

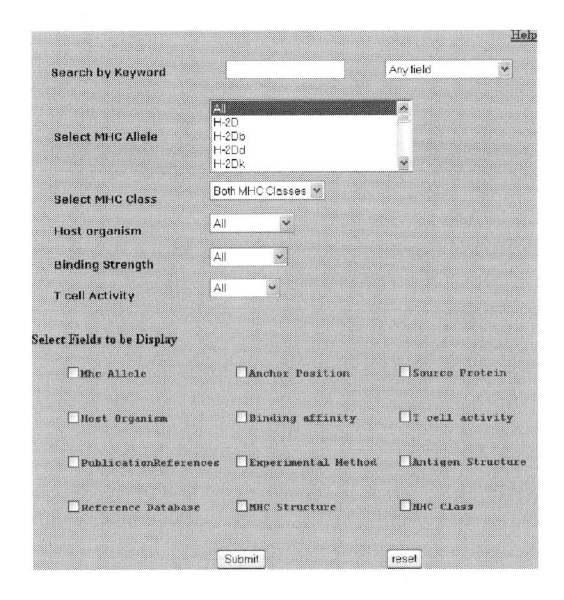

Fig. 3. Form for general query search.

- Any field: This option allows searching of a keyword in the selected field of entry only.
- Select MHC allele: One or more MHC alleles can be selected from this option box or else the default value "All alleles" is selected. Appropriate selection of the other options in the form, in combination, can prove to be highly useful. Otherwise, the default values of all the options would be considered.
- Select MHC class: Here, selection has to be made for the class of MHC to which the above-selected MHC allele belongs or to which the query peptide interacts, that is, MHC class I or MHC class II.
- Host organism: User may select the host organism harboring the MHC. For example, human, mouse, rat, and chimpanzee, or the default value "All" would be considered.
- Binding Affinity: The binding affinity of peptides to MHC is divided into four groups by using the parameter similar to MHCPEP database ***(6)***. The peptides having IC_{50} nm more than 50,000 were considered as nonbinders. In some entries, peptides are divided into high, moderate, low, and nonbinders according to the parameters mentioned by authors in papers. Depending on the need, users can select a high-affinity, moderate–affinity, or low-affinity binder, nonbinders, a peptide whose affinity is unknown, or peptides with any affinities.
- T-cell activity: This field contains a measure of immunogenic potency of a peptide. This is determined in whole database by using the same parameters as used in MHCPEP database ***(6)***. The T-cell activity is "YES?" where the peptide is immunogenic but its PD50 is unknown. One can select if the peptide required should show high, moderate, low, unknown, no T-cell activity, or every peptide shall be considered.
- Select fields to display: This option allows making choice of fields to be displayed in the result. Most of the fields in the result are linked locally or hyperlinked to other databases to obtain more detailed information. The result screen will display "No record found under specific condition" when no relevant data are found in the database.
- Output: After making appropriate selections in the form, one needs to click the submit button. This action returns the result page. On the top, a numerical figure is given that indicates the total number of entries found in the database as per to the selections made. The result is in a tabular form (Fig. 4) that contains the peptide sequences along with the options selected to be displayed in the general query search form.

7. Peptide search: This tool allows extracting of peptides from MHCBN, which are identical or have pattern similar to that of query peptide. The peptides with few mismatches can also be extracted by using this search tool. To narrow down the search conditions for more specific search, various options can be selected in the peptide search form (Fig. 5) as described below.

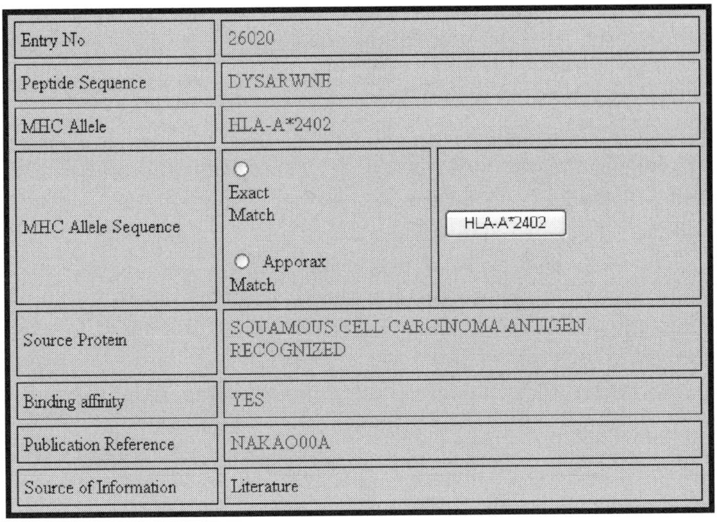

Entry No	26020	
Peptide Sequence	DYSARWNE	
MHC Allele	HLA-A*2402	
MHC Allele Sequence	Exact Match / Apporax Match	HLA-A*2402
Source Protein	SQUAMOUS CELL CARCINOMA ANTIGEN RECOGNIZED	
Binding affinity	YES	
Publication Reference	NAKAO00A	
Source of Information	Literature	

Fig. 4. Result of general query search when default parameters were selected.

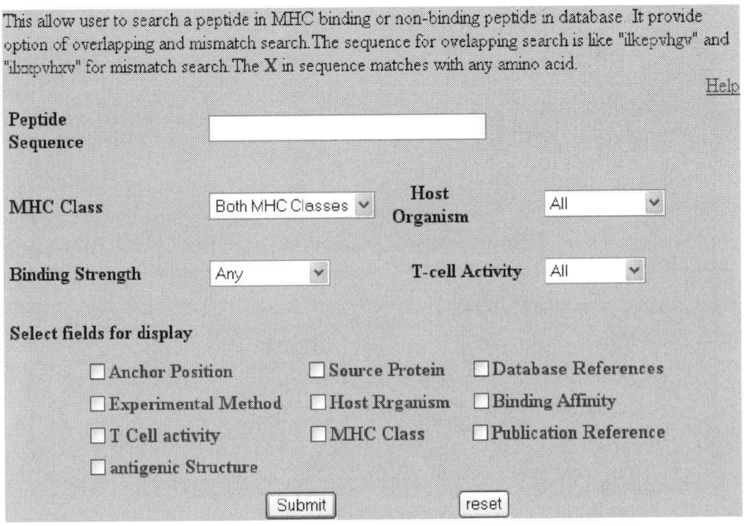

This allow user to search a peptide in MHC binding or non-binding peptide in database. It provide option of overlapping and mismatch search.The sequence for owelapping search is like "ilkepvhgv" and "ilxxpvhxv" for mismatch search.The X in sequence matches with any amino acid.

Help

Peptide Sequence

MHC Class Both MHC Classes **Host Organism** All

Binding Strength Any **T-cell Activity** All

Select fields for display

- Anchor Position
- Source Protein
- Database References
- Experimental Method
- Host Rrganism
- Binding Affinity
- T Cell activity
- MHC Class
- Publication Reference
- antigenic Structure

Submit reset

Fig. 5. Form for peptide query search.

- Peptide sequence: Paste or enter query peptide sequence as single amino acid code in the provided text box. All the nonstandard letters other than single amino acids code will be ignored. The "X" character in the sequence will match with any amino acid. The sequence of maximum 250 amino acids can be submitted.
- MHC Class: Refer Step 6.
- Host Organism: Refer Step 6.
- Binding Strength: Refer Step 6.
- T-cell Activity: Refer step 6.
- Select fields to display: Refer step 6.
- Output: The value of checked boxes will be displayed in result. The default output only has Entry no., and MHC allele corresponding to query peptide according to selected condition will be displayed. Every field displayed in result is further linked for obtaining more information (Fig. 6).

8. TAP search: This tool allows users to extract peptides interacting with TAP. The data about the peptides binding to TAP transporter will be useful in understanding the process of endogenous antigen processing. This data can be useful for the analysis of TAP-binding peptides and development of better method for prediction. Following are the options that can be selected from the TAP search form (Fig. 7):

   - Peptide Sequence: Refer Step 7.
   - Host Organism: Refer Step 6.
   - Select fields to display: Refer Step 6.

Entry No	29307
Peptide sequence	LMLGEFLKL
MHC Allele	HLA-A2
MHC Allele Sequence	⊙ Exact Match ○ Apporax Match [HLA-A2]
Source Protein	MUTATED SURVIVIN(M2)(96-104)
T-cell activity	YES HIGH
Source of Information	Literature
Comment	C50(uM)=1
Note	the c50 value is concentration of peptide required for half-maximal binding to HLA-A2

Fig. 6. Result for peptide query search for major histocompatibility (MHC) class I in human.

This options allows the search of peptides interacting with the Transporter Assiociated with Antigen Processing (TAP).The intearction of the peptide with TAP is expressed in term of IC50 or Relative IC50 Value.The user have to enter a minimum 6 amino acids.

Help

Peptide Sequence

Host Organism All

Select fields for display

☐ TAP Type ☐ Host Organism ☐ Experimental Method

☐ Publication Reference ☐ Database Reference ☐ Binding affinity(IC50)

☐ Antigen Structure

Submit reset

Fig. 7. transporter associated with antigen processing (TAP) search form.

- Output: The result displays, in a tabular format, quantitative measure of interaction between the query peptide and TAP molecule in terms of relative IC_{50} value along with the general information about the peptide (similar to the peptide search output).

9. Creation of Peptide data sets: One of the major goals behind the development of MHCBN is to provide a comprehensive source of data for the development of new and more accurate computational methods useful in subunit vaccine design. This interactive tool thus comes to aid for the creation of the allele-specific data set depending on the options selected in the data set creation form (Fig. 8):
   - Select MHC allele: Refer Step 6.
   - Select peptide-binding affinity: Refer Step 6.
   - Select peptide T-cell activity: Refer Step 6.
   - Optional field: The unique data set can be created for specific MHC allele by selecting this checkbox. Thus, no two peptides in the result would be identical.
   - Output: The action of clicking the submit button returns a table that contains the sequence of peptides for the options selected in the form. For example, on selecting the allele H-2d, high-binding affinity, and moderate T-cell activity, the output as shown in Fig. 9 was obtained.
10. Peptide mapping: This tool allows mapping of MHC binders, TAP binders and T-cell epitopes (available in MHCBN) on query protein sequence. Therefore,

User can access MHCBN database For creating datasets for Specific MHC Allele. Please restrict your query by following Field

Help

Select MHC Allele

H-2D
H-2Db
H-2Dd
H-2Dk
H-2Dq
H-2K
H-2Kb

Note:Select single MHC allele

Select peptide Binding affinity All[Binder/Non-binders

Select peptide T Cell Activity All

Optional Field

☐ **Unique Peptide Set for Specfic MHC Allele**

Submit reset

Fig. 8. Form for data set creation.

user can locate experimentally proven antigenic and nonantigenic regions in the query sequence. It is useful for the detection of immunodominant or promiscuous binding regions in query sequence. The results of the peptide mapping for specific antigenic sequence are shown in Fig. 5. All the mapped peptides are further linked to provide more detailed information such as MHC-binding specificity, binding

Total Number Of Unique Peptides In Dataset: **6**

CKGVNKEYL
KAVYNFATC
KAVYNFATCGI
SAINNYAQKL
SGPSNTPPEI
YIASNENMETMESSTLE

Fig. 9. Result for data set creation for H-2d with high-affinity and moderate T-cell activity.

affinity, T-cell activity, publication, and database reference. The user can map MHC binder, nonbinder, or T-cell epitopes of specific organism on the query sequence by selecting appropriate value of host organism and type of peptides for mapping. Users have to fill up the peptide search form (Fig. 10) and click submit button to obtain the results. The options to be selected in the peptide search form are:

- Paste your query antigen: Paste or enter your antigenic or protein sequence as single amino acid code in provided text area. All the nonstandard codes will be ignored from query sequence.
- Select MHC host organism: User can restrict search by selecting the host organism from this list.
- Select type of peptides to map: User can restrict search by selecting the value of the type of peptide mapping.
- Output: The graphical (frames) output is displayed in rows. The first row (black & bold) is amino acid sequence (single amino acid code) as submitted by the user. The remaining rows show the location of MHC binders or nonbinders or T-cell epitopes in the submitted sequence, which are found in MHCBN. Different rows or frames are taken in account for showing overlapping regions. A maximum of 20 frames are displayed for showing overlapping sequence (Fig. 11).

11. BLAST search against MHC/antigenic sequences: This tool allows BLAST search of query protein sequence against database of MHC alleles or antigenic peptide sequences. The BLAST search is useful in determining whether the query sequence belongs to MHC molecules or not. It allows to perform the following steps:

This allows the search of known MHC binders/non-binders,TAP binders/non-binders and T cell epitopes/non epitopes available in MHCBN database for query antigen or protein sequence. It display graphical mapping results.

Help

Paste your query Antigen/Protein sequence (Single letter amino acid code)

Select MHC host organism

All | Human | Mouse | Chimpanzee
Equine | Rat | Monkey | Other

Select type of Peptides to map

All | MHC Binders | Non-binders | T Epitope

Submit | Clear all

Fig. 10. Form for peptide mapping.

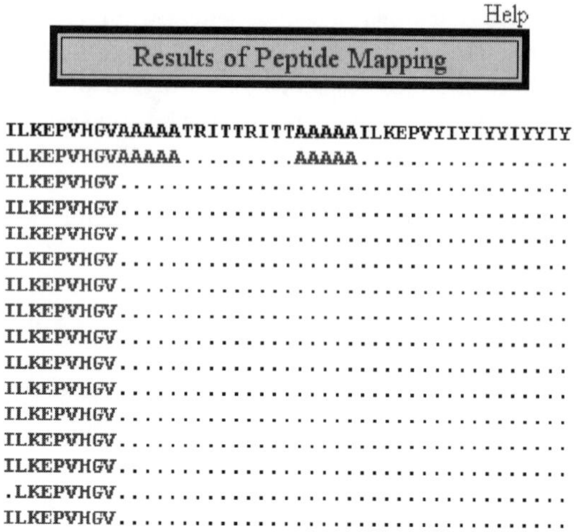

Help

Results of Peptide Mapping

```
ILKEPVHGVAAAAATRITTRITTAAAAAILKEPVYIYIYYIYYIY
ILKEPVHGVAAAAA.........AAAAA.................
ILKEPVHGV....................................
ILKEPVHGV....................................
ILKEPVHGV....................................
ILKEPVHGV....................................
ILKEPVHGV....................................
ILKEPVHGV....................................
ILKEPVHGV....................................
ILKEPVHGV....................................
ILKEPVHGV....................................
ILKEPVHGV....................................
ILKEPVHGV....................................
ILKEPVHGV....................................
.LKEPVHGV....................................
ILKEPVHGV....................................
```

Fig. 11. Sequence "ilkepvhgvaaaaatruitutruittaaaaailkepvyioyiyyiyoyiy," in human host and other default parameters.

(i) BLAST search of query sequence against MHC, (ii) extract full protein sequence of BLAST hits, (iii) multiple sequence alignment of sequences obtained from BLAST hits using CLUSTAL-W, and (iv) color view of multiple alignment by program Mview. Following fields need to be filled in BLAST search form (Fig. 12) in order to perform a BLAST search:

- Paste your Sequence: The protein sequence can be pasted into the text area, or a file containing amino acids sequence can be uploaded using this option.
- Your choice of BLAST: For example, BLASTN for nucleotide–nucleotide comparisons and BLASTP for protein–protein comparisons.
- Limit the expect value: the lower the *E* value, the more significant the score.
- Select region: Users can select a particular region from the input sequence data for similarity search. This option saves the user's job of continuously trimming and editing their sequence in case where they want to restrict their search to a particular region.
- Format type: Correct format type, that is, plain or standard (EMBL, FASTA, GENBANK, etc.) shall be selected.
- Weight matrix: BLAST uses different kinds of substitution matrices for similarity searches. It is well known that certain amino acids can easily substitute one another in related proteins, presumably because of their similar physico-chemical properties. These can be considered in calculating alignment scores in a flexible manner through the use of a substitution matrix, in which the score for any pair of amino acids can be easily looked up. Two most used matrices

Help

This option allow user to search similar sequence in database of MHC proteins in MHCBN, corresponding to their their query search. This search is performed by BLAST program at GWBLAST server at IMTECH, Chandigarh.

Paste your sequence (single letter code)
(The program will ignore all non-standard characters)

or Submit your sequence Browse...

BLAST Program blastp Expect Value (E) 1

Select region From To

Format Type Plain text (Single letter code)

Weight Matrix: BLOSUM62

Blast Against: Major Histocompatibility Complex (MHC)

Clear All Run Analysis!

Fig. 12. Form for MHC BLAST.

are Point accepted Mutation (PAM) and BLOCKS substitution (BLOSUM) matrices.

12. MHCBN statistics: This link would connect to the page having the statistical account of the total number of peptides, out of which how many are MHC binders, MHC nonbinders, Tap binders, Tap nonbinders, number of antigens, and the references.
13. MHC structure: The sequence and structure of different MHC alleles can be obtained from the database by clicking this link. Amino acid sequences of MHC alleles are stored in FASTA format with relevant hyperlinks to IMG/HLA-DB and GenBank databases ***(8,9)***. The sequence information is useful in evolutionary and variability analysis of MHC molecules. The database also maintains brief information about 3D structures of MHC molecules and MHC–peptide complexes. These structures are hyperlinked to PDB via OCA browser ***(10)*** to provide specific or detailed structural information. The structural information is useful in analysis of MHC-binding pocket and development of ab initio structure-based method for prediction.
14. MHC-linked disease search: The database also provides information about diseases associated with various MHC alleles (autoimmune disease). Users can search information by using either of the following options (Fig. 13). For example, MHC alleles responsible for rheumatoid arthritis can be easily obtained by specifying the name of disease or vice versa. This field is linked to OMIM database for more detailed information about a particular disease.

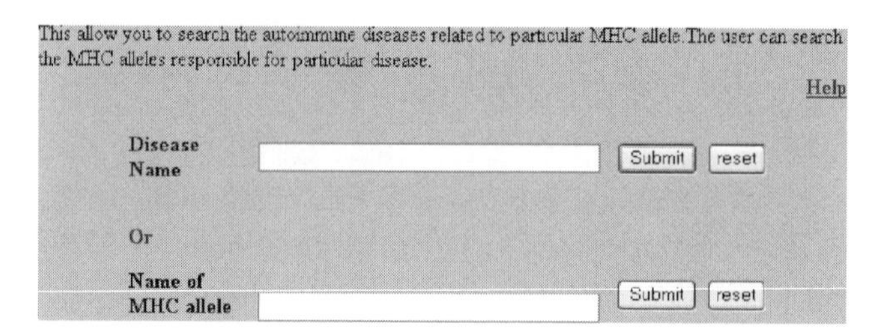
This allow you to search the autoimmune diseases related to particular MHC allele.The user can search the MHC alleles responsible for particular disease.

Help

Disease Name [ ] Submit reset

Or

Name of MHC allele [ ] Submit reset

Fig. 13. Form for MHC linked disease.

- Disease name: One can simply enter the name of the disease and click submit button to find out the alleles associated with that disease.
- Name of the allele: Users can also enter the name of the allele and find the diseases associated with that particular MHC allele.

15. Related Links: A click on this menu option leads to a page that has the names and web addresses of the various relevant immunological databases, servers, and web sites. These are also hyperlinked to the corresponding web addresses.
16. Online data submission: The MHCBN is compiled mainly from published literature reports and public databases and is regularly verified and updated. The database has a facility for online submission of MHCbinding, nonbinding peptides, and T-cell epitopes. The experimental biologist can submit the data of new MHC binders and T-cell epitopes. This will help us in maintaining the comprehensive database up-to-date. In order to maintain the quality, database team will

Please Submit your experimentally obtained MHC or TAP binding peptides here with whole information.Your sequence will included in queries after verification.

Name of MHC allele or alleles

Peptide Sequence [single AA Code]

Source Protein [e g HCV(1-14)]

Method For Obtaining Peptide

MHC Allele Host Organism [eg human]

Anchor Amino Acids Position eg 2,9

Comments

Select MHC Class: CLASS-1

Peptide Binding Affinity: High

T-cell Activity: High

Publication References

Submit reset

Fig. 14. Form for online submission of data.

crosscheck the submitted entries before inclusion in database. The submission has the following textboxes and options to be filled and selected, respectively (Fig. 14):

- Name of the MHC allele/alleles: Biologists must enter the MHC allele or alleles to which the newly found peptide is showing its binding Fig. 2. The entry in this field is mandatory (see the figure below).

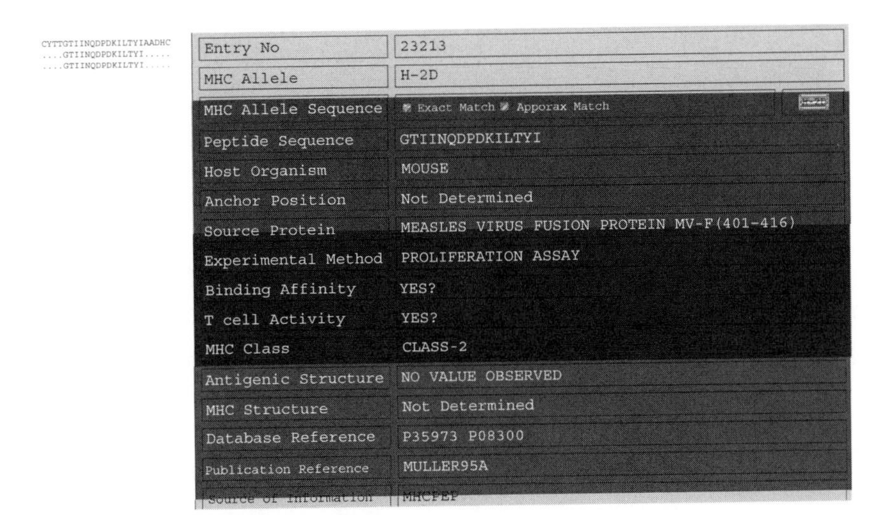

- Peptide sequence: One has to enter the sequence of a newly found peptide or protein sequence in single letter amino acid code. This field is also mandatory to fill in, and the combination of MHC allele and the sequence of the proteins is unique (see the figure below).

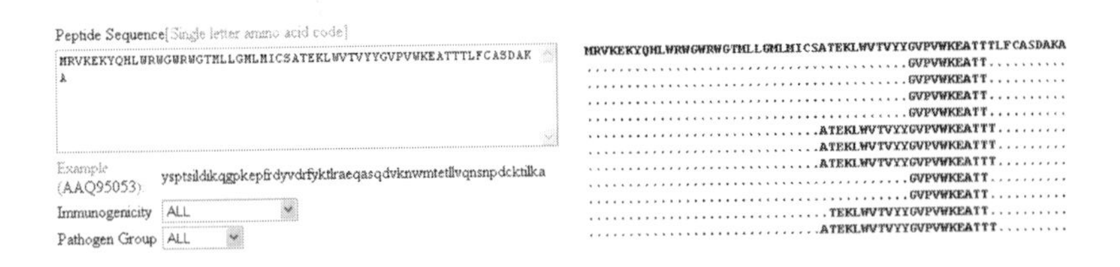

- Source protein: Users can specify the name of the protein from which the particular peptide was obtained.

- Method for obtaining the peptide: The experimental methods used for exploring peptide–MHC binding (e.g., competitive binding assay, peptide elution, MHC reconstitution assay, stabilization assay, and mass spectroscopy) and T-cell activity properties (e.g., cytotoxicity assay, proliferation assay, cytokine release assay, and ELISPOT assay) may be entered in this field.
- MHC allele host organism: Name of the host, which is harboring the specific MHC allele, may be selected from this list (e.g., human, mouse, monkey, and chimpanzee).
- Anchor amino acids position: The position of the anchor amino acids relative to N terminal can be entered in this text box. The value of field is "Not Determined" where anchor positions in peptide sequence are not analyzed.
- Comments: This text box is meant for the entry of special information or comment about the new entry. One can enter the IC_{50} and reference IC_{50} values to discriminate high, moderate, low, and non-MHC binders from each other.
- Select MHC class: Users can select the appropriate class of MHC from this list.
- Peptide-binding affinity: Users can select appropriate binding affinity based on parameters given in Step 6.
- T-cell activity: The immunogenic potency may be selected here depending upon the criteria defined in Step 6 (see the figure below).

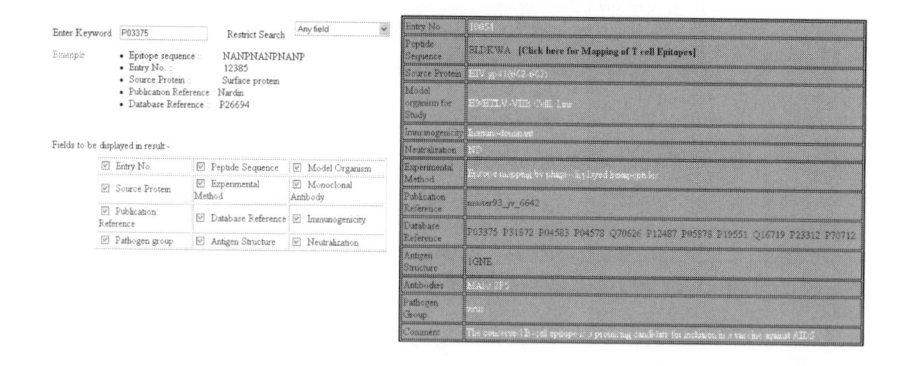

- Publication reference: This field contains the literature references of particular entry in coded form. The code has the surname of the first or corresponding author, the last two digits of the date of publication, and an identifying character in the end (if required). Each reference includes title of paper, authors, year of publication, and name of journal. The references are further linked to records of PubMed database at NCBI, which provides more comprehensive information (see the figure below).

Peptide Sequence CYTTGTIINQDPDKILTYIAADHC
Example NANPNANPNANP

Immunogenicity ALL
Pathogen Group ALL

Fields to be displayed in result -

Entry No.	Peptide Sequence	Model Organism
Source Protein	Experimental Method	Monoclonal Antibody
Publication Reference	Database Reference	Immunogenicity
Pathogen group	Antigen Structure	Neutralization

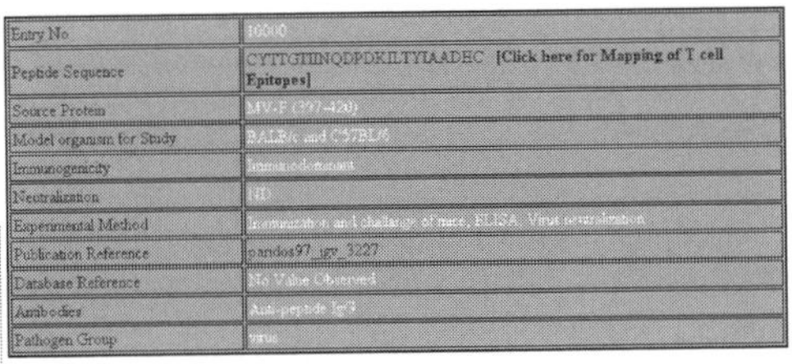

Entry No	10000
Peptide Sequence	CYTTGTIINQDPDKILTYIAADHC **[Click here for Mapping of T cell Epitopes]**
Source Protein	MV-F (397-420)
Model organism for Study	BALB/c and C57BL/6
Immunogenicity	Immunodominant
Neutralization	ND
Experimental Method	Immunization and challenge of mice, ELISA, Virus neutralization
Publication Reference	pandos97_jgv_3227
Database Reference	No Value Observed
Antibodies	Anti-peptide IgG
Pathogen Group	virus

- Submit: After all the information has been entered, users must click the submit button. On submission, a page in return will be displayed.

17. Supplementary Info: This page is linked to the supplementary information about the database.
18. MHCBN Team: This page has the name and addresses of the people involved in the development of the database.
19. Contact: This page contains the name and address of the concerned person who needs to be contacted in case of any query.

Acknowledgments

We acknowledge the financial support from the Council of Scientific and Industrial Research (CSIR) and Department of Biotechnology (DBT), Government of India.

Notes

1. To avoid the misuse of the site, the services are available for the registered users only. Users who are interested to use these servers are required to register themselves at http://www.imtech.res.in/errors/noauth.html. They need to fill up a registration form if they agree to the terms and conditions stated in the form. The user name and password is then sent by e-mail to the users.

References

1. Schirle, M., Weinschenk, T. and Stevanovic, S. (2001) Combining computer algorithms with experimental approaches permits the rapid and accurate identification of T cell epitopes from defined antigens. *J Immunol Methods*, **257**, 1–16.
2. Rammensee, H., Bachmann, J., Emmerich, N. P., Bachor, O.A. and Stevanovic, S. (1999) SYFPEITHI: database for MHC ligands and peptide motifs. *Immunogenetics*, **50**, 213–9.
3. Blythe, M.J., Doytchinova, I.A., and Flower, D. R. (2002) JenPep: a database of quantitative functional peptide data for immunology. *Bioinformatics,* **18**, 434–9.

4. Schonbach, C., Koh, J.L., Flower, D.R., Wong, L., and Brusic, V. (2002) FIMM, a database of functional molecular immunology: update 2002. *Nucleic Acids Res.*, **30**, 226–9.
5. Korber, T.M.B., Brander, C., Haynes, B.F., Koup, R., Kuiken, C., Moore, J.P., Walker, B.D., and Watkins, D.I. (2001) *Los Alamos National Laboratory, Theoretical Biology and Biophysics*, Los Alamos, New Mexico. LA-UR 02-4663.
6. Brusic, V., Rudy, G., and Harrison, L.C. (1998) MHCPEP, a database of MHC-binding peptides: update 1997. *Nucleic Acids Res.*, **26**, 368–71.
7. Wheeler, D.L., Church, D.M., Federhen, S., Lash, A.E., Madden, T.L., Pontius, J.U., Schuler, G.D., Schriml, L.M., Sequeira, E., Tatusova, T.A., and Wagner, L. (2003) Database resources of the National Center for Biotechnology Information. *Nucleic Acids Res.*, **31**, 28–33.
8. Robinson, J., Waller, M.J., Parham, P., Bodmer, J.G., and Marsh, S.G. (2001) IMGT/HLA Database—a sequence database for the human major histocompatibility complex. *Nucleic Acids Res.*, **29**, 210–3.
9. Benson, D.A., Karsch-Mizrachi, I., Lipman, D.J., Ostell, J., Rapp, B.A., and Wheeler, D.L. (2002) GenBank. *Nucleic Acids Res.*, **30**, 17–20.
10. Berman, H.M., Westbrook, J., Feng, Z., Gilliland, G., Bhat, T.N., Weissig, H., Shindyalov, I N. and Bourne, P.E. (2000) The protein data bank. *Nucleic Acids Res.*, **28**, 235–42.

7

Searching and Mapping of B-Cell Epitopes in Bcipep Database

Sudipto Saha and Gajendra P. S. Raghava

Summary

One of the major challenges in the field of subunit vaccine design is to identify the antigenic regions in an antigen, which can activate B cell. These antigenic regions are called B-cell epitopes. In this chapter, we describe how to use Bcipep, which is a database of experimentally determined linear B-cell epitopes of varying immunogenicity collected from literature and other publicly available databases. The current version of Bcipep database contains 3,031 entries that include 763 immunodominant, 1,797 immunogenic, and 471 null-immunogenic epitopes. The database provides a set of tools for analysis and extraction of data that includes keyword search, peptide mapping, and BLAST search. The database is available at http://www.imtech.res.in/raghava/bcipep/.

Key Words: B-cell epitope; immunodominant; immunogenic; neutralizing antibody; subunit vaccine design; pathogen

1. Introduction

Antibodies are the key component in the adaptive immune response of all higher vertebrates, and they recognize and bind to antigenic determinants or B-cell epitopes of an antigen. These epitopes in proteins are composed of hydrophilic amino acids, present on the protein surface, and composed of 5–30 residues. Those epitopes that produce a more pronounced immune response than others do under the same condition are termed immunodominant epitopes. B-cell epitopes can be classified into two categories: (i) conformational/discontinuous epitope, where residues are distantly separated in the sequence and brought into physical proximity by protein folding

From: *Methods in Molecular Biology, vol. 409: Immunoinformatics: Predicting Immunogenicity In Silico*
Edited by: D. R. Flower © Humana Press Inc., Totowa, NJ

and (ii) linear/continuous epitopes, which comprises a single continuous stretch of amino acids within a protein sequence that can react with anti-protein antibodies. Most of the B-cell epitopes were thought to be discontinuous. However, linear epitopes are easily identified through enzyme-linked immunoadsorbent assay (ELISA)-based epitope-mapping techniques (PEPSCAN), and experimental B-cell epitopes largely include linear epitopes. In the 1980s, it was shown that the conformational restriction is not a necessary condition for the production of protein-reactive anti-peptide antibodies. Thereafter, large numbers of linear B-cell epitopes have been reported in the literature. These epitopes can be exploited in the development of synthetic vaccines and disease diagnosis. Currently, a number of vaccines based on linear B-cell epitopes are under clinical phase trials against viruses, bacteria, and cancer. These epitopes are also important for allergy research and in determining cross-reactivity of immunoglobulin E (IgE)-type epitopes of allergens. This information of experimentally determined linear B-cell epitopes were scattered in the public domain and were collected and compiled in Bcipep database *(1,2)*. There are other databases that provide limited information on B-cell epitopes such as JenPep *(3)*, which has been superseded by AntiJen 2.0. In this chapter, we describe how to use Bcipep database efficiently.

2. Materials

2.1. Source of Information

Information about B-cell epitopes was collected from the literature (PubMed, http://www.ncbi.nlm.nih.gov/pubmed/; ScienceDirect, http://www.sciencedirect.com/). The information of the epitopes was curated manually and compiled. A large number of human immunodeficiency virus (HIV) B-cell epitopes were extracted from a book *HIV Molecular Immunology 2001* *(4)*. Statistics on pathogen group and on immunogenicity vice distribution of B-cell epitopes in the database has been shown in Table 1.

2.2. Description of Database

The Bcipep database provides (i) comprehensive information about B-cell epitopes, which includes source of protein, immunogenic potency of epitopes, model organism, monoclonal or polyclonal antibodies produced against an epitope, and neutralization potential of anti-peptide antibody; (ii) tools for extraction and analysis of this information such as keyword search and peptide mapping; and (iii) hyperlinks to MHCBN *(5)*, PUBMED *(6)*, Swiss-Prot *(7)*, and PDB *(8)* databases. The database also provides sequence retrieval system

Table 1
Statistics on pathogen group and immunogenicity vice distribution of B-cell epitopes in Bcipep database

Pathogen	Immunodominant	Immunogenic	Nonimmunogenic
Virus (2,046)	415	1474	157
Bacteria (539)	130	159	250
Protozoa (236)	139	57	40
Fungi (53)	17	23	15
Others (157)	62	84	9

(SRS) version to retrieve information (http://www.imtech.res.in/srs5bin/cgi-bin/wgetz?-fun+pagelibinfo+-info+BCIPEP).

2.3. Description of Web Interface

The Bcipep data have been maintained in "Relational Database Management System" (RDBMS) called PostgreSQL, which is a public domain software freely available. Full information about a peptide has been stored in a single table. Related information such as publication reference and source proteins reference has been stored in directories having internal links to main database. The web server was developed in a UNIX environment on SUN server 420E in Solaris 7.0. This server is designed to provide easy access to the users, based on a set of simple graphical user interface (GUI) forms. Methods for searching the databases and displaying the selected objects were built with HyperText Markup Language (HTML) and CGI-scripts in PERL 5.4. Requirements for accessing Bcipep are (i) Windows 95 and later version for personal computers and (ii) Internet Explorer 4.01 and later version or Netscape 3.01 and later version.

3. Methods

3.1. Description of Home Page

The Bcipep database is available at http://www.imtech.res.in/raghava/bcipep/. The menus are in the left side of the page and are interlinked. Following is a brief description of menus

3.1.1. Bcipep Home

It links to the home page of Bcipep, a database of B-cell epitopes. The snapshot of the home page of Bcipep has been shown in Fig. 1.

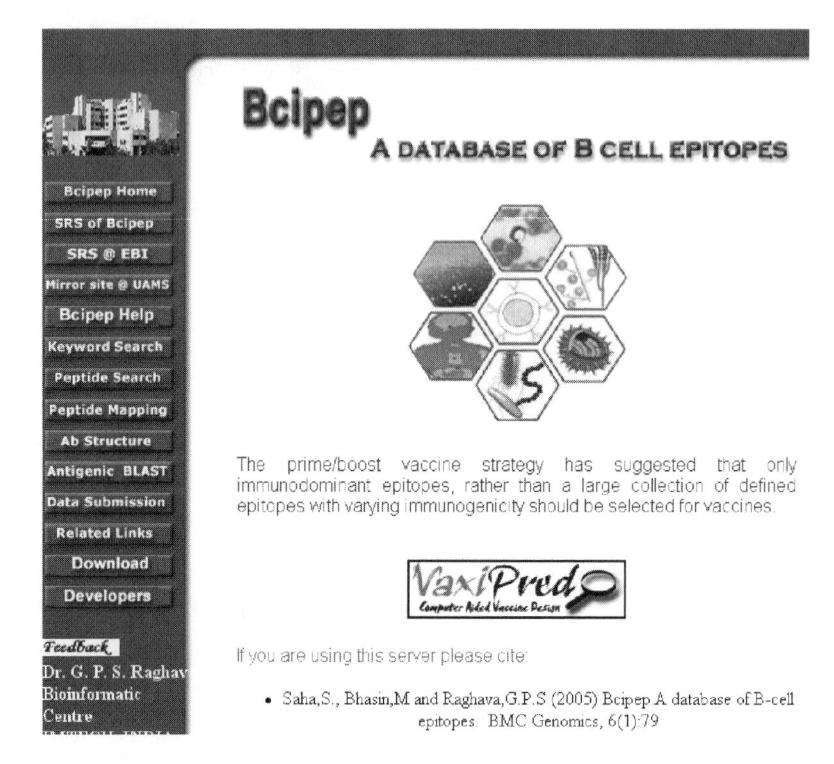

Fig. 1. The home page of Bcipep database.

3.1.2. SRS of Bcipep

It allows user to access SRS version of Bcipep from IMTECH site (http://www.imtech.res.in/srs5bin/cgi-bin/wgetz?-fun+pagelibinfo +-info+ BCIPEP).

3.1.3. SRS @ EBI

Users can access SRS version of Bcipep from European Bioinformatics Institute site, UK (http:// srs.ebi.ac.uk/srs6bin/cgi-bin/wgetz?-page+LibInfo+-id+1X2XW1JU5_L+-lib+BCIPEP).

3.1.4. Mirror Site @ UAMS

This menu allows user to access Bcipep from its mirror site which is available at University of Arkansas for Medical Sciences, Little Rock, USA (http:// bioinformatics.uams.edu/mirror/bcipep/).

3.1.5. Bcipep Help

This page contains general information about the database entries, schematic diagram of B-cell epitope, statistics on pathogen group and immunogenicity vice distribution of B-cell epitope in the database, and application of Bcipep data in prediction methods. There are other information related to specification and requirements for accessing the Bcipep.

3.1.6. Keyword Search

This menu allows users to perform keyword search. More information is available in Section 3.4.1.

3.1.7. Peptide Search

It allows users to search their peptide in Bcipep database (see Section 3.4.2).

3.1.8. Peptide Mapping

Users can map online B-cell epitopes in Bcipep database on their antigen sequence (see Section 3.4.4).

3.1.9. Ab Structure

Bcipep maintains structure of antibodies whose coordinates are available in Protein Data Bank; user can access these structures by clicking on *Ab Structure*.

3.1.10. Antigenic BLAST

This option allows users to search their antigen sequence against the antigen sequence in Bcipep using BLAST (see Section 3.4.5).

3.1.11. Data Submission

Users can submit new experimental B-cell epitope data using this link.

3.1.12. Related Links

It links to other related databases.

3.1.13. Download

This menu allows users to download B-cell epitope data (*see* **Note 1**).

3.1.14. Developers

It shows the addresses of Bcipep database developers.

3.2. Usage of Bcipep Database

The users are required to fill a request form available at http://www.imtech.res.in/errors/noauth.html for using web servers developed by raghava's group (http://www.imtech.res.in/raghava/). The user name (e-mail id) and password are provided through e-mail. The old users can directly access the database by providing the user name and password.

3.3. Description of Fields

Bcipep have more than 3,000 entries; each entry consists of 13 fields. Each field provides specific information related to B-cell epitopes. Following is a brief description of each field.

3.3.1. Entry Number

This is a five-digit unique identifier provided to each entry of the database. The entries can be searched from the database by using this unique identifier.

3.3.2. Peptide Sequence

This field has the primary amino acid sequence of the epitope. It has a link to MHCBN database *(5)* in order to identify the peptides that are B-cell as well as T-cell epitopes.

3.3.3. Source Protein

The information about the source protein from which the epitope is obtained is available in this field. The field also provides the specific information about the position of the peptide in the antigenic sequence.

3.3.4. Pathogen Group

The group of the organism to which the source protein or antigen belongs is provided. The major pathogenic groups are virus, bacteria, fungi, and protozoa.

3.3.5. Immunogenicity

This is a semiquantitative measure of immunogenic activity of the peptide. In Bcipep, it is divided into three categories: (i) immunodominant, if it increases twofold to threefold anti-peptide antibodies in comparison with reference or control (carrier protein, e.g., BSA or KLH); (ii) immunogenic, if it enhances anti-peptide antibodies by onefold in comparison to reference; and (iii) null-immunogenic, where no difference was observed when compared to reference. This information is very important for developing B-cell epitope prediction method.

3.3.6. Model Organism

The database also provides information about the experimental animal model used for immunization. Inbred mice, pigs, dogs, and monkeys were used as animal model.

3.3.7. Experimental Method

This field of each entry specifies the experimental methods, which were used for checking the immunogenic or antigenic properties of the peptide, such as ELISA.

3.3.8. Antibody

This field provides full information about monoclonal or polyclonal antibodies against an epitope. The information includes isotypes of Ig and name/number of monoclonal antibodies.

3.3.9. Neutralization

The database contains information about neutralization potential of anti-peptide antibody, which were crucial for considering a peptide for synthetic vaccine design. In cases where no data was available, it is marked as ND (not done by the author).

3.3.10. Antigen Structure

This field consists of PDB codes of protein structures having matching peptides. These PDB codes are linked to OCA browser (http://pdb.tau.ac.il/) in order to provide detailed structural information of these proteins. The database has structure of 1,216 antigenic proteins.

3.3.11. Database Reference

The Bcipep provides hyperlinks to various sequence databases in order to provide detailed information about peptides in other databases. The *database reference* field consists of name/code of protein available in Swiss-Prot.

3.3.12. Publication Reference

The field provides full information about related publications with link to PubMed ***(6)***.

3.3.13. Comments

More important information about the peptide used in the study is available in the *Comment* field (*see* **Note 2**).

3.4. Web Tools Integrated in Bcipep

3.4.1. Keyword Search

This option allows users to perform search on all major fields of the database (*Peptide Sequence*, *Source Protein*, *Publication Reference*, and *Database Reference*). One can restrict the keyword search on any specific field. It also allows users to select the fields to be displayed. An example of keyword search is shown in Fig. 2A, where key word 'P26694' is searched in any field of the database. The output/result of this keyword search is shown in Fig. 2B.

3.4.2. Peptide Search

The database provides option to search a peptide in Bcipep. The server permits users to search their query sequence in any pathogen group. Search can be restricted on the basis of immunogenicity (e.g., immunodominant, immunogenic, or null-immunogenic) and pathogen group (all, virus, bacteria, fungi, and protozoa). An example of input and output of peptide search is shown in Fig. 3A, B, respectively.

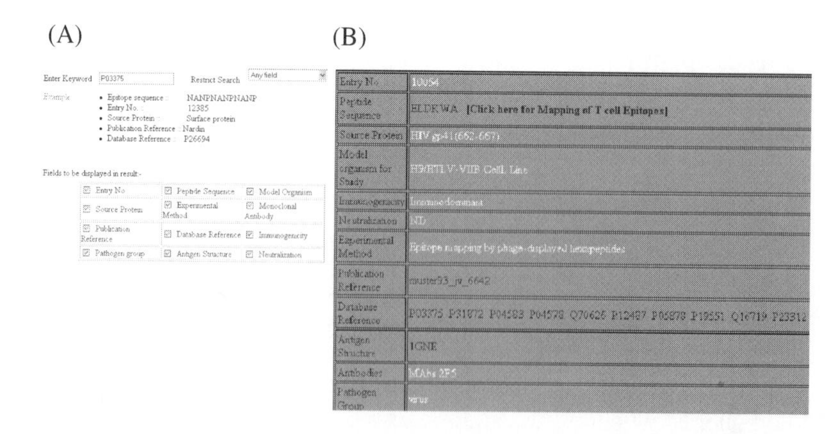

Fig. 2. The typical display of Bcipep database for keyword search, (**A**) input page of keyword search and (**B**) output of keyword search.

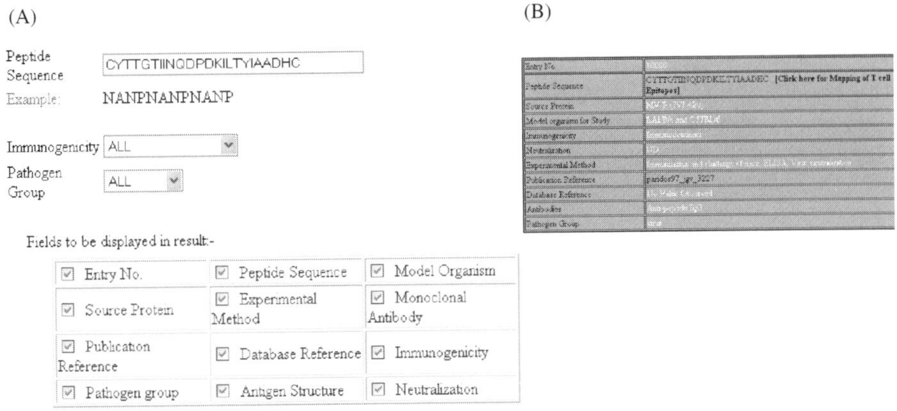

Fig. 3. An example that illustrates search on Bcipep, (**A**)peptide search page and (**B**) result of peptide search.

3.4.3. Mapping of T-cell Epitopes

It allows searching of peptide in Bcipep against MHCBN *(5)* database. By clicking on "mapping of T-cell epitopes" on *Peptide Sequence* field, it links to MHCBN database, which provides information about components of cell-mediated immunity such as MHC binders/nonbinders, T-cell epitopes, and TAP binders. The example of mapping of MHCBN peptides on B-cell epitope [by clicking on *Peptides Sequence* field (see Fig. 3B)] is shown in Fig. 4A. The full information of each map peptide can be obtained by clicking on the mapped sequence. One such example is shown in Fig. 4A, B. Thus, the server is useful in identifying the potential B-cell epitopes having T-cell epitopes (or MHC binders).

3.4.4. Peptide Mapping

The peptides of Bcipep can be mapped on query sequence using this option. The full information about mapped peptide can be obtained by clicking on it. The tool will assist the users in gaining knowledge about the known immunogenic or nonimmunogenic regions in target protein of interest. The users can specify the pathogen group and/or immunogenicity level of peptides to be mapped on query sequence. An example of input and output of peptide mapping is shown in Fig. 5A, B, respectively. The users can specify the pathogen group and/or immunogenicity level of peptides to be mapped on query sequence.

(A)

```
CYTTGTIINQDPDKILTYIAADHC
....GTIINQDPDKILTYI.....
....GTIINQDPDKILTYI.....
```

(B)

Entry No	23213
MHC Allele	H-2D
MHC Allele Sequence	⊙ Exact Match ○ Apporax Match
Peptide Sequence	GTIINQDPDKILTYI
Host Organism	MOUSE
Anchor Position	Not Determined
Source Protein	MEASLES VIRUS FUSION PROTEIN MV-F (401-416)
Experimental Method	PROLIFERATION ASSAY
Binding Affinity	YES ?
T cell Activity	YES ?
MHC Class	CLASS-2
Antigenic Structure	NO VALUE OBSERVED
MHC Structure	Not Determined
Database Reference	P35973 P08300
Publication Reference	MULLER95A
Source of Information	MHCPEP

Fig. 4. Mapping of peptide in MHCBN database, (**A**) mapping of MHCBN peptides on B-cell epitopes and (**B**) full information about a MHCBN peptide.

(A)

Peptide Sequence[Single letter amino acid code]

```
MRVKEKYQHLWRWGWRWGTMLLGMLMICSATEKLWVTVYYGVPVWKEATTTLFCASDAK
A
```

Example (AAQ95053): ysptsildikqgpkepfrdyvdrfyktlraeqasqdvknwmtetllvqnsnpdcktilka

Immunogenicity ALL

Pathogen Group ALL

(B)

```
MRVKEKYQHLWRWGWRWGTMLLGMLMICSATEKLWVTVYYGVPVWKEATTTLFCASDAKA
........................................GVPVWKEATT..........
........................................GVPVWKEATT..........
........................................GVPVWKEATT..........
........................................GVPVWKEATT..........
.............................ATEKLWVTVYYGVPVWKEATTT.........
.............................ATEKLWVTVYYGVPVWKEATTT.........
.............................ATEKLWVTVYYGVPVWKEATTT.........
........................................GVPVWKEATT..........
........................................GVPVWKEATT..........
..............................TEKLWVTVYYGVPVWKEATT..........
.............................ATEKLWVTVYYGVPVWKEATTT.........
```

Fig. 5. Mapping of B-cell epitopes on antigen sequence, (**A**) submission page of B-cell epitope mapping and (**B**) mapping results.

3.4.5. Antigenic BLAST

This tool allows users to search their query protein against antigenic proteins maintained at Bcipep. The sequence of 1,070 antigenic proteins has been obtained from Swiss-Prot. The similarity search is performed using the GWBLAST server (http://www.imtech.res.in/raghava/gwblast/). The

GWBLAST also allows users to analyze the BLAST output such as multiple alignments and phylogenetic analysis.

3.5. Link to Other Databases

The Bcipep provides hyperlinks to various sequence databases in order to provide detailed information about the peptides in other databases. The *database reference* field consists of name/code of protein available in Swiss-Prot. The *Antigenic structure* field consists of PDB codes of protein structures having matching peptides. These PDB codes are linked to OCA browser http://pdb.tau.ac.il/ in order to provide detailed structural information of these proteins. The *Publication reference* field provides full information about related publications with link to PubMed. Bcipep is also linked to MHCBN database in order to identify the peptides that are B-cell as well as T-cell epitopes.

3.6. Potential Applications

The identification of regions/stretches on an antigen from the data pool of known epitopes is an important step in vaccine design. The Bcipep database would be very useful as it consists of comprehensive information about experimentally verified linear B-cell epitopes and tools for mapping these epitopes on an antigen sequence. In case a query antigen contains known epitopes, this database might aid in the wet experimentation and lower the cost by reducing the overlapping repeats. This database also provides a link with MHCBN to search for overlapping regions of MHC binders and T-cell epitopes in the B-cell epitope. The epitopes in Bcipep can be used to derive rules for predicting B-cell epitopes.

Acknowledgments

We acknowledge the financial support from the Council of Scientific and Industrial Research (CSIR) and Department of Biotechnology (DBT), Government of India.

Notes

1. The users can download all the entries and also of specific pathogen group, after providing proper username and password.
2. Additional information is available in this field—about the mutagenic studies on particular residues, diagnosis study, and the development of vaccine candidate.

References

1. Saha, S., Bhasin, M. and Raghava, G.P.S. (2005) Bcipep: a database of B-cell epitopes. *BMC Genomics* **6**(1):79
2. Saha, S., Bhasin, M. and Raghava, G.P.S. (2005) Bcipep. *Nucleic Acids Res.* (Online; http://www3.oup.co.uk/nar/database/summary/642/).
3. Blythe, M.J., Doytchinova, I.A. and Flower, D.R. (2002) JenPep: a database of quantitative functional peptide data for immunology. *Bioinformatics* **18**:434–439.
4. Korber, B., Brander, C., Haynes, B., Koup, R., Kuiken, C., Moore, J., Walker, B. and Watkins, D. (2002) HIV monoclonal antibodies. In *HIV Molecular Immunology 2001*. Published by Theoretical Biology and Biophysics group T-10, Mail Stop K710 Los Alamos national Laboratory, Los Alamus, New Mexico, IV-B-1–IV-B-278.
5. Bhasin, M., Singh, H. and Raghava, G.P.S. (2003) MHCBN: a comprehensive database of MHC binding and non-binding peptides. *Bioinformatics* **19**(5):665–666.
6. Wheller, D.L., Church, D.M., Lash, A.E., Leipe, D.D., Madden, T.L., Pontius, J.U., Schuler, G.D., Schriml, L.M., Tatusova, T.A., Wagner, L. and Rapp, B.A. (2002) Database resources of National Center for Biotechnology Information: 2002 update. *Nucleic Acids Res.* **30**:13–16.
7. Bairoch, A. and Apweiler, R. (2000) The SWISS-PROT protein sequence database and its supplement TrEMBL in 2000. *Nucleic Acids Res.* **28**:45–48.
8. Westbrook, J., Feng, Z., Jain, S., Bhat, T.N., Thanki, N., Ravichandran, V., Gilliland, G.L., Bluhm, W.F., Weissig, H., Greer, D.S., Bourne, P.E. and Berman, H.M. (2002) The Protein Data Bank: unifying the archive. *Nucleic Acids Res.* **30**:245–248.

8

Searching Haptens, Carrier Proteins, and Anti-Hapten Antibodies

Shilpy Srivastava, Mahender Kumar Singh, Gajendra P. S. Raghava, and Grish C. Varshney

Summary

Haptens are small molecules that are usually nonimmunogenic unless coupled to some carrier proteins. The generation of anti-hapten antibodies is important for the development of immunodiagnostics and therapeutics. Recently, our group has developed a database called HaptenDB, which provides comprehensive information about 1,087 haptens. In this chapter, we describe following web tools integrated in HaptenDB: (i) keyword search facility allows search on major fields, (ii) browsing service, to display all haptens, carrier proteins and antibodies, and (iii) structure similarity search, which allows the users to search their structure against hapten structures.

Key Words: Carrier protein; database; hapten; haptenDB; pesticides

1. Introduction

Haptens are small molecules, such as pesticides, drugs, hormones, and toxins, which are usually nonimmunogenic unless coupled with some macromolecules such as proteins. These carrier molecules provide T lymphocyte help required for the induction of humoral (antibody) response. Direct coupling of hapten with carrier protein is possible where the target compound contains functional groups such as $-NH_2$ and –COOH. Alternatively, these functional groups can be introduced by derivatization of the hapten. Thus, the production of anti-hapten antibodies of desired specificity depends on hapten design (preserving the chemical structure and spatial conformation of target compound), selection of appropriate carrier protein, and the conjugation method ***(1)***. Antibodies once

From: *Methods in Molecular Biology, vol. 409: Immunoinformatics: Predicting Immunogenicity In Silico*
Edited by: D. R. Flower © Humana Press Inc., Totowa, NJ

generated can be exploited for multiple applications such as in serology, drug delivery, and development of immunodiagnostic kits.

Most of the contaminants in the environment including soil, water, air, and food are small molecules that are often nonimmunogenic (haptens). Moreover, the haptens can be altered structurally to raise the antibodies of defined specificity and affinities toward target analyte. Immunochemical techniques such as immunoassays, immunosensors, immunochromatography, and immunolabeling supplement traditional analytical methods in an ideal way because these are extremely sensitive, simple, and inexpensive. Standardized immunochemical methods for medicine, food, and environmental monitoring calls for the generation of antibodies of defined specificities and affinities against the analyte/hapten.

Immunology has followed the trend of molecular biology in the explosive generation of new data. The amount of data pertaining to haptens is overwhelmingly increasing because of its growing applied importance. Advances in database technology have enabled us to manage these data efficiently, while at the same time, bioinformatics have provided new tools for data analysis. Though there are number of immunological databases on protein sequences and peptides (epitopes) (KABAT, IMGT, FIMM, MHCBN, BCIPEP, and AntiJen 2.0) *(2–7)*, but there is only one database on haptens called HaptenDB *(8)*. HaptenDB is a comprehensive database comprising haptens, carrier molecules, and the antibodies where the information has been collected from the web sources and the standard literature *(8)*. HaptenDB, the first of its kind, aims at providing the information about chemical, physical, and structural properties of haptens to the user . Besides, it also contains information about the carrier molecules used to raise the antibodies against the particular hapten, together with the conjugation methods, immunization schedules, host organism, and the properties of the antibodies generated. The database further describes the assay method, which could be used to characterize the antibody, as well as the application of the antibody generated, e.g., in immunodiagnostics. The database is comprehensive in itself as it has integrated many aspects of the hapten that one would like to gather for research or application purpose. Furthermore, the database has some structure similarity tools that would enable the user to check against the query, whether the database has entries to similar/or related structures and respective antibodies. To collect the particular information, if not entered in the database, the reference and web link of each source is given. Although the database is made user-friendly by making each page self-explanatory, still one can go to Help, Information, and Related links options on Home Page (see **NOTE 3, 4 & 5**).

2. Materials

2.1. Web Server

The HaptenDB web server was developed in a UNIX environment on SUN server 420R in Solaris 7.0. This server is designed to provide easy access to the user, based on a set of simple graphical user interface (GUI) forms. Methods for searching the databases and displaying the selected objects were built with a combination of Java Scripts and CGI-scripts in PERL 5.4. One can access database and web tools via Internet from http://www.imtech.res.in/raghava/haptendb/ or http://www.imtech.ac.in/raghava/haptendb/ (see **NOTE 1**). In order to provide search on any field of database and to maintain standards, SRS version of HaptenDB (http://www.imtech.res.in/srs/) and its mirror sites have been launched on SGI origin server under IRIX environment, which is available from http://bioinformatics.uams.edu/.

2.2. Description of Data

The current version of the database has 2,021 entries for 1,087 haptens and 25 carrier proteins. Each entry provides comprehensive details about (i) nature of the hapten, (ii) information about carrier protein, (iii) coupling method, (iv) methods of anti-hapten antibody production, (v) assay method (used for characterization), and (vi) specificities of antibodies. Moreover, the haptens and the antibodies are categorized on the basis of their nature, for example, pesticides, herbicides, insecticides, drugs, toxins, steroids, and hormones. Tables 1 and 2 present the number of haptens and antibodies entered so far under different categories.

Table 1
Distribution of haptens (1,087)

Category	Number of entries
Pesticides, insecticides, fungicides, herbicides, etc.	225
Toxins	26
Drugs, antibiotics, analgesics, narcotics, etc.	120
Hormones, auxins, phytoestrogens, etc.	19
Synthetic and natural peptides	17
Vitamins and their analogs	18
Others (dyes, explosives, etc.)	99
Unclassified haptens or haptens belonging to smaller groups	563

Table 2
Distribution of anti-hapten antibodies entries (2,021)

Category	Number of entries
Pesticides, insecticides, fungicides, herbicides, etc.	650
Toxins	40
Drugs, antibiotics, analgesics, narcotics, etc.	200
Hormones, auxins, phytoestrogens, etc.	30
Synthetic and natural peptides	41
Vitamins and their analogs	50
Others (dyes, explosives, etc.)	210
Unclassified haptens or haptens belonging to smaller groups	800

3. Method

3.1. Browsing Tools

HaptenDB has number of browsing tools. To help the users, home page displays three options of Hapten, Carrier, and Antibodies browsers for direct search.

3.1.1. Browsing Haptens

This option allows users to browse haptens in database. The users can click the hapten link provided on the home page, which will provide brief information about each hapten. Figure 1 shows the example output of this option that includes haptens, their synonyms, and modifications.

3.1.1.1. Detailed description of hapten

One gets brief description about hapten by clicking on browsing option, HaptenDB. As shown in Fig. 1, each hapten record has clickable button 'Detail,' where user can get detailed information about a hapten. An example of hapten 2,4-dichlorophenoxyacetic acid is shown in Fig. 2, while Table 3 shows the name and description of field.

3.1.2. Browsing Carrier

Similarly, on clicking the carrier option on home page, one would receive the output (Fig. 3) as a list of 25 different carriers with their name, nature, and sequence distributed over two pages.

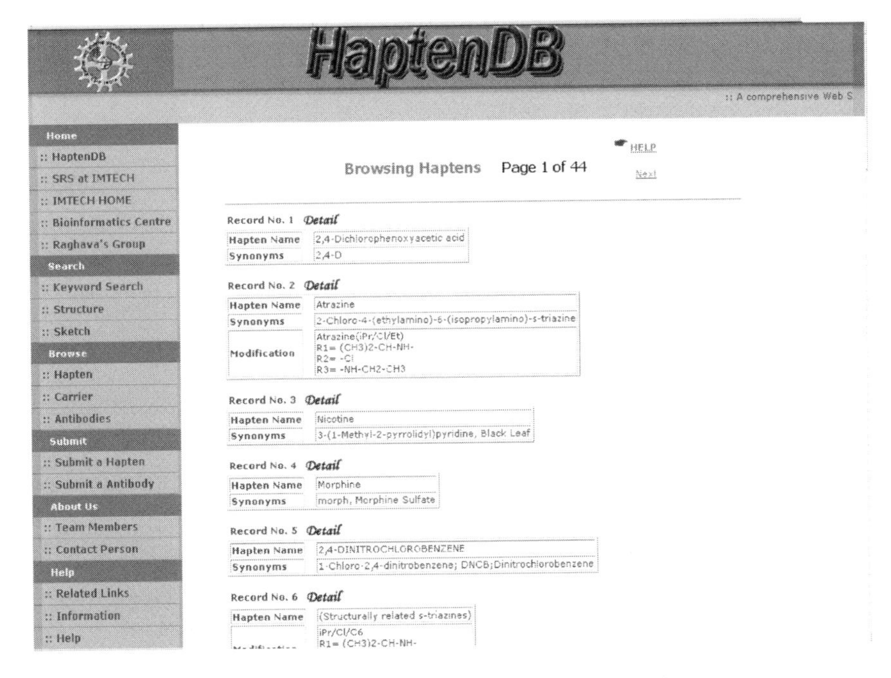

Fig. 1. Browsing of hapten molecules, an example output.

3.1.2.1. Detailed Description of Carrier Molecule

Each carrier record has clickable button 'Detail,' which provides detailed description of a carrier molecule. An example record of avidin is shown in Fig. 4. The carrier is usually a high molecular weight protein attached with the hapten to provide it immunogenicity. The brief description fields are given in Table 4.

3.1.3. Browsing Antibodies

The clicking of antibody browser on the home page will show output as a list of 238 records of different antibodies with their name and type (Fig. 5) distributed over ten pages, and clicking the detail of any record will show output as a list of the entries for a particular antibody raised against same, related, or different haptens, along with the type and cross-reactivity of the antibody (inlay in Fig. 5). Finally, clicking the detail of particular antibody against the particular hapten will show the output (Fig. 6) as a table describing the properties of the antibody.

Record No. 1

Download as A Text File

Hapten Name	2,4-Dichlorophenoxyacetic acid
Synonyms	2,4-D
Molecular Formulae	C8H6Cl2O3
Physical Properties	Colour:WHITE TO YELLOW CRYSTALLINE POWDER Odour:ODORLESS WHEN PURE Boiling Point: 160 DEG C AT 0.4 MM HG Melting Point: 138 DEG C Density: 1.416 @ 25 deg C
Nature	Pesticide (Herbicide)
Molecular Weight	221.04
Toxicity	Toxic
Area of Uses	AS A HERBICIDE FOR CONTROL OF BROADLEAF PLANTS & AS A PLANT-GROWTH REGULATOR. HERBICIDE USED ON GRASSES, WHEAT, BARLEY, OATS, SORGHUM, CORN, SUGARCANE, & NONCROP AREAS PASTURE AND RANGE LAND; It is used on tomatoes to cause all fruits to ripen at the same time for machine harvesting. /2,4-D free acid serves as the basic material from which the soluble esters & salts are produced. Used in forest management: Brush control; Conifer release; Tree injection. To increase latex output of old rubber trees. Fruit drop control

Download Structure in 2D/3D MOL Format

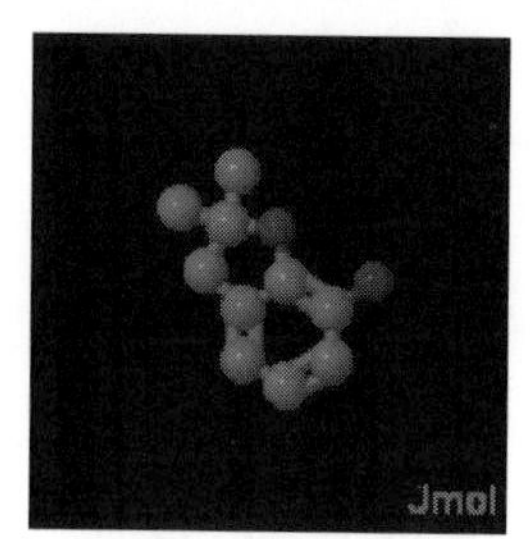

Fig. 2. The details of particular hapten (2,4-dichlorophenoxyacetic acid in this case).

The table showing antibody details is a comprehensive table to make one understand the major aspects covered in a particular paper completely. It starts with the name of the hapten, its synonym, modifications, if any, followed by the details of antibody generation, and its characterization. Following is the description of fields (see Fig. 6).

1. Hapten Name: Common name of haptenic compound.
2. Synonyms: Its chemical name or other commonly used names.
3. Modification: Modification in an existing well-known compound by introducing some groups or replacing one group with other.
4. Conjugation Method: The method used for the conjugation of hapten with the carrier molecules.

Table 3
Detail description of each field

Field name	Description
Hapten name	It displays the common name of haptenic compound.
Synonyms	This shows the chemical name or other commonly used names.
Modification	This specifies the modifications, if any, in an existing well-known compound by introducing some groups or replacing one group by the other.
Molecular formulae	Molecular formula of the hapten
Physical properties	This describes the physical properties in terms of its color, odor, boiling point, melting point, and density.
Nature	This gives nature or category of the haptenic compounds, e.g., pesticide, drug, peptide, hormone, and vitamins.
Molecular weight	Molecular weight of the compound.
Biological activity	It describes the effect of the compound in terms of toxicity on biological system.
Area of uses	This field contains information about the different uses of the hapten and their actions.
Structure	This field displays 2D (Fig. 3) and 3D (Fig. 4) structure of hapten. Jmol has been integrated into the database for the display and manipulation of 3D structures. Moreover, the structures could be downloaded in the form of Mol files.

5. Conjugation Method Details: They are well-defined protocols that are usually used with some modifications and are cited in literature, for example, active ester method and mixed anhydride method. Either the details or the reference of the paper is provided.
6. Spacer/Linkage Nature: The spacer arm, if any, attached to hapten before conjugation to carrier molecules. As regard to linkage nature, the nature of bond between the hapten and the carrier molecule, for example, amide linkage.
7. Hapten Carrier Ratio: It shows number of haptens attached per molecule of carrier.
8. Antibody Name: Name of the antibody that is raised against hapten.
9. Host organism: The host used to raise antibodies, that is, mouse, rabbit, goat, etc.
10. Type & Class: It is the type of antibody that is raised in the host organism, for example, monoclonal, polyclonal, or only antiserum. In case of monoclonal antibodies, the details of isotypes are also described.

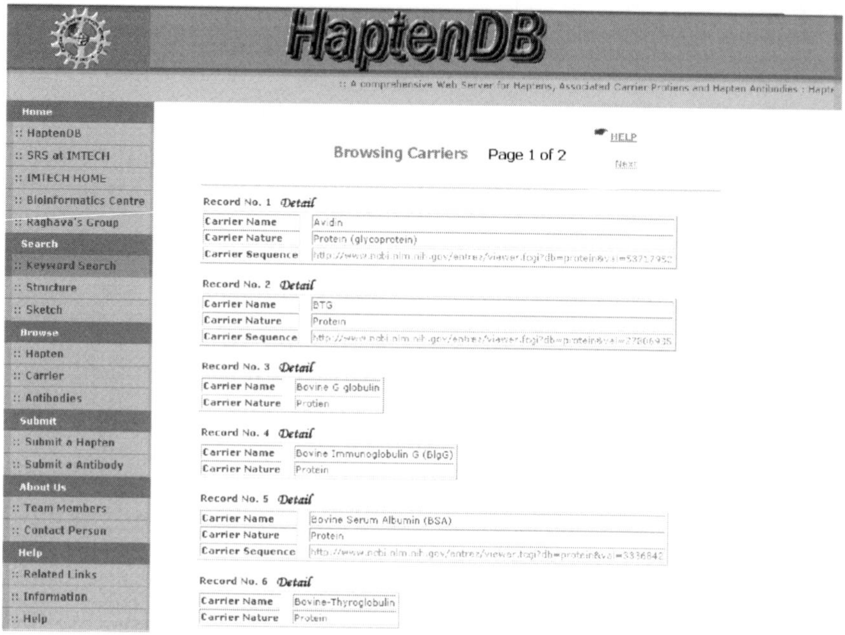

HaptenDB

:: A comprehensive Web Server for Haptens, Associated Carrier Protiens and Hapten Antibodies : Hapte

Home
:: HaptenDB
:: SRS at IMTECH
:: IMTECH HOME
:: Bioinformatics Centre
:: Raghava's Group
Search
:: Keyword Search
:: Structure
:: Sketch
Browse
:: Hapten
:: Carrier
:: Antibodies
Submit
:: Submit a Hapten
:: Submit a Antibody
About Us
:: Team Members
:: Contact Person
Help
:: Related Links
:: Information
:: Help

Browsing Carriers Page 1 of 2 HELP Next

Record No. 1 *Detail*

Carrier Name	Avidin
Carrier Nature	Protein (glycoprotein)
Carrier Sequence	http://www.ncbi.nlm.nih.gov/entrez/viewer.fcgi?db=protein&val=53717952

Record No. 2 *Detail*

Carrier Name	BTG
Carrier Nature	Protein
Carrier Sequence	http://www.ncbi.nlm.nih.gov/entrez/viewer.fcgi?db=protein&val=27806935

Record No. 3 *Detail*

Carrier Name	Bovine G globulin
Carrier Nature	Protien

Record No. 4 *Detail*

Carrier Name	Bovine Immunoglobulin G (BIgG)
Carrier Nature	Protein

Record No. 5 *Detail*

Carrier Name	Bovine Serum Albumin (BSA)
Carrier Nature	Protein
Carrier Sequence	http://www.ncbi.nlm.nih.gov/entrez/viewer.fcgi?db=protein&val=3336842

Record No. 6 *Detail*

Carrier Name	Bovine-Thyroglobulin
Carrier Nature	Protein

Fig. 3. Browsing of carrier molecules, an example output.

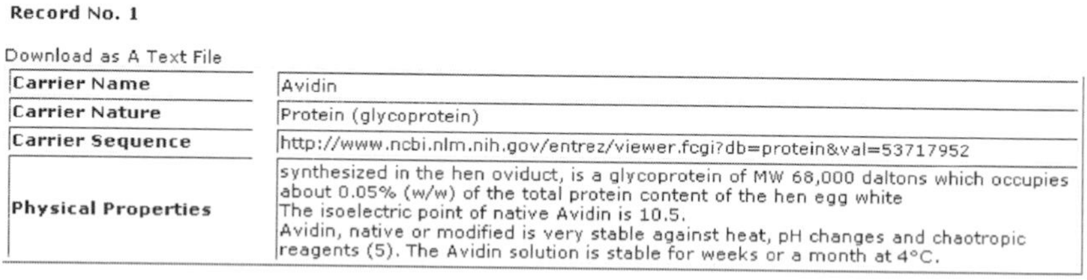

Record No. 1

Download as A Text File

Carrier Name	Avidin
Carrier Nature	Protein (glycoprotein)
Carrier Sequence	http://www.ncbi.nlm.nih.gov/entrez/viewer.fcgi?db=protein&val=53717952
Physical Properties	synthesized in the hen oviduct, is a glycoprotein of MW 68,000 daltons which occupies about 0.05% (w/w) of the total protein content of the hen egg white The isoelectric point of native Avidin is 10.5. Avidin, native or modified is very stable against heat, pH changes and chaotropic reagents (5). The Avidin solution is stable for weeks or a month at 4°C.

Fig. 4. The details of particular carrier molecule (avidin in this case).

11. Cross-reactivity: Cross-reactivity of the raised antibodies with other similar or related compounds has been mentioned as IC_{50} value, where IC_{50} is referred to the amount required for 50% inhibition of the antibody in the given set of conditions.
12. Sensitivity: This is also referred as limit of detection of the hapten with the raised antibody.
13. Assay System: The method used for characterizing the antibodies, for example, competitive ELISA, noncompetitive ELISA, and RIA.
14. Application: Likely application and future prospects of the ELISA method developed, antibody raised, etc.

Table 4
Description of fields of carrier record

Field name	Description
Carrier name	The name of the carrier
Nature	The nature of the protein such as, glycoprotein, and lipopeptide.
Sequence	The sequence of the protein. For this, the NCBI GENPEPT link is provided from where one can retrieve the information about the sequence, source, and origin of the carrier.
Physical properties	In terms of the molecular weight and any specific property for its advantage as carrier protein.

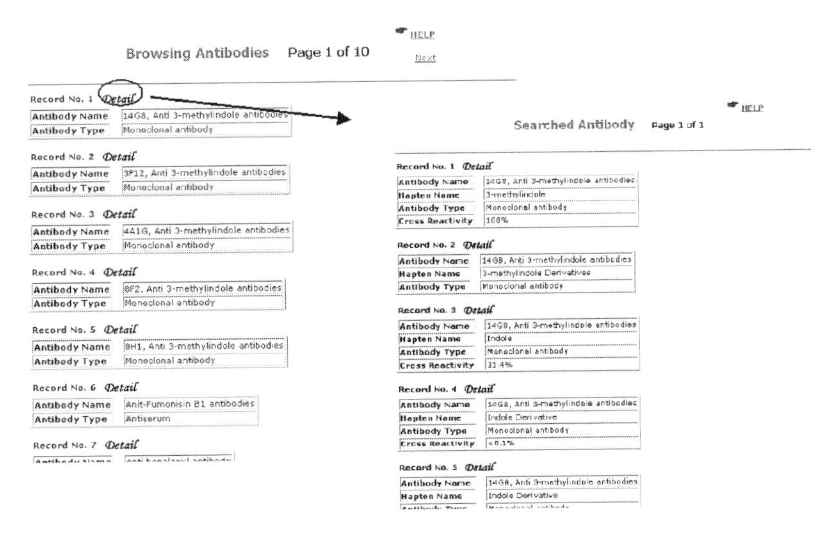

Fig. 5. Browsing of antibodies in HaptenDB, a screen shot from HaptenDB after clicking on "Browsing Antibodies."

15. Reference: This field has the details of the journal, author, title, volume, page numbers., and year of publication of the paper in which this information is reported.
16. Web link: This field contains the web link of the research paper that is cited in the reference field.
17. Comments: This field contains other relevant information that is not contained in all the above-mentioned fields such as immunization protocol, some other important properties of antibody, hapten, or carrier.

HELP

Record No. 1

Download as A Text File

Hapten Name	3-methylindole [Detail]
Antibody Name	14G8, Anti 3-methylindole antibodies [More Info]
Host Organism	BALB/c mice (4-9 month old)
Antibody Type	Monoclonal antibody
Cross-reactivity	100%
Assay System	Non-competitive time-resolved fluoroimmunoassays Competitive time-resolved fluoroimmunoassays
Reference	M Tuomola, R Harpio, H Mikola, P Knuuttila, M Lindstrom, V M Mukkala, M T Matikainen, T Lovgren : Production and characteris Immunological Methods : 240, 111-124 : 2000
Web Links	http://www.sciencedirect.com/science?_ob=ArticleURL&_udi=B6T2Y-40GHRX8-C&_user=529620&_coverDate=06%2F23% 2F2000&_alid=112964107&_rdoc=1&_fmt=summary&_orig=search&_cdi=4931&_sort=d&_st=4&_docanchor=&_acct=C00002
Comments	1) The antibody are produced against 3-Methylindole. 2) Cross Reactivity is calculated as a ratio of IC50 value of 3-Methylindole with IC50 value of cross reacting compound setting

Fig. 6. Table showing detailed information of an antibody, a screen shot.

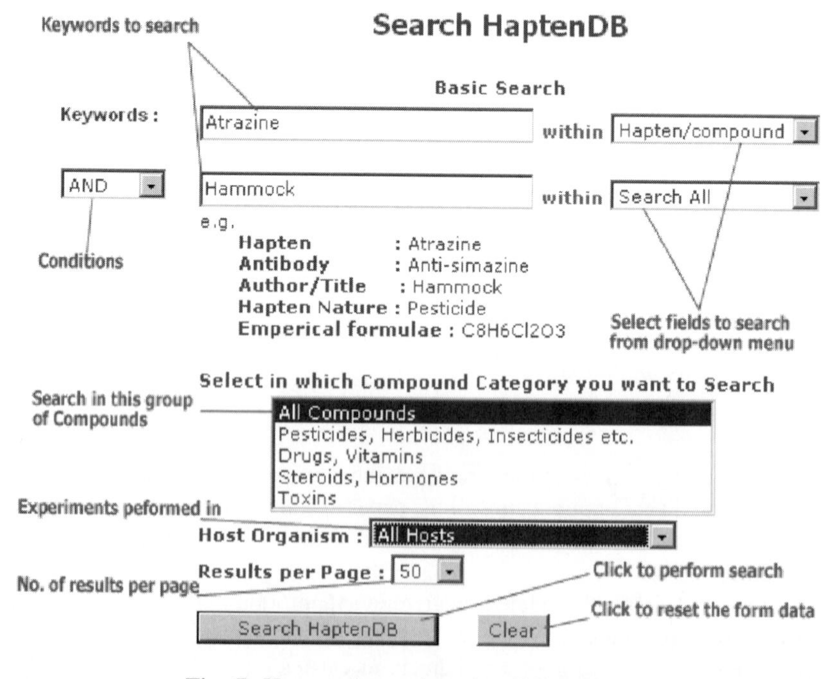

Fig. 7. Keyword search page of HaptenDB.

3.2. Searching Options

As browsing tools allow one to see the records as they were entered in the database. It is difficult to search a specific hapten or antibody or carrier using the browsing tool.HaptenDB also has searching facility, in order to assist the user in getting a specified hapten. The searching facility includes (i) keyword search and (ii) structure similarity search.

3.2.1. Keyword Search

Using this search engine, one can specify a search by giving keywords. The keyword, that is, input (Fig. 7), could be the name of the (i) hapten, (ii) antibody, (iii) author, (iv) title of the paper, (v) nature of the hapten, and (vi) empirical formula of the hapten. The users can also specify the category of the hapten: (i) all compounds; (ii) pesticides, herbicides, and insecticides; (iii) drug and vitamins; (iv) steroids and hormones; and (v) toxins and the host organism in which antibody is raised as (i) all hosts, (ii) mouse, (iii) sheep, and (iv) rabbit. Moreover, results per page can also be specified as desired: (i) 10, (ii) 25, (iii) 50, or (iv) 100 results per page. Figure 8 shows the keyword search, that is, output for the atrazine and the options to filter the search.

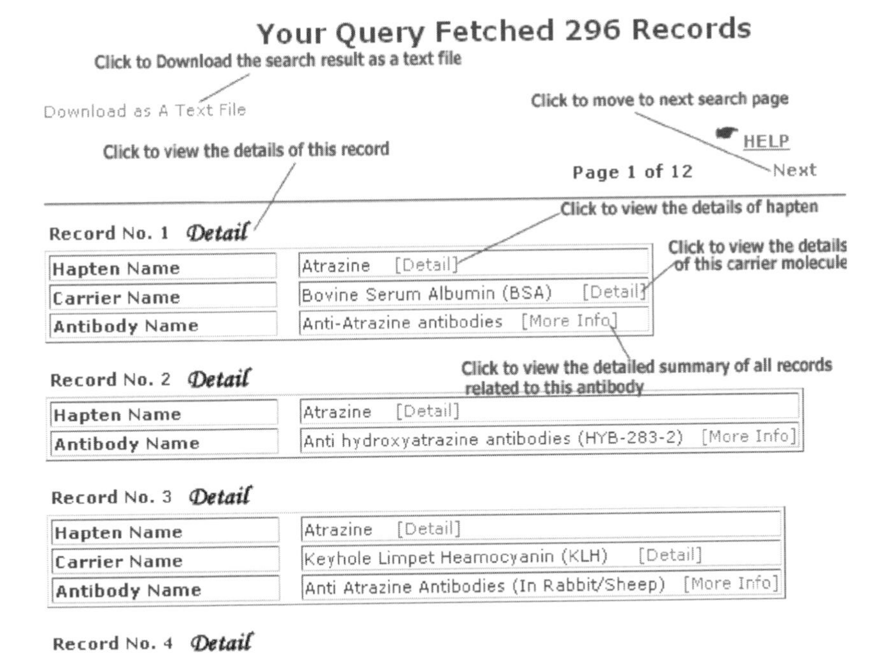

Fig. 8. An example output page of keyword search.

3.2.2. Structure Search

One of the powerful tools integrated in HaptenDB is structure similarity tool, which allows user to search similar hapten structures. The similarity search option can be divided into two categories: (i) upload and search structure and (ii) sketch and search the structure.

HELP

Search HaptenDB using Hapten Structure

Upload Structure [ ] Browse...

Similarity Type * substructure search

Reset Submit

Note : Users must submit the structure in chemical structure formats such as mol, pdb etc.[More] and not as images of structures.
**Similarity seach is done using Java based jcsearch command line utility of JChem for more information in this regard visit www.jchem.com.

Your Query Fetched 318 Records

Download as A Text File

HELP

Page 1 of 13 Next

Record No. 1 Detail

Hapten Name	2,4-Dichlorophenoxyacetic acid [Detail]
Antibody Name	Anti-2,4,5-Trichlorophenoxyacetic acid antibodies [More Info]

Record No. 2 Detail

Hapten Name	2,4-Dichlorophenoxyacetic acid [Detail]
Antibody Name	Anti Isoproturon Antibodies [More Info]

Record No. 3 Detail

Hapten Name	2,4-Dichlorophenoxyacetic acid [Detail]
Antibody Name	Anti-(Delor 103) antibodies [More Info]

Record No. 4 Detail

Hapten Name	2,4-Dichlorophenoxyacetic acid [Detail]
Carrier Name	Bovine Serum Albumin (BSA) [Detail]
Antibody Name	Anti (2,4-Dichlorphenoxyacetic acid) Antibodies [More Info]

Record No. 5 Detail

Hapten Name	2,4-Dichlorophenoxyacetic acid [Detail]
Carrier Name	Thyroglobulin (Tg) [Detail]
Antibody Name	Anti-2,4-Dichlorophenoxyacetic acid(MAb`s B5/C3, E2/B5, E2/G2, F6/C10, and F6/E5) [More Info]

Record No. 6 Detail

Fig. 9. Searching of similar structures, an example input and output screens of HaptenDB.

Upload and Search Structure

This option allows one to search their structure against hapten structure. In order to use this option, one needs to have structure in standard format readable by BABEL software. One needs to upload the structure file to be searched and to select appropriate options that include type of similarity search (e.g., substructure, superstructure, perfect, or exact search). The output (Fig. 9) will provide the list of haptens and the corresponding antibodies satisfying the criteria of the search (substructure search in this case), and again clicking the detail will lead to the detail of hapten or antibody.

Sketch Search

The database integrates JME molecular editor, using which one can sketch the structure of the query molecule instead of the uploading of the file. This option is very useful for creating and searching similar structure. Figure 10 shows the input for the chlorobenzene sketch search and submit for the

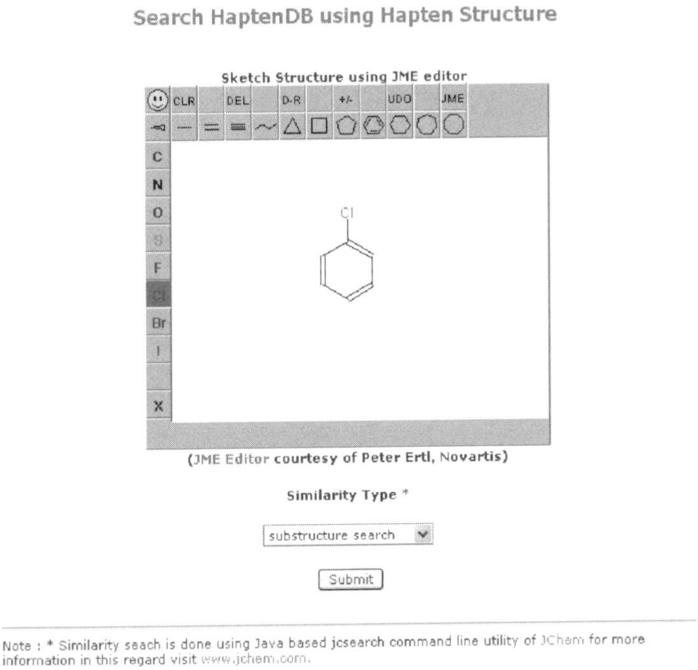

Search HaptenDB using Hapten Structure

Sketch Structure using JME editor

CLR DEL D-R +/- UDO JME

C N O S F Cl Br I X

Cl

(JME Editor courtesy of Peter Ertl, Novartis)

Similarity Type *

substructure search

Submit

Note : * Similarity seach is done using Java based jcsearch command line utility of JChem for more information in this regard visit www.jchem.com.

Fig. 10. A screen shot of sketching structure using JME editor.

similarity search as above; however, output would be the same (Fig. 9) as in case of structure search.

Acknowledgments

We acknowledge the financial support from the Council of Scientific and Industrial Research (CSIR) and Department of Biotechnology (DBT), Government of India.

Notes

1. The users are required to fill a request form available at http://www.imtech.res.in/errors/noauth.html for using web servers developed by Raghava's group (http://www.imtech.res.in/raghava/).
2. It is difficult for developers to maintain any database without the help of the scientific community. Users are requested to submit their new haptens.
3. Each page of the database is self-explanatory; still to help the user "Help" option is provided on the home page as well as individual pages.
4. Database have Related Links, which gives the web links of the sites either used for the construction of the database or could be useful for the browser in one or the other way.
5. Information option gives the information about the architecture of the database, category-wise analysis of database, data management of HaptenDB, system requirement to access, data submission and updates, and disclaimer and limitation of liability.

References

1. Suri, C.R., Raje, M. and Varshney, G.C. (2002) Immunosensors for the pesticide analysis: antibody production and sensor development. *Crit. Rev. Biotech.*, **22:** 15–32.
2. Bhasin, M., Singh, H. and Raghava, G.P.S. (2003) MHCBN: a comprehensive database of MHC binding and non-binding peptides. *Bioinformatics* **19:** 665–666.
3. Johnson, G. and Wu, T.T. (2000) Kabat database and its applications: 30 years after the first variability plot. *Nucleic Acids Res.* **28:** 214–218.
4. McSparron, H., Blythe, M.J., Zygouri, C., Doytchinova, I.A. and Flower, D.R. (2003) JenPep: a novel computational information resource for immunobiology and vaccinology. *J. Chem. Inf. Comput. Sci.* **43:** 1276–1287
5. Ruiz, M., Giudicelli, V., Ginestoux, C., Stoehr, P., Robinson, J., Bodmer, J., Marsh, S.G., Bontrop, R., Lemaitre, M., Lefranc, G., Chaume, D. and Lefranc, M.P. (2000) IMGT, the international ImMunoGeneTics database. *Nucleic Acids Res.* **28:** 219–221.
6. Saha, S., Bhasin, M., and Raghava, G.P.S. (2005) BCIPEP: a database of B-cell epitopes. *BMC Genomics* **6:** 79.

7. Schönbach, C., Koh, J.L.Y., Sheng, X., Wong, L. and Brusic, V. (2000) FIMM, a database of functional molecular immunology. *Nucleic Acids Res.* **28:** 222–224.
8. Singh, M.K., Srivastava, S., Raghava, G.P.S. and Varshney, G.C. (2006) Hapten DB: a comprehensive database of haptens, carrier proteins and anti-hapten antibodies. *Bioinformatics* **22:** 253–256.

II

Defining HLA Supertypes

9

The Classification of HLA Supertypes by GRID/CPCA and Hierarchical Clustering Methods

Pingping Guan, Irini A. Doytchinova, and Darren R. Flower

Summary

Biological experiments often produce enormous amount of data, which are usually analyzed by data clustering. Cluster analysis refers to statistical methods that are used to assign data with similar properties into several smaller, more meaningful groups. Two commonly used clustering techniques are introduced in the following section: principal component analysis (PCA) and hierarchical clustering. PCA calculates the variance between variables and groups them into a few uncorrelated groups or principal components (PCs) that are orthogonal to each other. Hierarchical clustering is carried out by separating data into many clusters and merging similar clusters together. Here, we use an example of human leukocyte antigen (HLA) supertype classification to demonstrate the usage of the two methods. Two programs, Generating Optimal Linear Partial Least Square Estimations (GOLPE) and Sybyl, are used for PCA and hierarchical clustering, respectively. However, the reader should bear in mind that the methods have been incorporated into other software as well, such as SIMCA, statistiXL, and R.

Key Words: HLA; MHC; supertype; principal component analysis; hierarchical clustering; GOLPE

1. Introduction

Human leukocyte antigen (HLA) is one of the most polymorphic proteins in human. There are more than 2,000 HLA sequences in the IMGT/HLA database, and the number is increasing yearly. Only a small percentage of HLA alleles have known binding motifs. Sette et al. *(1)* was the first to group class I HLA alleles with similar binding motifs into superfamilies. Several HLA supertypes were described—A2 *(1,2)*, A3 *(3)*, and B44 *(4)*. Later, the number of defined supertypes was extended to 9 *(5)*, which were A1 (A*0101, A*2501, A*2601,

From: *Methods in Molecular Biology, vol. 409: Immunoinformatics: Predicting Immunogenicity In Silico*
Edited by: D. R. Flower © Humana Press Inc., Totowa, NJ

A*2601, and A*3201), A2 (A*0201–07, A*6802, and A*6901), A24 (A*2301, A*2402–04, and A*3001–03), A3 (A*0301, A*1101, A*3101, A*3301, and A*6801), B7 (B*07, B*35, B*51, B*53, B*54, B*55, B*56, B*67, and B*78), B27 (B*1401–02, B*1503, B*1509, B*1510, B*1518, B*2701–08, B*3801, B*3802, B*3901–04, B*4801, B*4802, and B*7301), B44 (B*37, B*4001, B*4002, B*4006, B*41, B*44, B*45, B*47, B*49, and B*50), B58 (B*1516, B*1517, B*5701, B*5702, and B*58), and B62 (B*1301–02, B*1501, B*1502, B*1506, B*1512, B*1513, B*1514, B*1519, B*1521, B*4601, and B*52).

Sette's classification was a motif-based approach and required binding motifs for each allele. However, most of the 783 known class I HLA alleles have not been studied experimentally. To characterize all HLA alleles using experimental binding assays is both expensive and time-consuming; therefore, a chemometric strategy is applied to classify class I HLA molecules into supertypes, using information drawn solely from the protein sequences. The techniques used were GRID ***(6)*** and principal component analysis (PCA) ***(7,8)***. The molecular interaction fields (MIFs) between the chemical probes and the HLA molecules were calculated in GRID, and the MIFs were then used to build PCA/consensus PCA (CPCA) models. Results of the GRID/CPCA analysis were compared with the classification using hierarchical clustering analysis on CoMSIA fields; together, the results were used to classify HLA molecules and generate "supertype fingerprints," that is, the sequence features for supertype classification ***(9)***.

In chemical or pharmacological analysis, often many drug targets are studied in one experiment, and little information can be extracted from the data directly ***(10)***. PCA simplifies the data by replacing the large number of variables in the original data set with a few new, uncorrelated variables called principal components (PCs) ***(8)***. The PCs are calculated in the order of importance, and most of the variance in the data can be explained by the first few components. A variation of the PCA, CPCA, is also commonly used for calculations with multiple probes ***(11)***. CPCA divides values generated by each probe into blocks, and it is easier to see which property is the most important in the model ***(12,13)***.

2. Method Theory

2.1. GRID

The GRID program (version 21) finds the energetically favored or disfavored regions on molecules with known three-dimensional (3D) structures. Many molecules can be included in one calculation ***(6)***. A selection of chemical probes, which represent atoms or functional groups with different properties, is included in the program. GRID calculates the interaction energy between selected chemical probes and each of the molecules.

A GRID box is defined to include the interested molecular areas in the calculation (Fig. 1). GRID uses different probes placed at a regular interval throughout the grid box to calculate the interaction energy between the molecule and the probes. An example of the probes used in the HLA family calculation is listed in Table 1.

2.2. PCA

The PCA is commonly used in multivariate data analysis to reduce the number of variables. Data used in PCA are stored in a data matrix X (Fig. 2). There are N observations and K variables in the matrix. Each observation occupies one row; the variables are measurements of the observation and are stored in the columns.

PCA decomposes the matrix X into two smaller matrices: the scores matrix T and the loading matrix "P," which explain the overall variance of the X matrix. The scores matrix contains a few variables M (Fig. 2), that is, the PCs, which can be used to describe the observations. The loading matrix reveals the relationship between the variables in the original matrix and the PCs. Plots

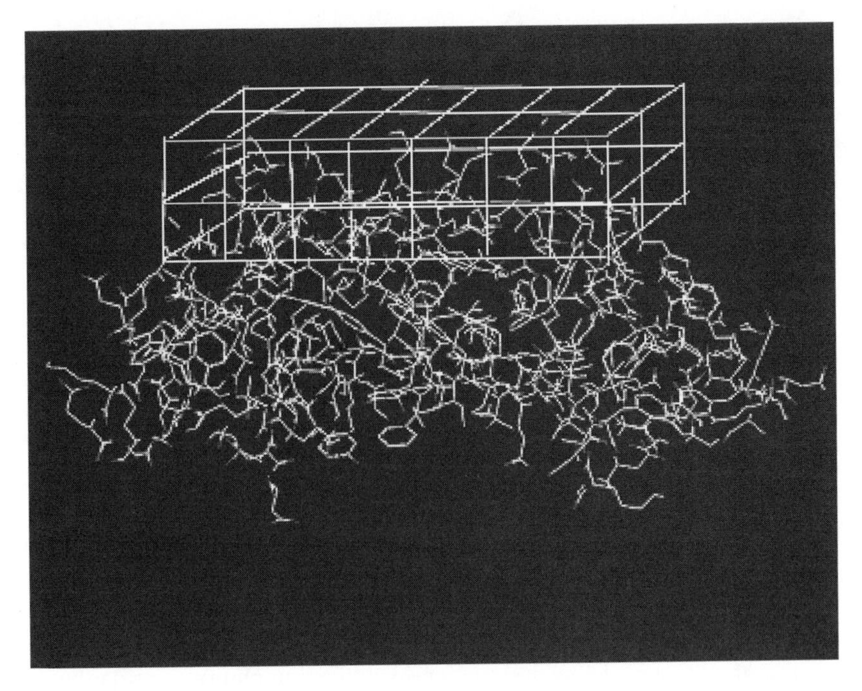

Fig. 1. The GRID box.

Table 1
Chemical probes used in the GRID/CPCA study

Probe	Chemical group	Represented amino acids
OH2	Water	Hydrophilic amino acids
Dry	The hydrophobic probe	Hydrophobic amino acids
H	Hydrogen	Hydrogen bond donor/accepter
C3	Methyl CH3 group	Aliphatic amino acids
C1=	sp2 CH aromatic or vinyl	Phe, Tyr, Trp, His
N:*	sp N with a lone pair	His
N:=	sp2 N with a lone pair	Asn Gln
N1	Neutral flat NH eg. Amide	Any amino acids
N2+	sp3 amine NH2 cation	Arg Lys
O1	Alkyl hydrox OH group	Ser Thr
OH	Phenol or carboxy OH	Tyr Asp Glu
O	sp2 carbonyl oxygen	Asp Asn Glu Gln
S1	Neutral SH	Cys Met

List of GRID probes used in the study. *A total 13 probes are selected from probes offered in GRID. These probes are chosen to represent different characteristics of the twenty amino acids.

of the observations in the multidimensional space are called the scores plots, which identify similarities and differences within the observations and groups them accordingly, whereas the loading plot relates the original variables with the PCs and identifies variables that are important in distinguishing groups of observations.

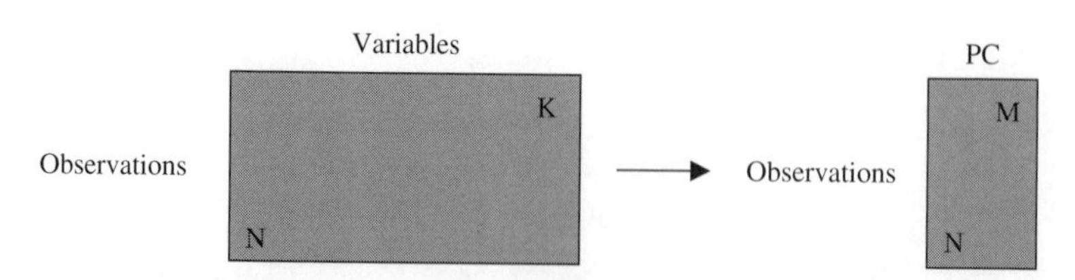

Fig. 2. The data in principal component analysis (PCA) are stored in a matrix, with N observations and K variables. The analysis builds a new model containing all the observations, and the variables in the original data set are replaced by a few new uncorrelated variables M, called principal components (PCs). By reducing the number of variables, the PCA model shows relationships between observations and variables and among observations themselves.

2.3. CPCA

Some multivariate data are organized in blocks; each block describes one molecular force. For example, in GRID, the interaction energy values are calculated using probes representing different chemical properties, and data are separated into the corresponding blocks. However, the variance can be very different among the data blocks. The energy values generated by probes representing weak nonbonded interactions such as van der Waals force and hydrophobic attractions will be masked by those generated by stronger interactions such as hydrogen bonds. Because the weak forces are equally important in molecular interaction, it is necessary that their effects are considered in the CPCA model. To overcome this problem, a scaling process is applied to the data to normalize their importance in the model. The scaling method used in Generating Optimal Linear Partial Least Square Estimations (GOLPE) is named block unscaled weights (BUW) scaling, in which data generated by each probe are organized into one block, and weighting coefficients are calculated for each block. The probes are scaled according to the weighting coefficients, which gives each probe the same importance in the model, whereas the relative scales of variables within the block do not change. Figure 3 illustrates the BUW scaling. Figure 3 A shows the initial variable distribution in each probe, and Fig. 3B shows the normalized variable distribution after the scaling.

2.4. GOLPE

GOLPE *(6)* improves the predictivity of the model by comparing the contributions of each variable and excluding those that make very small or no contributions. In this way, the model generated by GOLPE has a higher level of predictivity than the one generated by PLS alone *(6)*.

GOLPE also has one module for PCA calculation. The PCs are obtained by maximizing the variance of linear functions of the matrix. The results of the GRID field's calculations are stored in files with .kont extension and are imported into GOLPE. The data are pretreated before calculation; all the data with absolute values smaller than 0.03 or with standard deviation less than 0.03 are deleted. Positive interaction energy represented unfavorable steric repulsion between the probe and the molecule; therefore, it is removed by setting the maximum cutoff to 0 kcal/mol.

After calculating GRID energy fields using each probe, the probes that give the highest explained variance by the first three PCs are selected, and a GRID calculation is run using all these probes. The results are used to build a CPCA model.

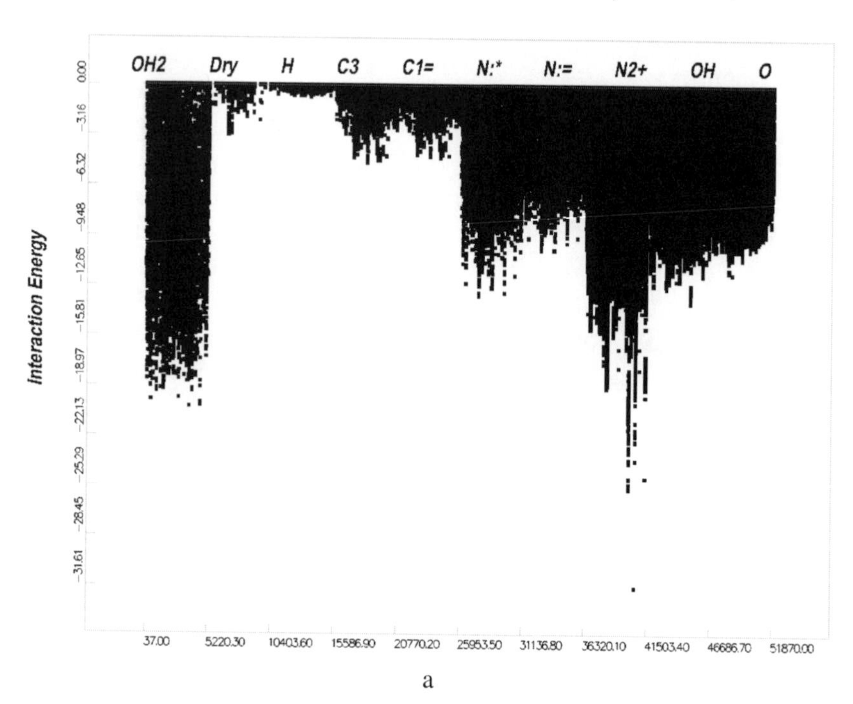
OH2
Dry
H
C3
C1=
N:*
N:=
N2+
OH
O
Interaction Energy
0.00
-3.16
-6.32
-9.48
-12.65
-15.81
-18.97
-22.13
-25.29
-28.45
-31.61
37.00
5220.30
10403.60
15586.90
20770.20
25953.50
31136.80
36320.10
41503.40
46686.70
51870.00

a

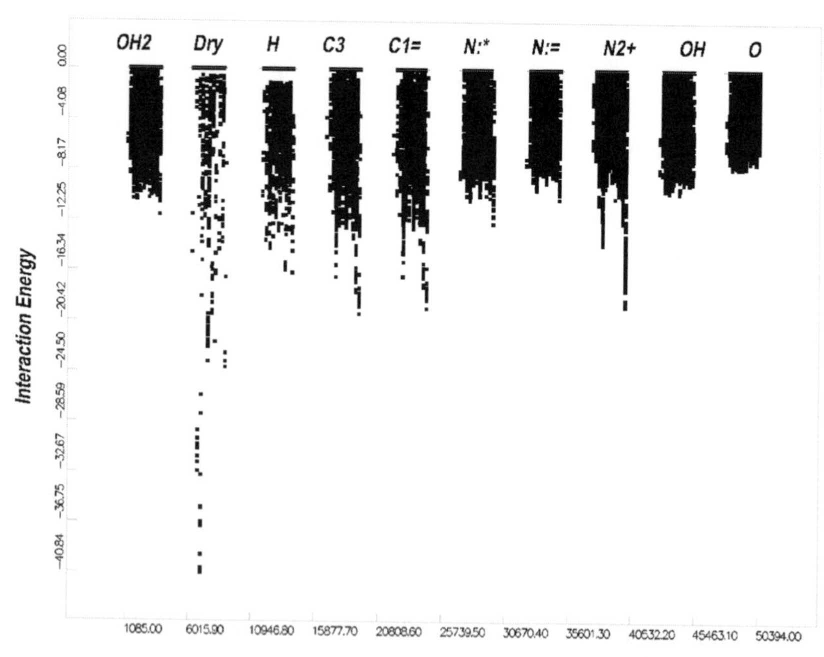
OH2
Dry
H
C3
C1=
N:*
N:=
N2+
OH
O
Interaction Energy
0.00
-4.08
-8.17
-12.25
-16.34
-20.42
-24.50
-28.59
-32.67
-36.75
-40.84
1085.00
6015.90
10946.80
15877.70
20808.60
25739.50
30670.40
35601.30
40532.20
45463.10
50394.00

b

When more than one probe is used in the GRID calculation, the data generated by different probes are grouped into blocks, and they are often analyzed by hierarchical PCA methods such as CPCA. The advantage of such methods over PCA is that they compare the relative importance of each block in the calculation and make a "consensus" clustering of the objects. CPCA uses the same principle as PCA: a CPCA model tries to explain the overall variance of the original data matrix. The algorithm used in CPCA is an adaptation of the NIPALS algorithm used in PCA ***(14)***. Like PCA, CPCA calculates the PCs and gives the scores and loading matrix. In addition, CPCA also calculates the importance of each data block. It calculates the scores and the loading matrix for each probe used and gives the weight matrix that illustrates the contribution of each probe in the overall scores.

2.5. CoMSIA

The CoMSIA calculations were performed on a Silicon Graphics workstation using Sybyl 6.9, as previously described ***(15)***. The structure of the HLA supertype A*0101 is used as a template to align all the HLA structures.

The HLA structures are evaluated using the five CoMSIA physicochemical properties included in the QSAR module of Sybyl 6.9: steric, electrostatic, hydrophobic, and hydrogen donor and hydrogen bond acceptor properties. The properties are evaluated using a probe atom placed at regular intervals within the grid. The probe has a radius of 1 Å, charge, hydrophobicity, and hydrogen bond donor and acceptor properties all equal to +1. Similarity indices are calculated using Gaussian-type distance dependence between the probe and the atoms of the peptides tested.

2.6. Hierarchical Clustering

Hierarchical clustering analysis is a statistical technique used in classifying large numbers of objects to reveal how closely the objects are related ***(16)***. A common form of hierarchical clustering is the agglomerative algorithm, in which the calculation of hierarchical clusters starts by separating each object into a separate cluster ***(17)***. The distance between two clusters is dependent on the similarities between the two objects. The clustering is then improved by merging clusters that have the shortest distance ***(17)***. The distance between the new clusters is recalculated. The steps are repeated until all clusters are

Fig. 3. The distribution of the variables for each probe. a. before the block unscaled weights (BUW) scaling, and b. after the block unscaled weights (BUW) scaling.

clustered into a single cluster *(18)*. The result of the clustering is a binary tree with a root and many leaves; each leaf represents one object *(19)*. The order of the leaves is arbitrary. An HLA classification is carried out using hierarchical clustering based on CoMSIA fields, in which the alleles were clustered by comparing the generated CoMSIA fields of each molecule *(9)*.

3. Methodology

3.1. GRID/CPCA Calculation

3.1.1. Protein Structures

GRID and CoMSIA calculation requires the 3D structures of the proteins. As there are only a few crystallized HLA protein structures available, the alternative approaches are to select a crystallized structure as a template and build the rest of the molecules by homology modeling. The template structure is selected from the RCSB protein data bank (http://www.rcsb.org/pdb/).

3.1.2. Computer Software

GRID calculation is carried out on the GRID software developed by Molecular Discoveries Ltd. The version of the program used is 21.

3.2. Calculating MIFs Using GRID

3.2.1. Import Structures

1. Molecules are imported into the program using one of the two options on the program tool bar “Add single” and “Add multiple,” which are for adding single structure file and multiple structure files, respectively. Because multiple structures are often used in GRID calculation, the option “Add multiple” is recommended. A file with the .lst extension containing the names of all the structure files is required for this option. A dialog is activated by clicking the “Add multiple” button. On the “List” panel, enter a .lst file in the text box and select “automatic” filtering level from the pull-down list. Click “OK,” and all the files included in the .lst file will start to be imported into the GRID. Depending on the numbers of structures, this process can take up to 10 min.
2. After importing the molecules, a table will appear in the main window of the GRID interface. Each row contains one structure. The status column shows whether the structures have been imported correctly. Usually it displays “ready,” which means the user can proceed. If there is an error, depending on the nature of the error, the row will appear in yellow or in red, with the words “GRIN error” in the status column. Note that to select “automatic filtering” in “Add multiple” or “PDB filtering” in “Add single” in step 2 will help to reduce the number of errors, particularly the water molecules in the PDB file, which may otherwise be recognized as errors.

3.2.2. Calculate MIFs

1. An interaction box is defined before calculation. The box is defined manually. On the Method list, click “define box size.” On the dialog box, deselect “automatic” and click the “interactive” button. On the window displaying the structure, right click and select “Toggle mode” to move the box around and modify the box size. Then select “OK.”

2. At least one probe is required for calculation. To select the probes, click “Probes” on the menu and select “Choose probes” from the drop-down list. After selecting the probes, click “OK.”

3. Several parameters are required to be set for calculation. This can be viewed from the keyword tree. Three parameters are set for calculating multiple structures and to generate output: NPLA $= 0.5$, MOVE $= 1$, and LIST $= -2$.

4. To run the program, select “Run” from the tool bar and choose “Joint” from “Advanced” panel and click “OK.” The status column in the main window should change from “ready” to “running.” When the calculation finishes, a dialog will appear to inform the user.

3.2.3. Build PCA/CPCA Models in GOLPE

The result file from GRID is saved in .kont extension. This file is imported into GOLPE to build PCA models.

1. Files are imported into GOLPE using the option “Import fields” on the pull-down list of “File,” where the user selects the .kont file they want to import and name the new golpe data file to which the program imports the data. The new .dat file is then opened by selecting “File” and “Open data file.”

2. Usually data is pretreated in GOLPE before further calculation. Select “Pretreatment” from the menu and “Advanced pretreatment.” From the dialog, select the following parameters:

a. “maximum cutoff” and set the values to be 0,
b. “zeroing values,” select both positive and negative values and set the value to be 0.03, and
c. “min SD cutoff,” and set the value to be 0.03, if multiple probes are used in GRID calculation, choose “BUW scaling” from the dialog and select “OK.”

3. Choose “Modeling” from the menu and select “Generate PCA model,” set dimensionality to be 5, click “OK.” The results of the PCA model are displayed in the main window.

4. To view the maps generated by the PCA model, choose “Plot” from the menu. For scores plot, select “2D-plot” or “3D-plot” from the drop-down list followed by “PCA scores.” To view the loading map, select “Grid plot” from the drop-down list and “PCA loadings.” Enter the component that the user wants to use and click “OK.” The maps are displayed in a separate pop-up window.

3.3. Hierarchical Clustering Based on CoMSIA Fields analysis

3.3.1. Evaluation of HLA Binding Site using CoMSIA Fields

3.3.1.1. Building a New Molecular Database

1. Select "File" from the menu and "database," choose "New database," and name the database. Select the "update" mode to enter molecules into the database.
2. Add all structures into the database. Select "Read" from "File" to read the structure file and center the molecule so that it is displayed in the center of the screen.
3. Select "Biopolymer" from the menu and "Add hydrogens." Select the molecule you just displayed and click "All" so that hydrogen atoms will be added to the whole molecule. Click "OK," which gives another dialog, choose "ALL" again and click "OK." All the hydrogen atoms should be added to the molecule now.
4. Select "Biopolymer" again from the menu and "Load charges." Select the molecule and choose "All" and click "OK." In the next dialog, select "KOLL_ALL" from the list followed by "OK." All the charges will then be added to the molecule.
5. All structures are aligned manually. To do this, select "Biopolymer" from the menu followed by "Align structures using homology." Highlight the template molecule to be the fixed molecule and select the new molecule to be movable. Click "OK."
6. Select "Build/Edit" from the menu and "Name molecule," give the newly aligned molecule a name, and press "OK." Now the new molecule is ready to be put into the database.
7. From "File" select "Database" and "Put molecule," select the newly named molecule from the list and click "OK." The molecule is now entered into the database. Steps 2–7 are repeated until all the molecules are entered into the database.

3.3.1.2. Calculating CoMSIA Fields and Hierarchical Clustering

1. From "File" choose "Spreadsheet" and "New". Select "Database" in the pop-up window and select the name of the newly developed database and press "Open." A new spreadsheet should be opened in a pop-up window with the name of the molecules listed in the first column.
2. From the spreadsheet window choose "QSAR" from the menu and select "manage QSAR" from the drop-down list. Select "Regions" and "Define" in the following dialog and choose "User specific" to define a grid manually. Enter values into the *X*, *Y*, and *Z* dimensions till the molecule in question is enclosed in the grid.
3. Select the second column from the spreadsheet and press the "Autofill" button. Select "CoMSIA" and choose "steric" from the list. The attenuation factor is set to 0.3. Name the column in the next dialog and press "OK." The steric interactions are automatically calculated and listed in the second column of the spreadsheet. Repeat this step for electrostatic, hydrophobic, and hydrogen bond donor and hydrogen bond acceptor interactions.

4. Choose "QSAR" from the spreadsheet menu and select "Hierarchical clustering" from the list. Use all default settings and press "Do." A dendrogram should be displayed in the window after calculation finishes.

Both GRID/CPCA and hierarchical clustering are useful chemometric techniques used in object classification and require some manual interpretation of the results. The classification by a CPCA model is reflected in the scores and loading plots. The scores plot gives a 2D or 3D overview of the objects in each cluster, but the size of the cluster and the outliers have to be defined by the user. Similarly, the energy levels of the loading plots also have to be adjusted manually to show the energetically favored and disfavored regions. Sybyl lists the molecules represented by each leaf of the hierarchical clustering, but the user needs to decide which level of the dendrogram to be used in the classification. Overall, there are no set rules, and the interpretation of data is largely dependent on individual user's experience.

References

1. Sidney, J, Grey, HM, Kubo, RT, and Sette, A, Practical, biochemical and evolutionary implications of the discovery of HLA class I supermotifs. Immunol Today, 1996. **17**(6): 261–6.
2. del Guercio, MF, Sidney, J, Hermanson, G, Perez, C, Grey, HM, Kubo, RT, and Sette, A, Binding of a peptide antigen to multiple HLA alleles allows definition of an A2-like supertype. J Immunol, 1995. **154**(2): 685–93.
3. Sidney, J, Grey, HM, Southwood, S, Celis, E, Wentworth, PA, del Guercio, MF, Kubo, RT, Chesnut, RW, and Sette, A, Definition of an HLA-A3-like supermotif demonstrates the overlapping peptide-binding repertoires of common HLA molecules. Hum Immunol, 1996. **45**(2): 79–93.
4. Sidney, J, Southwood, S, Pasquetto, V, and Sette, A, Simultaneous prediction of binding capacity for multiple molecules of the HLA B44 supertype. J Immunol, 2003. **171**(11): 5964–74.
5. Sette, A and Sidney, J, Nine major HLA class I supertypes account for the vast preponderance of HLA-A and -B polymorphism. Immunogenetics, 1999. **50**(3–4): 201–12.
6. Cruciani, G and Watson, KA, Comparative molecular field analysis using GRID force-field and GOLPE variable selection methods in a study of inhibitors of glycogen phosphorylase b. J Med Chem, 1994. **37**(16): 2589–601.
7. van der Voet, H and Franke, JP, A discussion of principal component analysis. J Anal Toxicol, 1985. **9**(4): 185–8.
8. Inoue, M and Kajiya, F, [Multivariate analysis in computer diagnosis. 3. Principal component analysis]. Iyodenshi To Seitai Kogaku, 1976. **14**(1): 52–7.
9. Doytchinova, IA, Guan, P, and Flower, DR, Identifying human MHC supertypes using bioinformatic methods. J Immunol, 2004. **172**(7): 4314–23.

10. Pate, ME, Turner, MK, Thornhill, NF, and Titchener-Hooker, NJ, Principal component analysis of nonlinear chromatography. Biotechnol Prog, 2004. **20**(1): 215–22.
11. Kastenholz, MA, Pastor, M, Cruciani, G, Haaksma, EE, and Fox, T, GRID/CPCA: a new computational tool to design selective ligands. J Med Chem, 2000. **43**(16): 3033–44.
12. Myshkin, E and Wang, B, Chemometrical classification of ephrin ligands and Eph kinases using GRID/CPCA approach. J Chem Inf Comput Sci, 2003. **43**(3): 1004–10.
13. Terp, GE, Cruciani, G, Christensen, IT, and Jorgensen, FS, Structural differences of matrix metalloproteinases with potential implications for inhibitor selectivity examined by the GRID/CPCA approach. J Med Chem, 2002. **45**(13): 2675–84.
14. Wold, S, Hellberg, S, Lundstedt, T, Sjostrom, M, and Wold, H, *Proc. Symp. on PLS Model Building: Theory and Application*. 1987, Germany: Frankfurt am Main.
15. Doytchinova, IA and Flower, DR, Toward the quantitative prediction of T-cell epitopes: coMFA and coMSIA studies of peptides with affinity for the class I MHC molecule HLA-A*0201. J Med Chem, 2001. **44**(22): 3572–81.
16. Johnson, SC, Hierarchical clustering schemes. Psychometrika, 1967. **32**(3): 241–54.
17. Guess, MJ and Wilson, SB, Introduction to hierarchical clustering. J Clin Neurophysiol, 2002. **19**(2): 144–51.
18. Glazko, GV and Mushegian, AR, Detection of evolutionarily stable fragments of cellular pathways by hierarchical clustering of phyletic patterns. Genome Biol, 2004. **5**(5): R32.
19. Levenstien, MA, Yang, Y, and Ott, J, Statistical significance for hierarchical clustering in genetic association and microarray expression studies. BMC Bioinformatics, 2003. **4**(1): 62.

10

Structural Basis for HLA-A2 Supertypes

Pandjassarame Kangueane and Meena Kishore Sakharkar

Summary

The human leukocyte antigen (HLA) alleles are extremely polymorphic among ethnic population, and the peptide-binding specificity varies for different alleles in a combinatorial manner. However, it has been suggested that majority of alleles can be covered within few HLA supertypes, where different members of a supertype bind similar peptides, yet exhibiting distinct repertoires. Nonetheless, the structural basis for HLA supertype-like function is not clearly known. Here, we use structural data to explain the molecular basis for HLA-A2 supertypes.

Key Words: HLA; alleles; peptide; binding; supertypes; structural basis

1. Introduction

The human leukocyte antigen (HLA) alleles are highly polymorphic among ethnic population. Today, more than 1,800 HLA alleles are known and about a 1,000 of them refer to the class 1 loci *(1)*. Class I alleles bind peptides of length 8–10 residues during T-cell-mediated immune response *(2)*. Therefore, the possible combination of specific HLA–peptide binding is large. However, it has been suggested that a majority of alleles can be grouped into few "HLA supertypes," where the members of a supertype bind similar peptides, yet exhibiting distinct binding repertoires *(3)*. The functional overlap between different alleles within defined supertypes will significantly reduce peptide-binding diversity. A catalog of functional overlap is critical for grouping alleles into supertypes from sequence information. In recent years, a number of supertypes have been defined by comparing peptide-binding data. Thus, HLA-A1 *(4)*, HLA-A2 *(3,5)*, HLA-A3 *(5)*, HLA-A24 *(4)*, HLA-B7 *(5)*, HLA-B27 *(4)*, HLA-B44 *(6)*, HLA-B58 *(4)*, and HLA-B62 *(4)* supertypes have been defined.

From: *Methods in Molecular Biology, vol. 409: Immunoinformatics: Predicting Immunogenicity In Silico*
Edited by: D. R. Flower © Humana Press Inc., Totowa, NJ

Alternatively, Chelvanayagam et al. *(7)*, Zhang et al. *(8)*, Zhao et al. *(9)*, and Doytchinova et al. *(10)* grouped HLA alleles into functionally overlapping clusters from sequence data. Chelvanayagam et al. *(7)* identified interaction pockets from HLA–peptide crystal structures; Zhang et al. *(8)* defined A–F structural binding pockets; Zhao et al. *(9)* defined functional pockets made of critical polymorphic functional residue positions (CPFRP); Doytchinova et al. *(10)* used molecular interaction fields (MIF), hierarchical clustering (HC), and principal component analysis (PCA); and Lund et al. *(11)* used clustering procedures for grouping HLA alleles into putative supertypes. However, the structural basis for "supertype-like" HLA function is not clearly known. Here, we use structural complexes of HLA–peptide structures to explain HLA supertypes.

2. Methodology

2.1. HLA Supertype Data

HLA-A2 supertype data are obtained from literature (Table 1). These data describe the binding/nonbinding of 25 peptides to A*0201, A*0202, A*0203, A*0206, and A*6802. Table 1 summarizes six peptides binding to all members of the A2 supertypes (A*0201, A*0202, A*0203, A*0206, and A*6802). The functional overlap between different members of the supertype is interesting. It also shows several peptides binding to some members but not all members of the A2 supertypes (Table 1).

2.2. HLA Sequences

The protein sequences of HLA-A (295 alleles) were obtained from IMGT/HLA (release 2.5) for this analysis *(1)*.

2.3. Functional Pockets in HLA Structures

The CPFRPs were used to define functional pockets in HLA structures *(9)*. HLA-A allele sequences are polymorphic but homologous among themselves. Hence, they have a similar 3D structure in space. However, the polymorphic residues are discontinuously distributed in structure. The residues at CPFRP demonstrated a change in solvent accessibility (ΔASA) of >0 Å^2 upon complex formation in a set of HLA–peptide structures *(9)* and at least one amino acid polymorphism among 295 HLA-A alleles (IMGT/HLA release 1.14). The 21 CPFRPs thus identified are then classified into virtual pockets for each peptide residue, as shown in Fig. 1.

Table 1
List of known HLA supertypes

Peptide	Supertypes	A*0201	A*0202	A*0203	A*0206	A*6802	Reference
LLFNILGGWV	A2	b	b	b	b	b	*(12)*
YLVAYQATV	A2	b	b	b	b	b	*(12)*
KVAELVHFL	A2	b	b	b	b	b	*(13)*
FLWGPRALV	A2	b	b	b	b	b	*(13)*
FLLLADARV	A2	b	b	b	b	b	*(12)*
IMIGVLVGV	A2	b	b	b	b	b	*(13)*
KIFGSLAFL	A2	b	b	b	b	nb	*(13)*
CLTSTVQLV	A2	b	b	b	b	nb	*(13)*
RLIVFPDLGV	A2	b	b	b	b	nb	*(12)*
YLQLVFGIEV	A2	b	b	b	b	nb	*(13)*
LLTFWNPPV	A2	b	b	b	b	nb	*(13)*
VLVGGVLAA	A2	b	b	b	b	nb	*(12)*
WMNRLIAFA	A2	b	b	b	nb	b	*(12)*
DLMGYIPLV	A2	b	nb	b	b	b	*(12)*
ILHNGAYSL	A2	b	b	b	nb	nb	*(13)*
YLSGANLNL	A2	b	b	b	nb	nb	*(13)*
VMAGVGSPYV	A2	b	b	b	nb	nb	*(13)*
ILAGYGAGV	A2	b	b	b	nb	nb	*(12)*
LMTFWNPPV	A2	b	nb	b	b	nb	*(13)*
YLVTRHADV	A2	b	nb	b	b	nb	*(12)*
HMWNFISGI	A2	b	nb	b	b	nb	*(13)*
YLLPRRGPRL	A2	b	nb	b	b	nb	*(12)*
LLFLLLADA	A2	b	b	nb	nb	nb	*(12)*
LLTFWNPPT	A2	b	nb	b	nb	nb	*(13)*
ALCRWGLLL	A2	b	nb	b	nb	nb	*(12)*

Peptides with known binding or nonbinding information are available for five HLA-A alleles. b, binder; nb, nonbinder.

2.4. HLA Supertypes and Virtual Binding Pockets of CPFRP

The HLA-A2 supertype data for 25 peptides covering A*0201, A*0202, A*0203, A*0206, and A*6802 are mapped manually to virtual pockets made of CPFRP in Fig. 1. Specific residue at CPFRP is assigned for each HLA allele with known supertype data. The visual representation of supertype-like function for known A2 supertypes along with structurally meaningful virtual pockets consisting of CPFRP provides structural insight into HLA supertype-like function.

G1	1	2	3	4	5	6	7	8	9	10
1	L	L	F	H	I	L	G	G	W	V
2	Y	L	V	A	Y	Q	A	T	V	
3	K	V	A	E	L	V	H	F	L	
4	F	L	W	G	P	R	A	L	V	
5	F	L	L	L	A	D	A	R	V	
6	I	M	I	G	V	L	V	G	V	

G2	1	2	3	4	5	6	7	8	9	10
1	K	I	F	G	S	L	A	F	L	
2	C	L	T	S	T	V	Q	L	V	
3	R	L	I	V	F	P	D	L	G	V
4	Y	L	C	L	V	F	G	I	E	V
5	L	L	T	F	W	N	P	P	V	
6	V	L	V	G	G	V	L	A	A	
7	W	M	N	R	L	I	A	F	A	
8	D	L	M	G	Y	I	P	L	V	

G3	1	2	3	4	5	6	7	8	9	10
1	I	L	H	N	G	A	Y	S	L	
2	Y	L	S	G	A	N	L	N	L	
3	V	M	A	G	V	G	S	P	Y	V
4	I	L	A	G	Y	G	A	G	V	
5	L	M	T	F	W	N	P	P	V	
6	Y	L	V	T	R	H	A	D	V	
7	H	M	W	N	F	I	S	G	I	
8	Y	L	L	P	R	R	G	P	R	L

G4	1	2	3	4	5	6	7	8	9	10
1	L	L	F	L	L	L	A	D	A	
2	L	L	T	F	W	N	P	P	T	
3	A	L	C	R	W	G	L	L	L	

1	2	3	4	5	6	7	8	9	10
7	7	7							
	9	9		9		9			
	24								
63	63								
66	66	66	66	66	66				
	70	70	70	70	70	70			
							73		
						74		74	
							76		
						77	77	77	
							80	80	
								81	
		97			97	97		97	
	99	99		99					
		114				114			
						116		116	
				152	152				
		156	156		156				
163									
167									

P	A*0201	A*0202	A*0203	A*0206	A*6802		
7	Y	Y	Y	Y	Y		1
9	F	F	Y	F	Y	1	2
24	A	A	A	A	A		3
35	R	R	R	R	R		4
63	E	E	E	E	N	2	5
66	K	K	K	K	N	3	6
70	H	H	H	H	Q	4	7
73	T	T	T	T	T		8
74	H	H	H	H	D	5	9
76	V	V	V	V	V		10
77	D	D	D	D	D		11
80	T	T	T	T	T		12
81	L	L	L	L	L		13
97	R	R	R	R	R		14
99	Y	Y	Y	Y	Y		15
114	H	H	H	H	H		16
116	Y	Y	Y	Y	Y		17
152	V	V	V	E	V	6	18
156	L	W	L	W	W	7	19
163	T	T	T	T	T		20
167	W	W	W	W	W		21
	A*0201	A*0202	A*0203	A*0206	A*6802		

Fig. 1. A graphical representation of human leukocyte antigen (HLA)-A2 supertypes with peptides binding to specific HLA alleles is mapped to critical polymorphic functional residue positions (CPFRP) for each of these alleles (Red, nonbinder; Green, binder).

3. Results

Table 1 summarizes 25 peptides with known binding/nonbinding data to A*0201, A*0202, A*0203, A*0206, and A*6802. These peptides bind to more than one HLA allele, and thus, they show overlapping function with two or more HLA alleles. Six of these peptides bind to all the five alleles, and eight of these peptides bind to any four of these alleles (Table 1). Table 1 also summarizes that eight peptides bind any three of these alleles and three peptides bind any two of these alleles. Overall, all the 25 peptides show overlapping function with at least two alleles.

Figure 1 shows the graphical representation of 25 peptides binding or nonbinding with A*0201, A*0202, A*0203, A*0206, and A*6802. Figure 1 also shows the mapping of each HLA allele to the 21 CPFRPs with their corresponding residues. Among alleles, A*0201, A*0202, A*0203, A*0206, and A*6802, 33% (seven) of CPFRP show variations. However, 66% (14) show no variation among A*0201, A*0202, A*0203, A*0206, and A*6802 at the CPFRP. The virtual pockets formed by the CPFRP are shown for each peptide residue position in Fig. 1. This comprehensive mapping between peptides, alleles, function, CPFRP, and virtual pockets is aimed at explaining the overlapping functional property in HLA-A2 supertypes.

4. Discussion

More than 1,800 HLA alleles have been defined *(1)*. Therefore, the number of theoretically possible combinations of HLA–peptide complexes is extremely large. However, the immune system maintains a homogenous balance by specific selection, degeneration, and discrimination (self/non-self) of short peptides using HLA molecules. Although, HLA molecules are polymorphic in ethnic population, they exhibit a substantial amount of functional overlap through the phenomenon of "HLA supertypes," where members bind similar peptides and yet display distinct repertoires. A number of "HLA supertypes" have already been defined using binding data (Table 1). Table 1 shows six peptides binding to all members of the A2 supertype (A*0201, A*0202, A*0203, A*0206, and A*6802). The functional overlap between different members of the supertype is interesting. These also show several peptides binding to some members but not all members of the A2 supertypes (Table 1).

The concept of HLA supertypes is that alleles belonging to supertypes bind a highly shared set of peptides; in principle, it should be possible to predict peptide binding of other members of a supertype using experimental results based on just one member of the type. However, as illustrated in Table 1, this promise does not hold true in the major supertype A2. Hence, the binding of peptides to different members of the A2 supertype is combinatorial in selection and degeneration. Moreover, this grouping is inconclusive, given the known number of HLA alleles. If the molecular basis for supertype-like function of

HLA molecules is known, extrapolation of supertype function to other HLA molecules will be trivial.

In Fig. 1, the A2 supertypes (A*0201, A*0202, A*0203, A*0206, and A*6802) are mapped to their corresponding CPFRP residues. This provides a graphical visualization of four groups of HLA supertypes with the CPFRP residues. For the five HLA alleles, the residues at seven CPFRP (9, 63, 66, 70, 74, 152, and 156) show residue-level changes. However, residues at 14 CPFRPs (7, 24, 35, 73, 76, 77, 80, 81, 97, 99, 114, 116, 163, and 167) are identical among A*0201, A*0202, A*0203, A*0206, and A*6802. The functional overlap among these alleles is partly due to the conservation at the 14 CPFRP. The difference in function among these alleles for some of the peptides given in Fig. 1 is due to the variations at the CPFRP.

Figure 1 also shows the virtual pockets defined for each peptide residue position using CPFRP residues. Data show that each of the eight virtual pockets (1, 2, 3, 4, 5, 6, 7, and 9) have at least one mutating residues at the CPFRP. This accounts for the subtle changes associated with the peptide-binding function. In an attempt to explain the difference in peptide-binding function of four groups of peptides with the five alleles, we mapped functional information with CPFRP residues and virtual pockets.

Six peptides in G1 (group 1) bind all the five alleles (Fig. 1). These peptides show functional overlap with these alleles, exhibiting supertype-like property. This implies that the residue-level changes at the seven CPFRPs are insensitive to peptide binding in these peptides. However, this is not strictly true for peptides in G2 (group 2), G3 (group 3), and G4 (group 4), as shown in Fig. 1.

HLA-A*0201 and HLA-A*0202 show 156L→156W mutation and residue 156 is involved in virtual pockets 3, 4, and 6. The involvement of 156 is deterministic for peptide binding in one peptide in G2, four peptides in G3, and two peptides in G4 (Fig. 1). Comparison of A*0202 and A*0203 with A*0201 shows 156L→156W between A*0201 and A*0202 and 9F→9Y between A*0201 and A*0203. Between A*0201 and A*0206, 152V→152E and 156L→156W changes are observed. These changes at 9 [AND, OR] 152 [AND, OR] 156 affect binding of peptides to A*0202, A*0203, and A*0206 despite their binding to A*0201. Comparison of A*6802 with A*0201 in Fig. 1 shows changes at six positions (9, 63, 66, 70, 74, and 156). Thus, 17 peptides that bind to A*0201 are nonbinders to A*6802. These data indicate that residues at CPFRP determine functional overlap between alleles in A2 supertype. However, it is important to generate mapping matrices incorporating functional overlap at multiple layers for gathering a more clear picture of "supertype-like" function in future investigations.

5. Conclusion

HLA–peptide binding is useful in the design of peptide vaccine candidates, immunotherapeutic targets, and diagnostics agents. The theoretically possible combinations are overwhelmingly large. However, the functional overlap between alleles occurs at the level of supertypes. An understanding of their structural principles has a significant role in generating supertypes from sequence. Here, we show that the 21 CPFRPs have a role to play in determining overlapping function between two or more alleles. The 14 conserved CPFRPs explain overlapping function, and the 7 nonconserved CPFRPs explain nonoverlapping function. We hope to create a much clear picture of this phenomenon in future studies.

References

1. Robinson J., M.J. Waller, P. Parham, N. de Groot, R. Bontrop, L.J. Kennedy, P. Stoehr & S.G.E. Marsh: IMGT/HLA and IMGT/MHC – sequence databases for the study of the major histocompatibility complex. *Nucleic Acids Res.* 31(1) 311–314 (2003).
2. Yewdell J.W., E. Reits & J. Neefjes: Making sense of mass destruction – quantitating MHC class I antigen presentation. *Nat. Rev. Immunol.* 3(12) 952–961 (2003).
3. Del Guercio M.F., J. Sidney, G. Hermanson, C. Perez, H.M. Grey, R.T. Kubo & A. Sette: Binding of a peptide antigen to multiple HLA alleles allows definition of an A2-like supertype. *J. Immunol.* 154(2) 685–693 (1995).
4. Sette A. & J. Sidney: Nine major HLA class I supertypes account for the vast preponderance of HLA-A and -B polymorphism. *Immunogenetics* 50(3–4) 201–212 (1999).
5. Sette A. & J. Sidney: HLA supertypes and supermotifs – a functional perspective on HLA polymorphism. *Curr. Opin. Immunol.* 10(4) 478–482 (1998).
6. Sidney J., S. Southwood, V. Pasquetto & A. Sette: Simultaneous prediction of binding capacity for multiple molecules of the HLA B44 supertype. *J. Immunol.* 171(11) 5964–5974 (2003).
7. Chelvanayagam G.: A roadmap for HLA-A, HLA-B, and HLA-C peptide binding specificities. *Immunogenetics* 45(1) 15–26 (1996).
8. Zhang C, A. Anderson & C. DeLisi: Structural principles that govern the peptide-binding motifs of class I MHC molecules. *J. Mol. Biol.* 281(5) 929–947 (1998).
9. Zhao B., A.E.H. Png, E.C. Ren, P.R. Kolatkar, V.S. Mathura, M.K. Sakharkar & P. Kangueane: Compression of functional space in HLA-A sequence diversity. *Hum. Immunol.* 64(7) 718–728 (2003).
10. Doytchinova I.A., P. Guan & D.R. Flower: Identifying human MHC supertypes using bioinformatics methods. *J. Immunol.* 172(7), 4314–4323 (2004).
11. Lund O., M. Nielsen, C. Kesmir, A.G. Petersen, C. Lundegaard, P. Worning, C. Sylvester-Hvid, K. Lamberth, G. Roder, S. Justesen, S. Buus & S. Brunak:

Definition of supertypes for HLA molecules using clustering of specificity matrices. *Immunogenetics* 55(12), 797–810 (2004).
12. Scognamiglio P., D. Accapezzato, M.A. Casciaro, A. Cacciani, M. Artini, G. Bruno, M.L. Chircu, J. Sidney, S. Southwood, S. Abrignani, A. Sette & V. Barnaba: Presence of effector CD8+ T cells in hepatitis C virus-exposed healthy seronegative donors. *J. Immunol.* 162(11), 6681–6689 (1999).
13. Kawashima I., S.J. Hudson, V. Tsai, S. Southwood, K. Takesako, E. Appella, A. Sette & E. Celis: The multi-epitope approach for immunotherapy for cancer: identification of several CTL epitopes from various tumor-associated antigens expressed on solid epithelial tumors. *Hum. Immunol.* 59(1), 1–14 (1998).

11

Definition of MHC Supertypes Through Clustering of MHC Peptide-Binding Repertoires

Pedro A. Reche* **and Ellis L. Reinherz**

Summary

Identification of peptides that can bind to major histocompatibility complex (MHC) molecules is important for anticipation of T-cell epitopes and for the design of epitope-based vaccines. Population coverage of epitope vaccines is, however, compromised by the extreme polymorphism of MHC molecules, which is in fact the basis for their differential peptide binding. Therefore, grouping of MHC molecules into supertypes according to peptide-binding specificity is relevant for optimizing the composition of epitope-based vaccines. Despite the fact that the peptide-binding specificity of MHC molecules is linked to their specific amino acid sequences, it is unclear how amino sequence differences correlate with peptide-binding specificities. In this chapter, we detail a method for defining MHC supertypes based on the analysis and subsequent clustering of their peptide-binding repertoires

Key Words: MHC; supertypes; clustering; peptide-binding repertoire

1. Introduction

Major histocompatibility complex (MHC) molecules play a key role in the immune system by capturing peptide antigens for display on cell surfaces. Subsequently, these peptide–MHC (pMHC) complexes are recognized by T cells through their T-cell receptors (TCRs). MHC molecules fall into two

* Address for correspondence: Department of Immunology, Faculated de Medicina, Universidad Complutense de Madrid, Ave Complutense S/N Madrid, 28040, SPAIN. TL: +34 91 394 7299; FX: +34 91 394 1641; Email: parecheg@med.ucm.es

From: *Methods in Molecular Biology, vol. 409: Immunoinformatics: Predicting Immunogenicity In Silico*
Edited by: D. R. Flower © Humana Press Inc., Totowa, NJ

major classes, MHC class I (MHCI) and MHC class II (MHCII). Antigens presented by MHCI and MHCII are recognized by two distinct sets of T cells, $CD8^+$ T and $CD4^+$ T cells, respectively ***(1)***. Because T-cell recognition is limited to those peptides presented by MHC molecules, prediction of peptides that can bind to MHC molecules is important for anticipating T-cell epitopes and designing epitope-based vaccines ***(2–4)***. Furthermore, the availability of computational methods that can readily identify potential epitopes from primary protein sequences has fueled a new epitope discovery-driven paradigm in vaccine development.

A major complication to the development of epitope-based vaccines is the extreme polymorphism of the MHC molecules. In the human, MHC molecules are known as human leukocyte antigens (HLAs), and there are hundreds of allelic variants of class I (HLA I) and class II (HLA II) molecules. These HLA allelic variants bind distinct sets of peptides ***(5)*** and are expressed at vastly variable frequencies in different ethnic groups ***(6)***. Consequently, the potential population coverage of epitope-based vaccines is greatly compromised. Interestingly, it has been noted that some HLA molecules can bind largely overlapping sets of peptides ***(7,8)***. Therefore, grouping of MHC molecules into supertypes according to peptide-binding specificity is of relevance for the formulation of epitope vaccines providing a wide population coverage.

The first supertypes were defined by Sidney, Sette, and co-workers ***(7,8)*** (hereafter Sidney–Sette et al.) upon inspection of the reported peptide-binding motifs of individual HLA alleles. However, the relationships between peptide-binding specificities of HLA molecules may be too subtle to be defined by visual inspection of these peptide-binding motifs. Furthermore, such sequence patterns have proven to be too simple to describe the binding ability of a peptide to a given MHC molecule ***(9,10)***. In view of these limitations, we developed an alternative method to define MHC supertypes by clustering the peptide-binding repertoire of MHC molecules. The core of the method consists of the generation of a distance matrix whose coefficients are inversely proportional to the peptide binders shared by any two MHC molecules. Subsequently, this distance matrix is fed to a phylogenic clustering algorithm to establish the kinship among the distinct MHC peptide-binding repertoires. The peptide-binding repertoire of any given MHC molecule is unknown, and thereby, defining supertypes through this method requires the estimation of the peptide-binding repertoire of MHC molecules. In this chapter, we will use position-specific scoring matrices (PSSMs) as the predictor of peptide–MHC binding ***(11,12)*** and describe in detail the generation of supertypes using, for example, a selection of HLA class I (HLA I) molecules for which PSSMs are readily available.

2. Materials

2.1. Prediction of Peptide–MHC Binding Repertoires

We consider the peptide-binding repertoire of any MHC molecule as the subset of peptides predicted to bind from a reference set consisting of a random protein of 1,000 amino acids. A selection of public online resources that can be used for the prediction of peptide–MHC binding is summarized in Table 1. In our study PSSMs derived from aligned MHC ligands as the predictors of peptide–MHC binding *(11,12)*. In this approach, the binding potential of any peptide sequence (query) to the MHC molecule is determined by its similarity to a set of known peptide–MHC binders and can be obtained by comparing the query to the PSSM. Prediction of peptide–MHC binding is threshold-dependent, and here we use the same threshold for all MHC molecules. Thus, the size of the peptide-binding repertoire of all MHC molecules is considered to be same (same number of peptides).

2.2. Supertype Construction

MHC supertypes are derived following the general scheme illustrated in Fig. 1 . First, the overlap between the predicted peptide-binding repertoires (see Section 2.1) of any two MHC molecules, $pMHC_i$ and $pMHC_j$, is computed as the number of peptide binders shared by the two molecules. Let that number be n_{ij}. Subsequently, a distance coefficient (d_{ij}) is defined as follows:

$$d_{ij} = N - n_{ij}, \quad (1)$$

where N is the size of the peptide-binding repertoire of the MHC molecule. Thus, if the peptide-binding repertoire between two MHC molecules is identical, then $d_{ij} = 0$. Alternatively, if they share no peptides in common, d_{ij} will match the size of the binding repertoire, N. Through the repetition of this process over all distinct pairs of MHC molecules, a quadratic distance matrix is derived containing the d_{ij} coefficients for all distinct pairs of MHC molecules. Once the distance matrix is obtained, we use the Phylogeny Inference Package (PHYLIP; http:// evolution.genetics.washington.edu/phylip.html) *(13)* to generate a phylogenic tree where the MHC molecules appear clustered according to their peptide-binding specificity. Specifically, within the PHYLIP package one must use applications such as *kitsch* and *neighbor* that take distance matrices as input. The *kitsch* application uses a Fitch–Margoliash criterion and assumes an evolutionary clock *(14)*. On the other hand, the *neighbor* application uses the popular neighbor-joining method to derive an unrooted tree without the assumption of a

Table 1
Public online resources for the prediction of peptide–major histocompatibility complex (MHC) binding

Name	URL	Method	Class	Reference
HLABIND	http://www.bimas.dcrt.nih.gov/molbio/hla_bind/	QM	I	*(22)*
MHCpred	http://www.wehih.wehi.edu.au/mhcpep	QM	I and II	*(23)*
PROPRED	http://www.imtech.res.in/raghava/propred/	QM	II	*(24)*
SVMHC	http://www.sbc.su.se/svmhc/information.html	SVM	I	*(25)*
SYFPEITHI	http://www.syfpeithi.de/Scripts/MHCServer.dll/EpitopePrediction.htm	Motif matrix	I and II	*(26)*
RANKPEP	http://www.mif.dfci.harvard.edu/Tools/rankpep.html	Motif PSSM	I and II	*(11)*
NetMHC	http://www.cbs.dtu.dk/services/NetMHC/	ANN	I	*(27)*
PREDEP	http://www.margalit.huji.ac.il/	Structure based	I	*(28)*

ANN, Artificial Neural Network; PSSM, position-specific scoring matrix; QM, quantitative matrix.

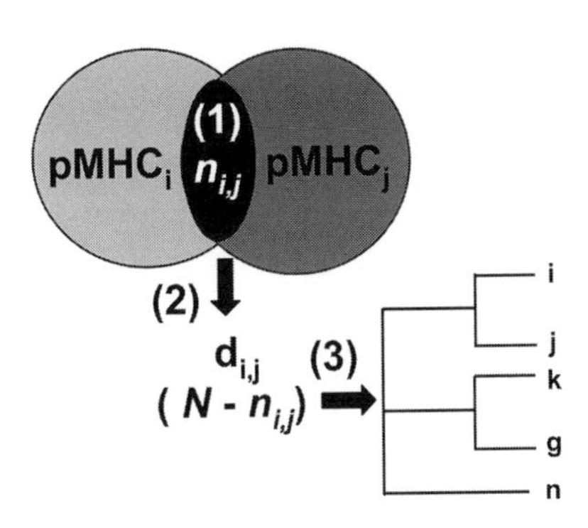

Fig. 1. Strategy to define major histocompatibility complex (MHC) supertypes. MHC supertypes are identified as follows: (1) estimate number of common peptides, n_{ij}, between the binding repertoires of any two MHC molecules, $pMHC_i$ and $pMHC_j$; (2) obtain a distance matrix whose coefficients, d_{ij}, are inversely proportional to the peptide-binding overlap between any pair of MHC molecules; and (3) derive a dendrogram using a phylogenic clustering algorithm to visualize MHC supertypes (groups of MHC molecules with similar peptide-binding specificity). N is the size of the peptide-binding repertoire of the MHC molecule.

clock *(15)*. For instance, to generate a tree using the neighbor-joining algorithm method one can use the command:

```
echo Y | neighbor > /dev/null.
```

This command will generate a tree from a distance matrix that must be named as *infile* using the default options of the *neighbor* application. Likewise, one may use similar commands to generate trees using other applications. In any case, these applications will generate two files, one named *outfile* displaying the tree and another named *treefile* describing the tree in NEWICK format, which can be used to visualize and manipulate the tree using third party applications such as TREEVIEW (http:// taxonomy.zoology.gla.ac.uk/rod/treeview.html).

3. Methods

3.1. HLA I Supertypes

Definition of MHC supertypes using the method described here requires the estimation of the peptide-binding repertoire of the MHC molecules using predictors of peptide–MHC binding. The prediction of peptide–MHCII binding is generally less reliable than that of peptide–MHCI binding *(12)*. Therefore, to illustrate the definition of MHC supertypes, we focused on 55 HLA I molecules (human MHCI) for which we can readily predict their peptide-binding repertoires using PSSMs (see Section 2). Given that MHCI ligands are usually nine residues in length, we selected PSSMs for the prediction of binders of that same size (nine residues). In previous studies we have shown that depending on the specific MHCI molecule, the accuracy of peptide–MHCI binding predictions is optimal by considering as binders the top 2–5% scoring peptides (2–5% threshold) within a protein query ***(12)***. Here we have estimated the peptide-binding repertoire of the selected HLA I molecules using a 2% threshold. Thus, following the method described above with a Fitch and Margoliash clustering algorithm ***(14)*** (Section 2.2; *kitsch* application), we generated the phylogenic tree, which is shown in Fig. 2. In this tree, HLA I molecules with similar peptide-binding specificity (large overlap in their peptide-binding repertoires) branch together in groups or supertypes. The relationship between the peptide-binding specificities of HLA I molecules is extensive, and although affinities are mostly confined to alleles belonging to the same gene, they also reach to alleles belonging to different genes (Fig. 2, B15 cluster; B*4002 and A*2902; and A*2402 and B*3801). We clearly identified the classic A2, A3, B7, B27, and B44 supertypes previously defined by Sidney–Sette et al., as well as three new potential supertypes, BX, AB, and B57 (Fig. 2). Furthermore, this analysis indicates that classic HLA I supertypes may be larger than that previously thought. For instance, the A2 supertype would also include the A0207, A0209, and A0214, and the A3 supertype will also include A*6601.

3.2. Combined Phenotypic Frequency of HLA I Supertypes

HLA I-restricted peptides are the targets of $CD8^+$ cytotoxic T lymphocytes (CTLs). The population protection coverage (PPC) of a vaccine composed of CTL epitopes is given by the combined phenotypic frequency (CPF) of the HLA I molecules restricting the epitopes, and it can be computed from the gene and haplotype frequencies ***(16)***. Using the allelic and haplotype frequencies reported by Cao et al. ***(17)*** corresponding to five major American ethnic groups (Black, Caucasian, Hispanic, Native American, and Asian), we have computed the CPF for the HLA I supertypes defined in the previous section (Section 3.1), and the values are tabulated in Table 2 . Targeting HLA I supertypes for the prediction of promiscuous peptide binders allows to minimize the total number of predicted epitopes without compromising the population coverage required in the design of multi-epitope vaccines. However, including many distantly related HLA I molecules in the supertypes may result in too few or no epitopes predicted to bind to all the alleles included in the supertype. Therefore, for the CPF calculations, we have limited the composition of HLA I supertypes to

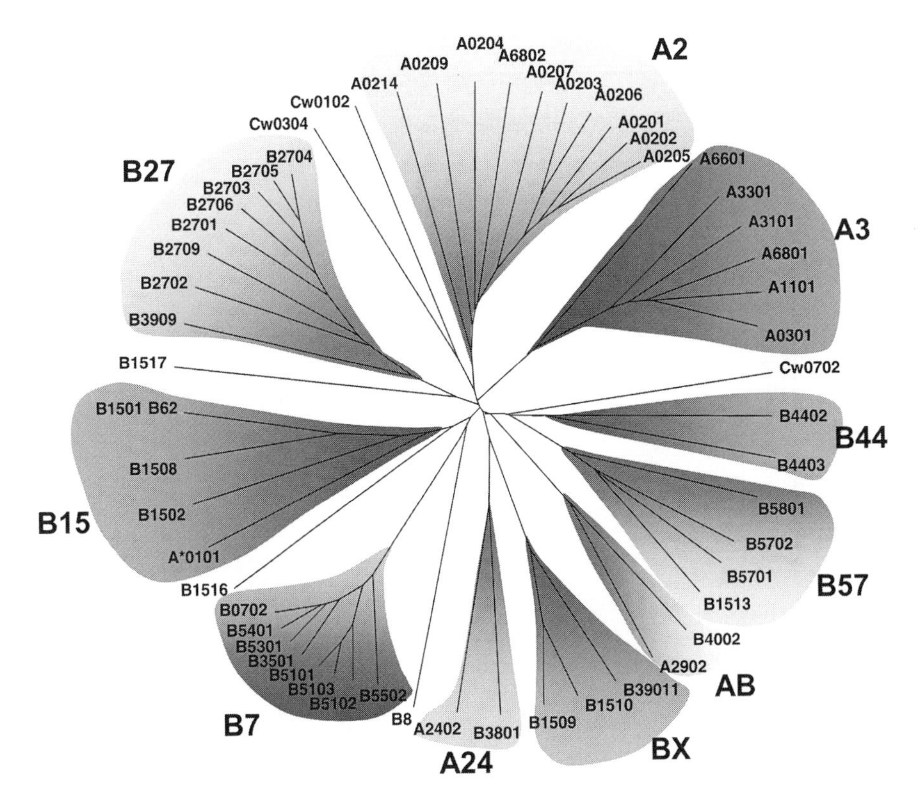

Fig. 2. Human leukocyte antigen (HLA) I supertypes. This figure shows an unroot dendrogram reflecting the relationships between the peptide-binding specificities of HLA I molecules. The closer the HLA I alleles branch, the larger the overlap between their peptide-binding repertoires. Groups of HLA I alleles with similar peptide-binding specificities branch together defining supertypes (shaded groups).

include only those HLA I alleles with $\geq$20% peptide-binding overlap (pairwise between any pair of alleles).

The A2, A3, and B7 supertypes have the largest CPF in the five studied ethnic groups, providing a CPF close to 90%, regardless of ethnicity. To increase the CPF to 95% in all ethnicities, it is necessary to include at least two more supertypes. Specifically, the supertypes A2, A3, B7, B15, and A24 or B44 represent the minimal supertypic combination providing a CPF $\geq$95%. These results indicate that as few as five epitopes restricted by the mentioned HLA I supertypes may be enough to develop a vaccine eliciting CTL responses in the whole population, regardless of ethnicity.

Table 2
Cumulative phenotype frequency of defined supertypes

Supertype	Alleles	Blacks (%)	Caucasians (%)	Hispanics (%)	N.A. Natives (%)	Asians (%)
A2	A*0201–7, A*6802	43.7	49.9	51.8	52.4	44.7
A3	A*0301, A*1101, A*3101, A*3301, A*6801, A*6601	35.4	46.9	41.5	40.7	47.9
B7	B*0702, B*3501, B*5101–02, B*5301, B*5401	45.9	42.2	40.5	52.0	31.3
B15	A*0101, B*1501_B62, B1502	13.06	37.80	16.75	27.26	21.04
A24	A*2402, B*3801	15.5	17.28	25.85	41.94	35.0
B44	B*4402, B*4403	10.4	27.7	17.15	14.4	10.1
B57	B5701–02, B5801, B*1503	19.2	10.3	5.9	5.8	16.5
ABX	A*2902, B*4002	7.4	11.3	19.1	16.3	16.3
B27	B*2701–06, B*2709, B*3909	2.3	4.8	5.1	16.9	4.7
BX	B*1509, B*1510, B*39011	3.1	0.7	4.2	7.8	4.1
AB	A*2902, B4002	7.4	0.11	0.19	0.16	0.07

N.A., North American.

4. Conclusions

HLA molecules are represented by hundreds of allelic variants displaying distinct peptide-binding specificities, and grouping them into supertypes is relevant for developing epitope-based vaccines with a wide PPC. The peptide-binding specificity of HLA molecules stems from the specific amino acids lining their binding groove, and consequently, supertypes may be defined from structural analysis ***(18–20)***. However, it is not always clear how amino acid sequence differences among HLA molecules translate into distinct peptide-binding specificities. Indeed, structure-based methods for the prediction of peptide–MHC binding are still in their infancy. Therefore, in thischapter, we described a method for defining HLA supertypes based on the analysis and subsequent clustering of their predicted peptide-binding repertoires. Furthermore, we have shown that the method can identify experimentally defined HLA I supertypes, suggesting in addition new potential relationships between the peptide-binding specificity of HLA I molecules. When the predictor of peptide–MHC binding is a specificity matrix such as a PSSM, clustering of the HLA molecules according to peptide-binding specificity may alternatively be achieved by comparison of the matrix coefficients ***(21)***. However, it is important to stress that the clustering method described here to derive supertypes can be applied in combination with any predictor of peptide–MHC binding. Although, not indicated in this chapter, minor differences in the defined supertypes appear depending on the phylogenic algorithm used to cluster the HLA I molecules. There are also two other limitations to the method described here. First, the method is limited by both the quality and availability of the peptide–MHC binding predictors. Thus, we do not discard the possibility that the fine structure of the supertypes may suffer some changes as new and better predictors of peptide–MHC binding develop. The second limitation is that we have considered the size of the peptide-binding repertoire of all MHC molecules to be the same. However, that might not always be the case. Indeed, it has been noted that, for instance, the A*0201 appears to be quite promiscuous, binding larger sets of peptides than the other HLA I molecules (Azouz, Reinhold, and Reinherz, unpublished results).

References

1. Margulies, D.H. 1997. Interactions of TCRs with MHC-peptide complexes: a quantitative basis for mechanistic models. *Curr Opin Immunol* 9:390–395.
2. Yu, K., Petrovsky, N., Schonbach, C., Koh, J.Y., and Brusic, V. 2002. Methods for prediction of peptide binding to MHC molecules: a comparative study. *Mol Med* 8:137–148.

3. Flower, D. 2003. Towards in silico prediction of immunogenic epitopes. *Trends Immunol* 24:667–674.
4. Flower, D., and Doytchinova, I.A. 2002. Immunoinformatics and the prediction of immunogenicity. *Appl Bioinformatics* 1:167–176.
5. Reche, P.A., and Reinherz, E.L. 2003. Sequence variability analysis of human class I and class II MHC molecules: functional and structural correlates of amino acid polymorphisms. *J Mol Biol* 331:623–641.
6. David W. Gjertson, and Paul I. Terasaki, E. (Eds) 1998. *HLA 1998*. American Society for Histocompatibility and Immunogenetics, Lenexa.
7. Sette, A., and Sidney, J. 1999. Nine major HLA class I supertypes account for the vast preponderance of HLA-A and -B polymorphism. *Immunogenetics* 50:201–212.
8. Sette, A., and Sidney, J. 1998. HLA supertypes and supermotifs: a functional perspective on HLA polymorphism. *Curr Opin Immunol* 10:478–482.
9. Bouvier, M., and Wiley, D.C. 1994. Importance of peptide amino acid and carboxyl termini to the stability of MHC class I molecules. *Science* 265:398–402.
10. Ruppert, J., Sidney, J., Celis, E., Kubo, T., Grey, H.M., and Sette, A. 1993. Prominent role of secondary anchor residues in peptide binding to HLA-A2.1 molecules. *Cell* 74:929–937.
11. Reche, P.A., Glutting, J.-P., and Reinherz, E.L. 2002. Prediction of MHC class I binding peptides using profile motifs. *Hum Immunol* 63:701–709.
12. Reche, P.A., Glutting, J.-P, Zhang, H., and Reinherz, E.L. 2004. Enhancement to the RANKPEP resource for the prediction of peptide binding to MHC molecules using profiles. *Immunogenetics* 56:405–419
13. Retief, J.D. 2000. Phylogenetic analysis using PHYLIP. *Methods Mol Biol* 132:243–258.
14. Fitch, W.M., and Margoliash, E. 1967. Construction of phylogenetic trees. *Science* 155:279–284.
15. Saitou, N., and Nei, M. 1987. The neighbor-joining method: a new method for reconstructing phylogenetic trees. *Mol Biol Evol* 4:406–425.
16. Dawson, D.V., Ozgur, M., Sari, K., Ghanayem, M., and Kostyu, D.D. 2001. Ramifications of HLA class I polymorphism and population genetics for vaccine development. *Genet Epidemiol* 20:87–106.
17. Cao, K., Hollenbach, J., Shi, X., Shi, W., Chopek, M., and Fernandez-Vina, M.A. 2001. Analysis of the frequencies of HLA-A, B, and C alleles and haplotypes in the five major ethnic groups of the United States reveals high levels of diversity in these loci and contrasting distribution patterns in these populations. *Hum Immunol* 62:1009–1030.
18. Doytchinova, I.A., Guan, P., and Flower, D.R. 2004. Quantitative structure-activity relationships and the prediction of MHC supermotifs. *Methods* 34:444–453.
19. Doytchinova, I.A., and Flower, D.R. 2005. In silico identification of supertypes for class II MHCs. *J Immunol* 174:7085–7095.

20. Doytchinova, I.A., Guan, P., and Flower, D.R. 2004. Identifying human MHC supertypes using bioinformatic methods. *J Immunol* 172:4314–4323.
21. Lund, O., Nielsen, M., Kesmir, C., Petersen, A.G., Lundegaard, C., Worning, P., Sylvester-Hvid, C., Lamberth, K., Roder, G., Justesen, S., Buus, S., and Brunak, S. 2004. Definition of supertypes for HLA molecules using clustering of specificity-matrices. *Immunogenetics* 55:797–810.
22. Parker, K.C., Bednarek, M.A., and Coligan, J.E. 1994. Scheme for ranking potential HLA-A2 binding peptides based on independent binding of individual peptide side chains. *J Immunol* 152:163–175.
23. Guan, P., Doytchinova, I.A., Zygouri, C., and Flower, D. 2003. MHCPred: a server for quantitative prediction of peptide-MHC binding. *Nucleic Acids Res* 31:3621–3624.
24. Singh, H., and Raghava, G.P. 2001. ProPred: prediction of HLA-DR binding sites. *Bioinformatics* 17:1236–1237.
25. Donnes, P., and Elofsson, A. 2002. Prediction of MHC class I binding peptides, using SVMHC. *BMC Bioinformatics* 3:25.
26. Rammensee, H.G., Bachmann, J., Emmerich, N.P.N., Bacho, O.A., and Stevanovic, S. 1999. SYFPEITHI: database for MHC ligands and peptide motifs. *Immunogenetics* 50:213–219.
27. Buus, S., Lauemoller, S.L., Worning, P., Kesmir, C., Frimurer, T., Corbet, S., Fomsgaard, A., Hilden, J., Holm, A., and Brunak, S. 2003. Sensitive quantitative predictions of peptide-MHC binding by a 'Query by Committee' artificial neural network approach. *Tissue Antigens* 62:378–384.
28. Altuvia, Y., Sette, A., Sidney, J., Southwood, S., and Margalit, H. 1997. A structure based algorithm to predict potential binding peptides to MHC molecules with hydrophobic binding pockets. *Hum Immunol* 58:1–11.

12

Grouping of Class I HLA Alleles Using Electrostatic Distribution Maps of the Peptide Binding Grooves

Pandjassarame Kangueane and Meena Kishore Sakharkar

Summary

Human leukocyte antigen (HLA) molecules involved in immune function by binding to short peptides (8–20 residues) have different sequences in different individuals belonging to distinct ethnic population. Hence, the peptide-binding function of HLA alleles is specific. Class I HLA alleles (alternative forms of a gene) are associated with CD8+ T cells, and their allele-specific sequence information is available at the IMGT/HLA database. The available sequences are one-dimensional (1D), and the peptide-binding functional inference often requires 3-dimensional (3D) structural models of respective alleles. Hence, 3D structures were constructed for 1,000 class I HLA alleles (310 A, 570 B, and 120 C) using MODELLER (a comparative protein modeling program for modeling protein structures). The electrostatic distribution maps were generated for each modeled structure using Deep View (Swiss PDB Viewer Version 3.7). The 1,000 models were then grouped into different categories by visual inspection of their electrostatic distribution maps in the peptide binding grooves. The distribution of the models based on electrostatic distribution was 30% negative (300), 1% positive (12), 8% neutral (84), and 60% (604) mixed (random mixture of negative, positive, and neutral). This grouping provides insight toward the inference for functional overlap among HLA alleles.

Key Words: HLA; alleles; grouping; peptide binding groove; electrostatic potential; negative; positive; neutral

1. Introduction

Human leukocyte antigen (HLA) proteins are involved in T-cell-mediated immune response by binding to short peptides of 8–20 residues long ***(1,2)***. The binding of HLA molecules to peptides is highly specific. However, HLA alleles

From: *Methods in Molecular Biology, vol. 409: Immunoinformatics: Predicting Immunogenicity In Silico*
Edited by: D. R. Flower © Humana Press Inc., Totowa, NJ

are highly polymorphic among different ethnic groups, and more than 1,500 HLA alleles are known *(3)*. Therefore, HLA–peptide binding is combinational in nature. Nonetheless, peptide binding is determined by groove geometry and chemistry.

Zhang et al. *(4)* grouped HLA alleles using structural pockets. Chelvanayagam *(5,6)* classified class I and DR alleles using pocket information. Recently, Zhao et al. *(7)*, Kangueane et al. *(8)*, and Guan et al *(9)* developed A–F pocket systems for grouping HLA molecules. Doytchinova et al. *(10)* used molecular interaction fields (MIF), hierarchical clustering (HC), and principal component analysis (PCA), and Lund et al. *(11)* used clustering procedures for grouping HLA alleles into putative supertypes (where different members bind similar peptides, yet exhibiting distinct repertoires). Here, we describe a novel methodology to group HLA alleles using electrostatic distribution map of the peptide binding groove in HLA molecules.

2. Methodology

The procedure used in this analysis is outlined in Fig. 1 using a work flow diagram. The work flow diagram describes the different steps involved in HLA modeling, electrostatic calculation for the model, and grouping of HLA alleles.

2.1. HLA modeling

HLA modeling consists of (1) identification of suitable templates from Protein databank (PDB), (2) selection of structural templates, (3) target-to-template alignment, (4) model building, and (5) generation of 3D models. Structural templates were searched in the PDB using PSI-BLAST (http://www.ncbi.nih.gov/blast) at an *E* value cut-off <0.01. Template structures with the highest similarity to the query sequence were chosen for homology modeling using MODELLER (a comparative protein modeling program for modeling protein structures). The models thus obtained were further checked for errors using PROSA-II (using energy calculations—http://www.came.sbg.ac.at/Services/prosa.html), PROCHEK (a program to check the stereochemical quality of protein structures), and WHATIF (http://swift.cmbi.kun.nl/whatif/).

2.2. Electrostatic distribution maps

The electrostatic distribution maps were calculated for each modeled structure using Coulomb's Law (the force between two point charges is directly proportional to the product of the charges and inversely proportional to the

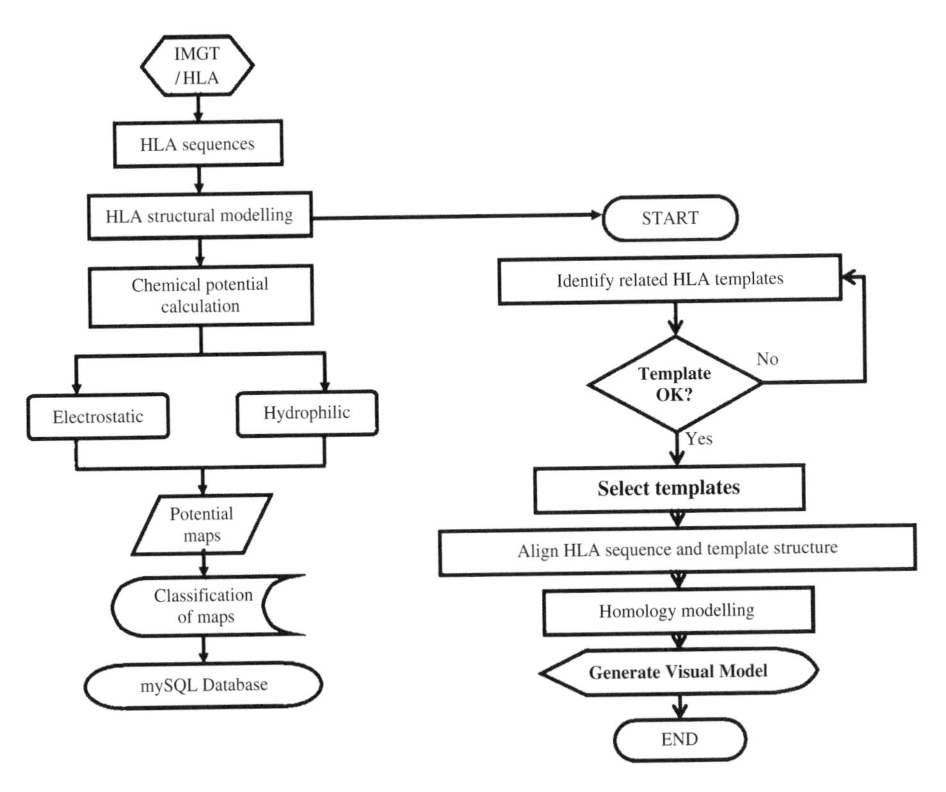

Fig. 1. Work flow diagram showing human leukocyte antigen (HLA) modeling and electrostatic calculation.

square of the distance between them) as implemented in Deep View (Swiss PDB Viewer Version 3.7).

2.3. *Grouping of models*

The models were then grouped by visual inspection based on the nature of color in the binding groove. Red color refers to electronegative groove, Blue to electropositive groove, White to neutral groove, and Mixed for a mixture of red, blue, and white groove (Table 1).

3. Results

Table 1 summarizes the grouping of 1,000 class I HLA models (310 HLA-A, 570 HLA-B, and 120 HLA-C) using the type of electrostatic potential in the peptide binding groove. We used negative (red color), positive (blue color), and

Table 1
Distribution of human leukocyte antigen (HLA)-A, HLA-B, and HLA-C alleles using electrostatic distribution of the peptide binding grooves

	Combination	HLA-A	HLA-B	HLA-C	Total
G1	R-R-R	106 (34%)	183 (32%)	11 (9.0%)	300 (30%)
G2	B-B-B	1 (0.5%)	9 (1.5%)	2 (2.0%)	12 (1.2%)
G3	W-W-W	51 (16%)	26 (4.5%)	7 (5.5%)	84 (8.4%)
G4	R-B-W	71 (23%)	57 (10%)	1 (1.0%)	129 (12.9%)
G5	B-R-W	24 (8.0%)	11 (2.0%)	2 (2.0%)	37 (3.7%)
G6	W-R-B	5 (1.5%)	26 (4.5%)	22 (18%)	57 (5.7%)
G7	R-W-B	27 (9.0%)	230 (40%)	70 (58%)	327 (33%)
G8	B-W-R	20 (6.5%)	26 (4.5%)	2 (2.0%)	48 (4.8%)
G9	W-B-R	5 (1.5%)	2 (1.0%)	3 (2.5%)	10 (1%)
Total		310 (100%)	570 (100%)	120 (100%)	1000 (100%)

G1 to G9 = groups 1 to groups 9; R, Red; B, Blue; W, White; Red means negative; Blue means positive; and W means neutral. The peptide binding groove is divided into three regions with respect to the binding peptide (N-terminal, C-terminal, and central) by visual inspection of their potential maps. The configuration W-W-W denotes white in the entire groove signifying neutral potential map in the groove. The configuration R-B-W denotes N-terminal red, C-terminal white and central blue in the entire groove signifying negative, neutral, and positive, respectively.

neutral (white color) in this study for grouping. Red is denoted by "R," blue by "B," and white by "W" in Table 1. Each peptide binding groove is divided into three regions with reference to the N-terminal, C-terminal, and middle region of the peptide binding the groove (Fig. 2). The group G1 is classified as R-R-R, where the groove is completely red and electronegative. Table 1 summarizes that 30% of 1,000 class I models have electronegative binding grooves, and this group is predominant among HLA-A alleles. Interestingly, 33% of HLA models belong to G7 with R-W-B configuration in the groove, and majority of the B and C alleles are of this type. Thus, 1,000 class I models are grouped into nine groups using the type of electrostatic maps at the binding groove. This grouping is useful in understanding the overlapping peptide-binding function exhibited by HLA alleles.

4. Discussion

HLA alleles have distinct sequences. These alleles demonstrate specific peptide-binding function. However, functional overlaps are also seen among them *(8)*. Homology models of 1,000 HLA class I molecules were constructed

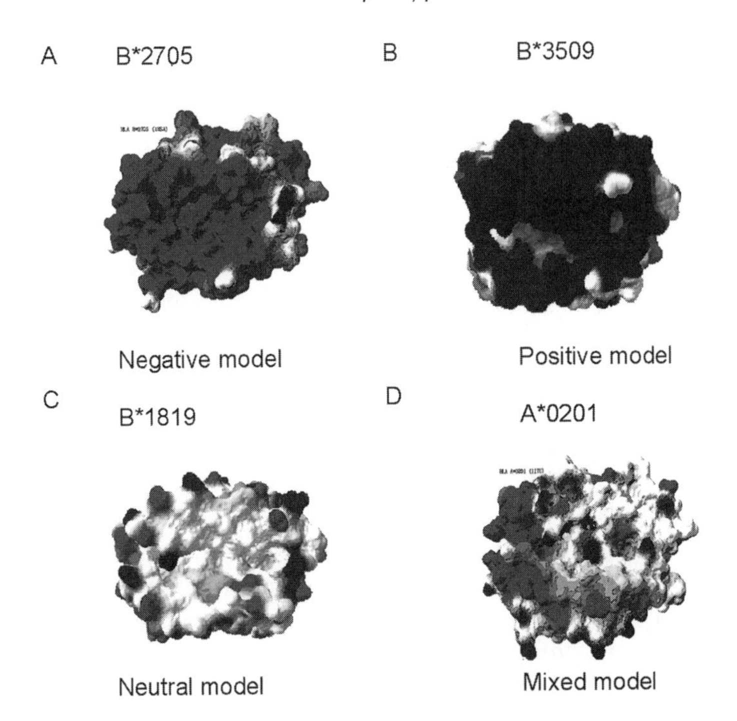

Fig. 2. Grouping of human leukocyte antigen (HLA) alleles into negative (red), positive (blue), neutral (white), and mixed (red, blue, and white) models.

using MODELLER. Electrostatic potentials for each model were then calculated using Deep View (Swiss PDB Viewer Version 3.7), and the electrostatic distribution maps of the peptide binding groove in HLA alleles were generated as shown in Fig. 2. Some HLA alleles have electronegative distribution, and some others have electropositive distribution. Interestingly, some also have neutral and others have a mixture of negative, positive, and neutral distribution in the groove. The grouping of alleles based on the electrostatic distribution of the HLA binding groove shows 300 negative (30%), 12 positive (1%), 84 neutral (8%), and 604 mixed (random mixture of negative, positive, and neutral) (60%) potential maps (Table 1). These models are thereafter referred as negative, positive, neutral, and mixed models. The distribution of HLA alleles with 30% negative, 1% positive, and 8% neutral models is interesting. This provides insight toward functional overlap in peptide binding to HLA alleles.

Table 1 summarizes that 34% HLA-A alleles and 32% HLA-B alleles are negative models. We divided the 604 (60%) mixed models into six groups (Table 1) with respect to the peptide binding in the groove (N-terminal,

C-terminal, and central regions of the peptide). The 60% mixed models among A, B, and C alleles show extensive variation in the peptide binding groove. In this category, 40% of B alleles have R-W-B configuration [N–terminal red (negative)' C–terminal blue (positive); central white (neutral)] in the groove.

The grouping of 1,000 HLA alleles into nine functional groups based on the electrostatic potential maps of the peptide binding grooves finds utility in the understanding of functional overlaps between alleles. This study complements the work done by Zhang et al. ***(4)***, Chelvanayagam ***(5,6)***, Zhao et al. ***(7)***, Kangueane et al. ***(8)***, Guan et al. ***(9)***, Doytchinova et al. ***(10)***, and Lund et al. ***(11)***.

5. Conclusion

HLA alleles are polymorphic. Hence, the peptide-binding specificity varies in a combinatorial manner. However, functional overlap between alleles is shown for several HLA supertypes. Previous studies have shown HLA functional overlap between alleles using several distinct methods ***(4–11)***. Here, we group 1,000 HLA class I alleles into different categories by visual inspection of their electrostatic distribution in the peptide binding grooves. The distribution is 300 negative (30%), 12 positive (1%), 84 neutral (8%), and 604 mixed (random mixture of negative, positive and neutral) models (60%). The groupings find utility in T-cell epitope design for peptide vaccines, immunotherapeutics, and diagnostics agents.

References

1. Govindarajan KR, Kangueane P, Tan TW, Ranganathan S. MPID: MHC-Peptide interaction database for sequence-structure-function information on peptides binding to MHC molecules. **Bioinformatics** (Oxford, England). 19(2): 309–310, 2003.
2. Kangueane P, Sakharkar MK, Kolatkar PR, Ren EC: Towards the MHC-peptide combinatorics. **Hum Immunol** 62(5):539–556, 2001.
3. Robinson J, Waller MJ, Parham P, de Groot N, Bontrop R, Kennedy LJ, Stoehr P, Marsh SGE IMGT/HLA and IMGT/MHC: sequence databases for the study of the major histocompatibility complex Nucleic Acids Research. 31:311–314, 2003. http://www.ebi.ac.uk/imgt/hla/intro.html
4. Zhang C, Anderson A, DeLisi C: Structural principles that govern the peptide binding motifs of class I MHC molecules. **J Mol Biol** 281:929, 1998.
5. Chelvanayagam G: A roadmap for HLA-A, HLA-B and HLA-C peptide binding specificities. **Immunogenetics** 45:15, 1996.
6. Chelvanayagam G: A roadmap for HLA-DR peptide binding specificities. **Hum Immunol** 58:61, 1997.

7. Zhao B, Png AEH, Ren EC, Kolatkar PR, Mathura VS, Sakharkar MK, Kangueane P: Compression of functional space in HLA-A sequence diversity. **Hum Immunol** 64:718, 2003.
8. Kangueane P, Sakharkar MK, Rajaseger G, Bolisetty S, Sivasekari B, Zhao B, Ravichandran M, Shapshak P, Subbiah S: A framework to sub-type HLA supertypes, **Frontiers in Bioscience** 10:879–886, 2005.
9. Guan PP, Doytchinova IA, Flower DR: HLA-A3 super-motif defined by quantitative structure-activity relationship analysis. **Protein Eng** 16:11, 2003.
10. Doytchinova IA, Guan P, Flower DR: Identifying human MHC supertypes using bioinformatics methods. **J Immunol** 172(7):4314–4323, 2004.
11. Lund O, Nielsen M, Kesmir C, Petersen AG, Lundegaard C, Worning P, Sylvester-Hvid C, Lamberth K, Roder G, Justesen S, Buus S, Brunak S: Definition of supertypes for HLA molecules using clustering of specificity matrices. **Immunogenetics** 55(12):797–810, 2004.

III

Predicting Peptide-MHC Binding

13

Prediction of Peptide–MHC Binding Using Profiles

Pedro A. Reche[*] and Ellis L. Reinherz

Summary

Prediction of peptide binding to major histocompatibility complex (MHC) molecules is a basis for anticipating T-cell epitopes. Peptides that bind to a given MHC molecule are related by sequence similarity. Therefore, a position-specific scoring matrix (PSSM)—also known as profile—derived from a set of aligned peptides known to bind to a given MHC molecule can be used as a predictor of both peptide–MHC binding and T-cell epitopes. In this approach, the binding potential of any peptide sequence (query) to the MHC molecule is determined by its similarity to a set of known peptide–MHC binders and can be obtained by comparing the query to the PSSM. Following structural considerations of the peptide–MHC interaction, we will describe here how to derive alignments and PSSMs that are suitable for the prediction of peptide–MHC binding.

Key Words: PSSM; MHC; binding; epitopes; profile; prediction

1. Introduction

T-cell immune responses are triggered by the recognition of foreign peptide antigens bound to cell membrane-expressed major histocompatibility complex (MHC) molecules ***(1–3)***. Because T-cell recognition is limited to those peptides presented by MHC molecules, prediction of peptides that can bind to MHC molecules is the basis for the anticipation of T-cell epitopes ***(4–6)***. Peptides binding to MHC molecules must fit into a specific chemical and physical

[*]Address for correspondence: Department of Immunology, Facultad de Medicina, Universidad Complutense de Madrid, Ave Complutense S/N, Madrid, 28040, SPAIN. TL: +34 91 394 7229; FX: +34 91 394 1641; Email: parecheg@med.ucm.es

From: *Methods in Molecular Biology, vol. 409: Immunoinformatics: Predicting Immunogenicity In Silico*
Edited by: D. R. Flower © Humana Press Inc., Totowa, NJ

environment conditioned by polymorphic residues in the MHC molecule ***(7–9)***. Consequently, distinct MHC molecules have distinct peptide-binding specificities ***(9)***. In addition, the peptides that bind to the same MHC molecule are related by sequence similarity. Sequence patterns reflecting amino acid preferences in peptide–MHC binders (anchor residues) are routinely used for defining peptide–MHC binding motifs and prediction of peptide–MHC binding ***(10,11)***. For example, the binding motif of the human MHC class I (MHCI) molecule A*0201 may be described by the following sequence pattern:

$$X\text{-}[AILMVT]\text{-}X_6\text{-}[AILMVT].$$

This motif indicates that A*0201 will preferentially bind peptides of nine residues having an Ala, Ile, Leu, Met, Val, or Thr residue at positions 3 and 9, which act as anchor positions. However, the binding ability of a peptide to a given MHC molecule cannot be explained by the presence of a few anchor residues, and indeed, non-anchor residues contribute to peptide–MHC binding ***(12,13)***. Instead, a position-specific scoring matrix (PSSM) or profile created from a set of aligned sequences of peptide–MHC binders provides a better alternative for capturing the complexity of peptide–MHC binding motifs. These PSSMs can be also used to quantify the relatedness of any peptide to the known peptide–MHC binders, thus serving as predictors of peptide–MHC binding.

PSSMs were first introduced by Gribskov et al. ***(14)*** for the detection of distantly related proteins and are now widely used for the representation and identification of sequence motifs ***(15,16)***. In essence, a PSSM consists of a table containing a form of frequency count of each one of the 20 amino acids observed in every column of an alignment divided by the corresponding expected frequency of that amino acid in the background (usually the frequency of the amino acid in a reference database). In addition, methods for the derivation of profiles also provide corrections for missing data and sequence redundancy in the alignments, which are essential to increase the detection limits of PSSMs ***(17,18)***. Missing and/or low counts in the alignments are corrected using pseudo-counts estimated from substitution matrices ***(17)***, whereas sequence redundancy is corrected by applying sequence weights before the estimation of the amino acid counts.

A PSSM is a good descriptor of the peptide–MHC binding motif, only if the peptide–MHC binders are aligned by structural and/or sequence similarity. There are two types of MHC molecules, class I (MHCI) and class II (MHCII), which actually present peptide antigens for recognition by two distinct sets of T cells, $CD8^+$ and $CD4^+$, respectively ***(7)***. MHCI and MHCII molecules bind peptides in a different mode, and thus, for aligning MHCI and MHCII ligands, we devised two distinct procedures that are compatible with the structural and molecular basis of the peptide–MHCI and peptide–MHCII interactions. In thischapter, we will describe these two procedures, and we will illustrate the prediction of peptide–MHC binding through the use of PSSMs.

2. Materials

2.1. Databases

Prediction of peptide–MHC binding using profiles require the availability of the sequence of peptides known to bind to MHC molecules. These sequences can be retrieved from any of the available public databases of MHC ligands (Table 1). However, in this study, we used the EPIMHC database ***(19)*** as the only source of MHC ligands (Table 1). All peptides in EPIMHC are MHC binders, and their binding strength is reported as unknown, low, moderate, or high. Importantly, the EPIMHC database (http://bio.dfci.harvard.edu/epimhc/) has been designed to facilitate the query, extraction, and analysis of data by third parties. To illustrate the prediction of peptide–MHC binding using PSSMs, we selected from the EPIMHC the sequences of 178 and 80 peptides annotated to bind with high affinity to A*0201 (human MHCI molecule) and DRB1*0401 (human MHCII molecule), respectively. The protein sources of the peptides were also retrieved from the EPIMHC database. All A*0201 peptide binders had a length of nine residues (9 mers), whereas the DRB1*0401 peptide binders were variable in length with at least nine residues. These sets of peptides are available as supplemental data from the site http://bio.med.ucm.es/methods/.

2.2. Software

The applications used in this study are indicated in Table 2. All these packages are freely available for academia users and were compiled and/or under the LINUX operating system. The core applications used for deriving alignments and profiles from MHC ligands are PROFILEWEIGHT ***(18)***, BLIMPS ***(20)***, and MEME ***(21)***. In addition to these applications, we used a set of Perl scripts to format data and/or handle the applications described above. These scripts are summarized in Table 2, and their use will be described elsewhere in Methods.

2.3. Leave-One Out Cross-Validation

Performance of PSSMs predicting peptide–MHC binding was evaluated using a leave-one out cross-validation (LOOCV). Briefly, for a set of peptides n known to bind to a given MHC molecule, a PSSM is generated from $n - 1$ peptides and used to test the binding of the remaining peptide (target peptide). This process is repeated n times until the binding of each peptide is tested.

Table 1
Selected public databases of MHC ligands

	Availability	Description	Reference
MHCPEP	http://wehih.wehi.edu.au/mhcpep	Database of MHC–binding peptides	***(44)***
MHCBN	http://www.imtech.res.in/raghava/mhcbn	Database of MHC-binding and nonbinding peptides	***(45)***
ANTIGEN	http://www.jenner.ac.uk/antijen/	Database of quantitative functional peptide data for immunology	***(46)***
FIMM	http://research.i2r.a-star.edu.sg/fimm/	A database of functional molecular immunology	***(47)***
EPIMHC	http://bio.dfci.harvard.edu/epimhc/	Curated database of MHC Ligands	***(19)***

Table 2
Computer applications used in this study

	Download	Description
BLIMPS	ftp://ftp.ncbi.nih.gov/repository/blocks/unix/	Motif discovery program
MEME	ftp://ftp.sdsc.edu/pub/sdsc/biology/meme/	Motif discovery program
PROFILEWEIGHT	ftp://ftp.ebi.ac.uk/pub/software/unix/profile.tar.Z	Program to create a PSSM from a GCG/MSF alignment
READSEQ	ftp://iubio.bio.indiana.edu/molbio/redseq/classic/	Program for sequence format conversion
epimhc.pl	http://bio.med.ucm.es/software/	Perl script to get peptides into a FASTA file from EPIMHC
meme2fasta.pl	http://bio.med.ucm.es/software/	Perl script to format block alignment in MEME output in FASTA
Mkmatrix.pl	http://bio.med.ucm.es/software/	Perl script to generate PSSMs using BLIMPS and PROFILEWEIGHT
rankpep.pl	http://bio.med.ucm.es/software/	Perl script to score and rank peptides using PSSMs

PSSM, position-specific scoring matrix.

3. Methods

3.1. Structural Alignments of MHCI and MHCII ligands

Capturing the complexity of the peptide–MHC binding motif in the form of a PSSM that can be used for the prediction of peptide–MHC binding requires the alignment of known MHC ligands by structural and/or sequence similarity. Peptides bound to MHCI molecules are in an extended conformation with several side chains accommodated in the binding pockets of the MHCI binding groove, and the N-terminal and C-terminal pinned into the groove *(7,8)* (Fig. 1A). As a consequence, MHCI ligands are of short length (8–11 residues), and proper structural alignment can be best accomplished by piling up peptides that have the same length *(22)*. In contrast, the peptide binding groove of MHCII molecules is open, allowing both the N-terminal and C-terminal of a peptide to extend beyond the binding groove *(7,8)* (Fig. 1B). Consequently, peptides bound to MHCII molecules display a great variability in length (9–22 residues). Nevertheless, only a peptide core of nine residues fits into the MHCII binding groove per se and is responsible for anchoring the peptide to the MHCII molecule *(3)*. This peptide core of nine residues binds in a conserved mode across the different peptide–MHCII complexes, sitting in the groove in an extended conformation connected through a network of hydrogen bonds between its backbone and conserved residues in the MHCII molecule *(3,7,8,23)*. As a result, the peptide-binding repertoire of MHCII molecules is broader than that of MHCI molecules, and MHCII ligands share less sequence similarity than MHCI ligands. Poor amino acid sequence similarity between MHCII ligands together with their great variability in sequence length makes their alignment difficult, hampering the use of global alignment algorithms such as CLUSTALW *(24)*. Because alignment of the MHCII ligands requires the identification of their binding core, we use the motif discovery program MEME *(21)* for aligning them. MEME uses an expectation-maximization algorithm in combination with a priori information to identify sequence motifs. The a priori information we use for aligning MHCI ligands is consistent with the interaction of peptides and MHCII molecules, namely, the existence of a single peptide-binding register per se MHCII ligand stretching nine residues.

3.2. Generation of Alignments and PSSMs from MHCI and MHCII Ligands

The strategy to derive alignments and profiles from known MHCI and MHCII ligands for the prediction of peptide–MHC binding consists of three basic steps: (i) peptide collection and subsequent subsetting by their MHC-binding

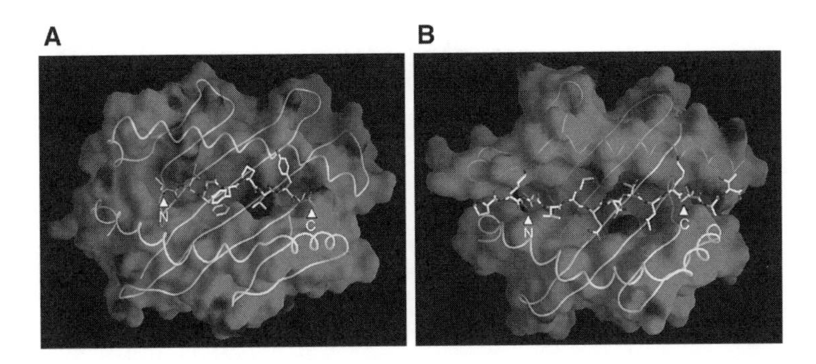

Fig. 1. Binding of peptide ligands to major histocompatibility complex class I (MHCI) and MHCII molecules. The figure shows the top of the molecular surface of the antigen-presenting platform of representative human MHCI (**A**) and MHCII (**B**) molecules as viewed by the T-cell receptor. The MHCI molecule corresponds to HLA-A*0201 in complex with a peptide LLFGYPVYV from HTLV-1 TAX protein [PDB: 1HHK;*(41)*]. The MHCII molecule corresponds to HLA-DR1 in complex with peptide PKYVKQNTLKLAT from influenza hemagglutinin protein [PDB:1FYT *(42)*]. Peptides bound to these molecules are represented by sticks to highlight the contours of the binding groove. Note how the peptide binding groove of the MHCI molecule is closed, and peptides bind in a manner such that both the N-terminal and C-terminal ends of the peptide (indicated by arrows) are nested into the MHCI binding groove, restricting their lengths to 8–11 residues. In contrast, the peptide binding groove of the MHCII molecule is open, thereby imposing no limitation to the size of ligands, whose N-terminal and C-terminal ends can extend beyond the binding grove. The side chains of N-terminal and C-terminal ends of the 9-mer peptide core fitting into the MHCII binding groove are indicated. The figure was prepared using GRASP *(43)*.

specificity and length in the case of MHCI ligands; (ii) generation of ungapped alignments; and (iii) generation of PSSMs from alignments. An outline of this strategy is shown in Fig. 2 and the detailed description is as follows.

1. Peptide collection and subsetting: In the case of MHCI ligands, the sequences must be subgrouped into files according to their MHCI-binding specificity and subsequently by sequence length. Peptides with 12 or more amino acids bind to MHCI molecules only exceptionally, and therefore, alignments and profiles should only be made from subsets of peptides of length 8, 9, 10, and 11. Furthermore, given that most of the known MHCI-restricted peptides are 9 mers (~90%) (data not shown), we suggest to preferentially make/use profiles from peptides of nine residues (9 mers). In the case of MHCII ligands, sequences must be subgrouped into distinct files only by their MHCII-binding specificity, and peptides with less than nine residues must be discarded. MHC

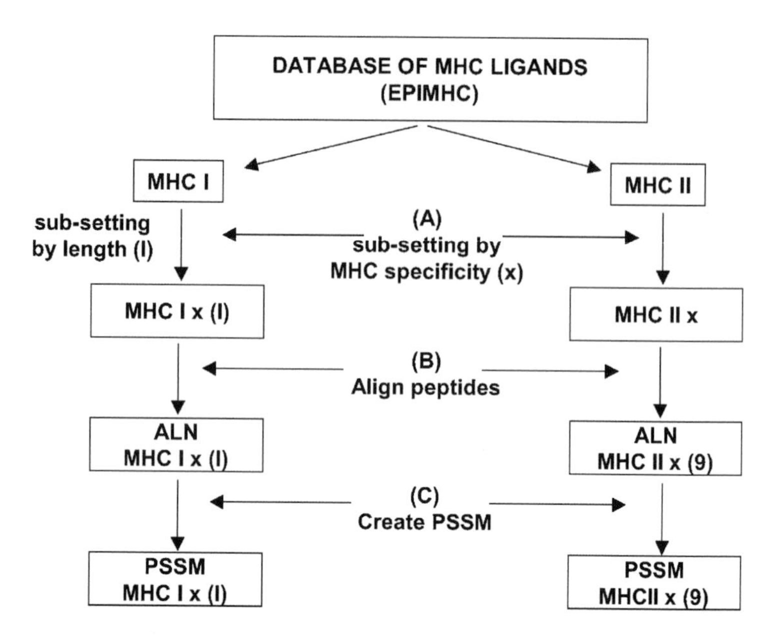

Fig. 2. Overview of the strategy for defining position-specific scoring matrices (PSSMs) from major histocompatibility complex class I (MHCI) and MHCII ligands. The basic steps for defining PSSMs are (A) peptide collection and subsetting of peptides by their MHC-binding specificity (x) and length (l) in the case of MHCI ligands; (B) generation of ungapped alignments; and (C) generation of PSSMs from alignments.

ligands meeting the above criteria can be obtained using the web interface of EPIMHC. Peptides should be saved as plain TEXT and in FASTA format. Alternatively, the Perl script *epimhc.pl* can be used to retrieve peptides from EPIMHC on the command line. For example, to create a FASTA file with all peptides in EPIMHC binding with high affinity to HLA-A*0201, one can use the following command:

*epimhc.pl -m 'HLA-A*0201' -s 9 -b high*

Likewise, the command:

*epimhc.pl -m 'HLA-DRB1*0401' -b high*

will generate a FASTA file with all peptides in EPIMCH binding to DRB1*0401 with high affinity. Peptides with less that seven residues will not be included in this file.

2. Generation of ungapped motif alignments: MHCI ligands of the same length in the FASTA files generated in the step above are already aligned. For aligning MHCII ligands, we use MEME with the following command:

```
meme mhcii_lig.fasta -protein -mod oops -nmotifs 1 -minsites 4\
  -maxsites 300 -minw 9 -maxw 9 -evt 10000 > mhcii_lig.meme
```

where *mhcii_lig.fasta -protein* corresponds to each of the MHCII-specific subsets of peptide sequences in FASTA format; *-mod oops* indicates that each sequence has a binding site; *-minsites 4 -maxsites 500* indicates that the motif should contain between 4 and 500 sequences; *-min 9 -maxw 9* indicates that the size of the motif is exactly 9; and finally *-evt 10000* is the expected threshold value for a sequence to be included in the motif. The output of MEME (*mhcii_lig.meme*) will contain a log-odd and a probability PSSM of the MHCII ligands' binding core which can readily be used for the prediction of peptide–MHCII binding. However, for consistency with the profiles derived from MHCI ligands, we obtain instead the motif alignment in the MEME output using the Perl script *meme2fasta.pl* and built the PSSM in the next step. The use of the script *meme2fasta.pl* is as follows:

```
meme2fasta.pl -i mhcii_lig.meme
```

This command will format the motif alignment in the output of MEME into FASTA format, discarding repeated sequences. The alignment obtained with MEME encompasses the binding core of the MHCII ligands.

3. Generation of PSSMs from alignments of MHC ligands: There are many methods to derive profiles from alignments that differ in the sequence weighting and in the computation of amino acid counts and pseudocounts. Here, we will describe the generation of profiles using PROFILEWEIGHT ***(18)*** and the applications included in the BLIMPS package ***(17,25)***. In both cases, pseudocounts are estimated using the BLOMUS62 substitution matrix-derived protein blocks ***(26)***. To learn about the actual equations used in these packages see Thompson et al. ***(18)*** and Henikoff and Henikoff ***(27)***. PROFILEWEIGHT uses a branch-proportional weighting method and requires an alignment in GCG/MSF format as input. BLIMPS PSSMs are generated through the sequential use of the following three applications: *mablock*, to translate alignments from FASTA format to BLOCK format; *blweight*, to apply weights to the sequences in the alignment, and *blk2pssm*, to generate the actual PSSM. The application *blweight* supports four distinct weighting methods: P, position-based method ***(28)***; A, pairwise distance method ***(29)***; V, Voroni method ***(30)***, and Cn,

clustering method *(26)*. The generation of matrices with PROFILEWEIGHT and BLIMPS can be facilitated using the Perl script *mkmatrix.pl*. For example, the command:

mkmatrix.pl -i peptides.tfa -w pw

will convert the alignment *peptides.tfa* into GSF/MSF format and create a PSSM using PROFILEWEIGHT. The PSSM will be saved under the name *peptides.pw.mtx*. Likewise the command:

mkmatrix.pl -i peptides.tfa -w p

will generate a PSSM under the name *peptides.p.mtx* using BLIMPS and position-based weights.

3.3. Scoring Peptide–MHC Binding Using PSSMs

PSSMs can be used to provide scores indicating the similarity (and hence binding potential) of any peptide to the set of aligned peptides known to bind to a given MHC molecule. These scores are computed by aligning the PSSM with the protein segments with the same length than the width of the PSSM (length of the alignment) and adding up the appropriate profile coefficients matching the residue type and position in the protein segment. Scoring all peptides in an entire protein sequence requires a dynamic algorithm that starts scoring at the beginning of the sequence and then moves the PSSM over the entire sequence one residue at a time to score the remaining peptides. Here, we provide the Perl script *rankpep.pl* as an example of dynamic scoring algorithm. The use of the script is as follows:

rankpep.pl -i sequence.fasta -m file.mtx

where *sequence.fasta* is the sequence query in FASTA format and *file.mtx* is the PSSM. The output of the program is a list of all peptides in the input sequence ranked by their score. Rank per se may, however, be insufficient to assess whether a peptide is a potential binder. Consequently, to better address whether a peptide might bind or not to a given MHC molecule, one should consider scoring all the peptides in the alignment from which the PSSM was obtained. Then, any given peptide can be considered a binder if it has a score within the range of scores of the peptides known to bind to the relevant MHC molecule.

3.4. Performance of PSSMs Predicting Peptide Binding to MHCI and MHCII Molecules

Only peptides that bind to MHC with an affinity above a necessary threshold are able to elicit a T-cell response. Therefore, determining whether known peptide–MHC binders can be identified among the high-scoring peptides within their protein sources is the best way to check whether prediction of peptide–MHC binding using PSSMs is of practical utility. Here, we have tested this notion for two sets of peptide ligands, one consisting of high-affinity binders to the human MHCI molecule A*0201 (Fig. 3A) and another of high-affinity binders to the human MHCII molecule DRB1*0401 (DR4) (Fig. 3B). These MHC ligands were aligned as indicated in section 3.2, and the binding of each of the peptides in the resulting alignments to the relevant MHC molecule was tested at different thresholds (0.5%, 1%, 2%, 3%, 4%, 5%, 10%, and 20%) under a LOOCV (see Section 2.3). At a given threshold, a peptide is computed as "to bind" if it is among the top scoring peptides from its protein source at that threshold. It is known that sequence weighting increases the sensitivity of profiles. Therefore, we carried out these prediction tests using PSSMs generated with PROFILEWEIGHT which applies branch-proportional weights (empty bars) and BLIMPS with position-based weights (black bars). The results

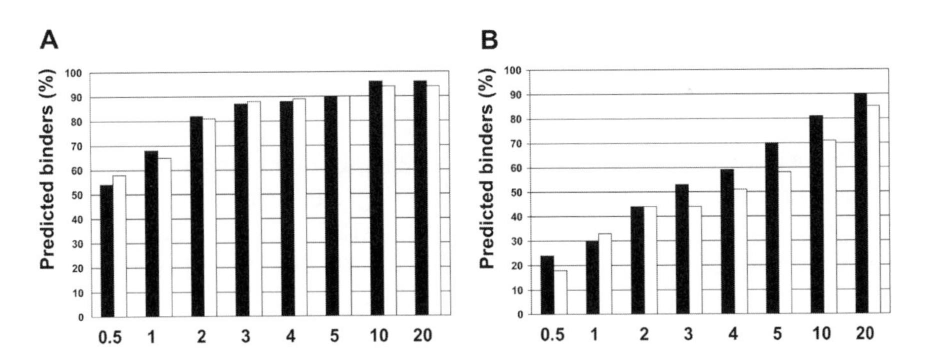

Fig. 3. Performance of position-specific scoring matrices (PSSMs) predicting peptide–major histocompatibility complex (MHC) binding. Performance of PSSMs predicting the binding of 178 peptides to A*0201 (**A**) and 66 peptides to DR4 (**B**) was evaluated by testing whether the peptides are predicted from their protein sources under a leave-one out cross-validation (LOOCV). Predictions were carried out at different thresholds (abscissa), and the percentage of correctly predicted peptides is plotted in the figure (ordinate). PSSMs were derived using PROFILEWEIGHT (empty bars) and BLIMPS with position-based weights (black bars).

indicated that ≥80% of the A*0201 peptide binders are predicted at a 2% threshold (Fig. 3A), whereas ~45% of the DR4 peptide binders are predicted at this threshold (Fig. 3B). As previously reported ***(22,31)***, PSSMs derived with BLIMPS and PROFILEWEIGHT are comparable predicting peptide binding to A*0201. However, for the prediction of peptide binding to DR4, PSSMs obtained with BLIMPS applying position-based weights were significantly better than those obtained using PROFILEWEIGHT (Fig. 3B).

4. Concluding Remarks

PSSMs are powerful tools to detect new and diverse sequences that are functionally related to those included in the original alignment (peptides binding to MHC molecules) and can be used to identify those peptides that can bind to MHC molecules. Prediction of peptide–MHC binding using PSSMs appears to be more accurate for MHCI molecules. (Fig. 3). This observation does not necessarily indicate that the MHCII-specific PSSMs were derived from incorrect alignments but rather could reflect the greater structurally inherent peptide-binding promiscuity of MHCII molecules (see Section 3.1).

Prediction of peptide–MHC binding has been approached by a large array of methods including quantitative matrices ***(32–34)***, machine learning algorithms (MLAs) ***(35,36)***, and peptide threading ***(37,38)***. Despite the fact that direct comparison between the various methods is not straightforward, some reports have indicated that MLAs such as artificial neural networks yield the best predictors of peptide–MHC binding, and it has been linked to the fact that MLAs can model binding interferences between peptide side chains, whereas the remaining methods, including PSSMs, assume independent binding of each side chain. Nevertheless, independent binding is generally supported by experimental evidence ***(32,39)***, and furthermore, considering side chain pair interactions only results in marginal improvement peptide–MHC binding predictions ***(40)***. Likewise, in a recent study, we have shown that PSSMs give similar or better results than those reported for MLAs ***(31)***. Thus, there may be more disadvantages than benefits when applying MLAs to the prediction of peptide–MHC binding. Thus, unlike PSSMs, MLAs are very prone to overfit data and are very sensitive to "dirty data." Consequently, much care and time has to be taken in preprocessing the data before training. Also, MLAs, as well as most data-driven methods used for predicting peptide–MHC binding, do not account for unseen data, instead only fitting the data they are provided with. Not surprisingly, it has also been shown that simple motif matrices outperform MLAs predicting peptide–MHC binding when the training sets are composed of a reduced set of samples (≤100 peptides) which is by large the most frequent scenario ***(4)***.

Prediction of peptide–MHC binding using PSSMs is also available at the RANKPEP web site (http://bio.dfci.harvard.edu/Tools/rankpep.html). Currently, 88 and 50 different MHCI and MHCII molecules, respectively, can be targeted for peptide-binding predictions in RANKPEP. This server is very versatile providing a framework for the prediction of MHC–peptide binding using profiles provided by the user.

References

1. Margulies, D.H. 1997. Interactions of TCRs with MHC-peptide complexes: a quantitative basis for mechanistic models. *Curr Opin Immunol* 9:390–395.
2. Garcia, K.C., Teyton, L., and Wilson, I.A. 1999. Structural basis of T cell recognition. *Annu Rev Immunol* 17:369–397.
3. Wang, J.-H., and Reinherz, E.L. 2001. Structural basis of T cell recognition of peptides bound to MHC molecules. *Mol Immunol* 38:1039–1049.
4. Yu, K., Petrovsky, N., Schonbach, C., Koh, J.Y., and Brusic, V. 2002. Methods for prediction of peptide binding to MHC molecules: a comparative study. *Mol Med* 8:137–148.
5. Flower, D. 2003. Towards in silico prediction of immunogenic epitopes. *Trends Immunol* 24:667–674.
6. Flower, D., and Doytchinova, I.A. 2002. Immunoinformatics and the prediction of immunogenicity. *Appl Bioinformatics* 1(4):167–176.
7. Stern, L.J., and Wiley, D.C. 1994. Antigen peptide binding by class I and class II histocompatibility proteins. *Structure* 2:245–251.
8. Madden, D. 1995. The three-dimensional structure of peptide-MHC complexes. *Annu Rev Immunol* 13:587–622.
9. Reche, P.A., and Reinherz, E.L. 2003. Sequence variability analysis of human class I and class II MHC molecules: functional and structural correlates of amino acid polymorphisms. *J Mol Biol* 331:623–641.
10. D'Amaro, J., Houbiers, J.G., Drijfhout, J.W., Brandt, R.M., Schipper, R., Bavinck, J.N., Melief, C.J., and Kast, W.M. 1995. A computer program for predicting possible cytotoxic T lymphocyte epitopes based on HLA class I peptide binding motifs. *Hum Immunol* 43:13–18.
11. Rammensee, H.G., Bachmann, J., Emmerich, N.P.N., Bacho, O.A., and Stevanovic, S. 1999. SYFPEITHI: database for MHC ligands and peptide motifs. *Immunogenetics* 50:213–219.
12. Bouvier, M., and Wiley, D.C. 1994. Importance of peptide amino acid and carboxyl termini to the stability of MHC class I molecules. *Science* 265:398–402.
13. Ruppert, J., Sidney, J., Celis, E., Kubo, T., Grey, H.M., and Sette, A. 1993. Prominent role of secondary anchor residues in peptide binding to HLA-A2.1 molecules. *Cell* 74:929–937.
14. Gribskov, M., McLachlan, A.D., and Eisenberg, D. 1987. Profile analysis: detection of distantly related proteins. *Proc Natl Acad Sci USA* 84:4355–4358.

15. Gribskov, M., and Veretnik, S. 1996. Identification of sequence pattern with profile analysis. *Methods Enzymol* 266:198–212.
16. Pearson, W. 1997. Identifying distantly related protein sequences. *Comput Appl Biosci* 13:325–332.
17. Henikoff, J.G., and Henikoff, S. 1996. Using substitution probabilities to improve position-specific scoring matrices. *Comput Appl Biosci* 12:135–143.
18. Thompson, J.D., Higgins, D.G., and Gibson, T.J. 1994. Improved sensitivity of profile searches through the use of sequence weights and gap excision. *Comput Appl Biosci* 10:19–29.
19. Reche, P.A., Zhang, H., Glutting, J.-P., and Reinherz, E.L. 2005. EPIMHC: a curated database of MHC-binding peptides for customized computational vaccinology. *Bioinformatics* 21:2140–2141.
20. Henikoff, S., Henikoff, J.G., Alford, W.J., and Pietrokovski, S. 1995. Automated construction and graphical presentation of protein blocks from unaligned sequences. *Gene* 163:17–26.
21. Bailey, T.L., and Elkan, C. 1995. The value of prior knowledge in discovering motifs with MEME. *Proc Int Conf Intell Syst Mol Biol* 3:21–29.
22. Reche, P.A., Glutting, J.-P., and Reinherz, E.L. 2002. Prediction of MHC class I binding peptides using profile motifs. *Hum Immunol* 63:701–709.
23. Barber, L.D., and Parham, P. 1993. Peptide binding to major histocompatibility complex molecules. *Annu Rev Cell Biol* 9:163–206.
24. Thompson, J.D., Higgins, D.G., and Gibson, T.J. 1994. CLUSTAL W: improving the sensitivity of progressive multiple sequence alignment through sequence weighting, position-specific gap penalties and weigh matrix choice. *Nucleic Acids Res* 2:4673–4680.
25. Henikoff, S., Henikoff, J.G., and Pietrokovski, S. 1999. Blocks+: a non-redundant database of protein alignment blocks derived from multiple compilations. *Bioinformatics* 15:471–479.
26. Henikoff, S., and Henikoff, J.G. 1992. Amino acid substitution matrices from protein blocks. *Proc Natl Acad Sci USA* 89:10915–10919.
27. Henikoff, J.G., and Henikoff, S. 1996. Substitution probabilities to improve position-specific scoring matrices. *Comput Appl Biosci* 12:135–143.
28. Henikoff, S., and Henikoff, J.G. 1994. Position-based sequence weights. *J Mol Biol* 243:574–578.
29. Vingron, M., and Sibbald, P. 1993. Weighting in sequence space: a comparison of methods in terms of generalized sequences. *Proc Natl Acad Sci USA* 90: 8777–8781.
30. Sibbald, P., and Argos, P. 1990. Weighting aligned protein or nucleic acid sequences to correct for unequal representation. *J Mol Biol* 216:813–818.
31. Reche, P.A., Glutting, J.-P., and Reinherz, E.L. 2004. Enhancement to the RANKPEP resource for the prediction of peptide binding to MHC molecules using profiles. *Immunogenetics* 56:405–419.

32. Parker, K.C., Bednarek, M.A., and Coligan, J.E. 1994. Scheme for ranking potential HLA-A2 binding peptides based on independent binding of individual peptide side chains. *J Immunol* 152:163–175.
33. Stryhn, A., Pederson, L.O., Romme, T., Holm, A., and Buus, S. 1996. Peptide binding specificity of major histocompatibility complex class I resolved into an array of apparently independent subspecificities: quantitation by peptide libraries and improved prediction of binding. *Eur J Immunol* 26: 1911–1918.
34. Udaka, K., Wiesmuller, K.H., Kienle, S., Jung, G., Tamamura, H., Yamigishi, H., Okumura, K., Walden, P., Suto, T., and Kawasaki, T. 2000. An automated prediction of MHC class I-binding peptides based on positional scanning with peptide libraries. *Immunogenetics* 51:816–828.
35. Adams, H.P., and Koziol, J.A. 1995. Prediction of binding to MHC class I molecules. *J Immunol Methods* 185:181–190.
36. Gulukota, K., Sidney, J., Sette, A., and DeLisi, C. 1997. Two complementary methods for predicting peptides binding major histocompatibility complex molecules. *J Mol Biol* 267:1258–1267.
37. Altuvia, Y., Sette, A., Sidney, J., Southwood, S., and Margalit, H. 1997. A structure based algorithm to predict potential binding peptides to MHC molecules with hydrophobic binding pockets. *Hum Immunol* 58:1–11.
38. Schueler-Furman, O., Altuvia, Y., Sette, A., and Margalit, H. 2000. Structure-based prediction of binding peptides to MHC class I molecules: application to a broad range of MHC alleles. *Protein Sci* 9:1838–1846.
39. Sturniolo, T., Bono, E., Ding, J., Raddrizzani, L., Tuereci, O., Sahin, U., Sinigaglia, F., and Hammer, J. 1999. Generation of tissue-specific and promiscuous HLA ligand databases using DNA microarrays and virtual HLA class II matrices. *Nat Biotechnol* 17:555–561.
40. Peters, B., Tong, W., Sidney, J., Sette, A., and Weng, Z. 2003. Examining the independent binding assumption for binding of peptide epitopes to MHC-I molecules. *Bioinformatics* 19:1765–1772.
41. Madden, D., Garboczi, D.N., and Wiley, D.C. 1993. The antigenic identity of peptide-MHC complexes: a comparison of the conformations of five viral peptides presented by HLA-A2. *Cell* 75:693–708.
42. Hennecke, J., Carfi, A., and Wiley, D.C. 2000. Structure of a covalently stabilized complex of a human alpha beta T-cell receptor, influenza HA peptide and MHC class II molecule, HLA-DR1. *EMBO J* 19:5611–5624.
43. Nicholls, A., Sharp, K., and Honig, B. 1991. Protein folding and association insights from the interfacial and thermodynamic properties of hydrocarbons. *Proteins* 11:281–296.
44. Brusic, V., Rudy, G., Kyne, A.P., and Harrison, L.C. 1998. MHCPEP, a database of MHC-binding peptides: update 1997. *Nucleic Acids Res* 26:368–371.

45. Bhasin, M., Singh, H., and Raghava, G.P. 2003. MHCBN: a comprehensive database of MHC binding and non-binding peptides. *Bioinformatics* 19:665–666.
46. Blythe, M.J., Doytchinova, I.A., and Flower, D. 2002. JenPep: a database of quantitative functional peptide data for immunology. *Bioinformatics* 18: 434–439.
47. Schonbach, C., Koh, J.L., Flower, D., Wong, L., and Brusic, V. 2002. FIMM, a database of functional molecular immunology: update 2002. *Nucleic Acids Res* 30: 226–229.

14

Application of Machine Learning Techniques in Predicting MHC Binders

Sneh Lata, Manoj Bhasin, and Gajendra P. S. Raghava

Summary

The machine learning techniques are playing a vital role in the field of immunoinformatics. In the past, a number of methods have been developed for predicting major histocompatibility complex (MHC)-binding peptides using machine learning techniques. These methods allow predicting MHC-binding peptides with high accuracy. In this chapter, we describe two machine learning technique-based methods, nHLAPred and MHC2Pred, developed for predicting MHC binders for class I and class II alleles, respectively. nHLAPred is a web server developed for predicting binders for 67 MHC class I alleles. This sever has two methods: ANNPred and ComPred. ComPred allows predicting binders for 67 MHC class I alleles, using the combined method [artificial neural network (ANN) and quantitative matrix] for 30 alleles and quantitative matrix-based method for 37 alleles. ANNPred allows prediction of binders for only 30 alleles purely based on the ANN. MHC2Pred is a support vector machine (SVM)-based method for prediction of promiscuous binders for 42 MHC class II alleles.

Key Words: Artificial neural network; machine learning techniques; MHC binders; support vector machine; T-cell epitopes

1. Introduction

It is well established in literature that binding of a peptide to major histocompatibility complex (MHC) molecules is a prerequisite for its recognition by T lymphocytes. Therefore, for a peptide to be a T-cell epitope, it is mandatory that it shall first form a complex with an MHC molecule. One of the ways to fish out such peptides (from a protein sequence) that bind to a particular MHC molecule is to get the experimental validation done, that is, to chop off

From: *Methods in Molecular Biology, vol. 409: Immunoinformatics: Predicting Immunogenicity In Silico*
Edited by: D. R. Flower © Humana Press Inc., Totowa, NJ

the antigenic protein in overlapping fragments and to determine the binding characteristics of each and every peptide with the specific MHC molecule. But the experimental identification of MHC binders is arduous, time-consuming, and economically not feasible. Therefore, development of reliable computational methods for MHC binders prediction may reduce cost and number of wet laboratory experiments to identify these binders. These methods are well known as indirect methods as they predict MHC binders rather than T-cell epitopes.

In the past, a number of methods have been developed to predict MHC binders, broadly divided into two categories: (i) knowledge-based methods where discrimination model are derived from known binders and (ii) ab inito methods where discrimination models are based on the rules of MHC–peptide interaction in three dimensions. Knowledge-based methods can be divided into the following categories: (i) binding motifs, (ii) quantitative matrices, and (iii) machine learning methods ***(1–4)***. In motif-based algorithm, the binding of a peptide to an allele is examined on the basis of occurrence of specific residues at specific position. These residues are known as anchor residues, and positions are known as anchor positions. The prediction of motif-based method is low because all MHC binders do not have binding motifs ***(5)***. The quantitative-based methods consider the contribution of each residue at each position in peptide instead of anchor positions/residues. Thus, matrix-based methods are more sound than motif-based methods. The quantitative matrix (*see* **Note 1**) methods, though reasonably good, fail in dealing with the nonlinearity in data, which may result in missing of distantly related set of binders. In order to handle nonlinearity in the data and to adapt self-training, machine learning techniques such as artificial neural network (ANN) and support vector machine (SVM) are used to predict MHC binders. In this chapter, we describe two web servers nHLAPred and MHC2Pred for predicting MHC class I and MHC class II binders, respectively.

2. Materials

2.1. nHLAPred

nHLAPred is a web server developed for predicting binders for 67 MHC class I alleles. This server has two methods: ANNPred and ComPred. The ANNPred allows the user to predict binders for 30 MHC class I alleles in an antigen sequence. This is based on ANN and allows the user to predict binders at different cutoff. The ComPred is a comprehensive method, which allows the user to predict binders in an antigen sequence for 67 MHC class I alleles. The ComPred predicts binders for 30 MHC alleles using hybrid approach, which is a combination of ANN and quantitative matrices. It predicts binders for the remaining 37 MHC class I alleles using quantitative matrix *(3)* only.

nHLAPred is a user-friendly web server developed and launched on SUN server 420R under Solaris environment. Stuttgart Neural Network Simulator, SNNS 4.2 *(6)*, was used to develop ANN-based method ANNPred in nHLAPred server. The web server was launched using public domain software package Apache. All web pages are written in hypertext markup language (HTML), and CGI scripts are written in PERL and JavaScript. ReadSeq (developed by Dr Don Gilbert) has been integrated in the server, which allows the user to submit their sequence in any standard formats. The server nHLAPred is accessible from http://www.imtech.res.in/raghava/nhlapred/ or http://www.imtech.ac.in/raghava/nhlapred/. In addition, nHLAPred has been mirrored at University of Arkansas for Medical Sciences, Little Rock, USA on SGI origin server under IRIX environment (http://bioinformatics.uams.edu/raghava/nhlapred/).

2.2. MHC2Pred

MHC2Pred is an SVM-based method for prediction of promiscuous MHC class II binders. The method is developed for the prediction of binders for 42 MHC class II alleles. MHC class II binding peptides are 10–30 amino acids long *(7)* with a binding core of nine amino acids containing primary anchor residues *(8)*. Therefore, in case of MHC2Pred prediction method, an additional method for finding the nine amino acids binding core from ligands of variable length is used. The data for training have been extracted from MHCBN *(9)* and JenPep databases *(10)*.

MHC2Pred is a user-friendly web server developed and launched on SUN server 420R under Solaris environment. SVM was implemented using the freely downloadable software SVM_light *(11)*. The web servers were launched using public domain software package Apache. All web pages are written in HTML, and CGI scripts are written in PERL and JavaScript. ReadSeq (developed by Dr Don Gilbert) has been integrated in the server, which allows the user to submit their sequence in any standard format. This server is accessible from http://www.imtech.res.in/raghava/mhc2pred/ or http://www.imtech.ac.in/raghava/mhc2pred/. This server has also been mirrored at University of Arkansas for Medical Sciences, Little Rock, USA on SGI origin server under IRIX environment (http://bioinformatics.uams.edu/raghava/mhc2pred/).

3. Method

3.1. nHLAPred

Accessing the above web address leads to the home page of the server nHLAPred. (Fig. 1). Certain menu options appear on the menu bar that are further linked to relevant sections. These options are:

- Home: This link leads to the home page of the nHLAPred server.
- ComPred: One is directed to the submission form of method ComPred on clicking the ComPred option.
- ANNPred: When a user clicks this option, he is directed to the submission form of the method ANNPred.
- Reference: A click at this option leads to the list of all publication references consulted in the development of the method.
- Help: This link would lead to a section of the server that provides complete information about the method and stepwise guidance to use it in order to predict MHC binders and CTL epitopes.
- Matrices: This option is linked to a table that contains list of all the matrices used in the method.
- Algorithm: This link directs to the page containing description of the steps involved in the development of the server one by one.

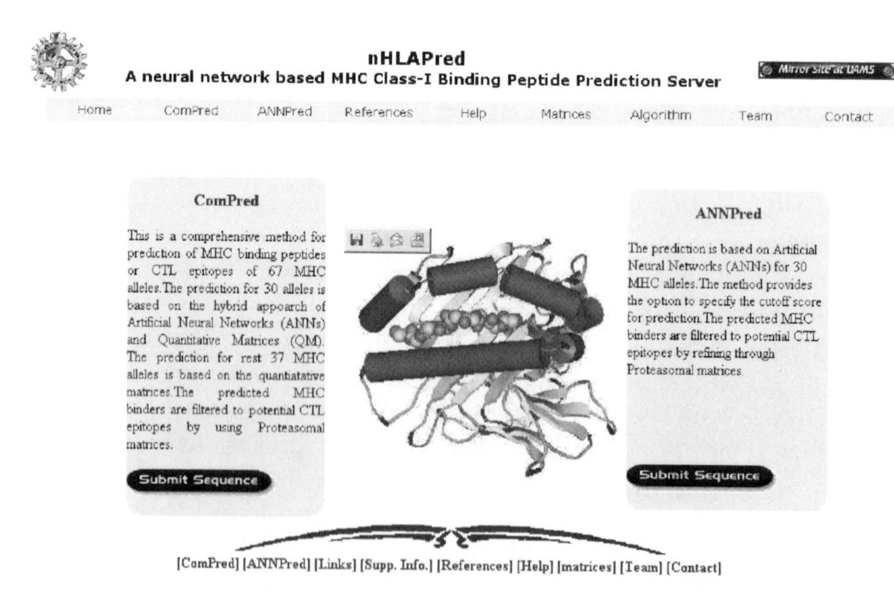

Fig. 1. Snapshot of home page of nHLApred web server.

- Team: This option is linked to the page containing information about the people involved in the development of the server.
- Contact: This option is linked to the page having the address and e-mail ID of concerned person to be contacted in case of any query.

3.1.1. ANNPred

The prediction of MHC binders in ANNPred is solely based on ANN. The major constraints of neural prediction are that it requires large amount of data of MHC binders and nonbinders for prediction. Thus, it is available only for 30 MHC alleles (Table 1) for which sufficient MHC binder was available in MHCBN *(9)*.

Instructions to use ANNPred server for predictions

The submit button at the right bottom of the column ANNPred or the ANNPred option in the menu bar must be clicked, on the appearance of the home page (Fig. 1) of the server nHLAPred, if the user wants to use ANNPred server. This action displays the sequence submission form of this method (Fig. 2).

Submission Form

A sequence submission form is a web interface wherein users can paste their query sequence, select among the choices provided, parameters of their choice, and submit it to the server that returns the result of this query. The fields that are required to be filled in the submission form are as follows:

- Sequence: The server accepts, as input, a protein or peptide sequence (at least of nine amino acid residues) in single-letter amino acid code or in any of the standard sequence format. The sequence can be pasted in the text box

Table 1
Major histocompatibility complex (MHC) class I alleles for which ANNpred allows to predict binders

HLA-A1	HLA-A2	HLA-A*0201	HLA-A*0202	HLA-A*0203
HLA-A*0206	HLA-A2.1	HLA-A3	HLA-A*0301	HLA-A11
HLA-A*1101	HLA-A*2402	HLA-A31	HLA-A*6801	HLA-A*6802
HLA-B7	HLA-B*0702	HLA-B8	HLA-B14	HLA-B27
HLA-B*2705	HLA-B*3501	HLA-B*51	HLA-B*5101	HLA-B*5301
HLA-B*5401	H2-Db	H2-Ld	H2-Kb	H2-Kd

ANNPred

The Prediction is based on the artificial neural networks.The server can run prediction for **30** MHC alleles.

Sequence Submission Form

Sequence	Paste your sequence	
	Or Submit Sequence	Browse...
Format	Input sequence format	Amino acids in single letter code (plain text) Standard sequence format (PIR/FASTA/EMBL etc.)
Allele selection	For multiple selection use the Alt or Meta key	ALL HLA-A1 HLA-A2 HLA-A*0201 HLA-A*0202
Result	Output Display mode, select one	HTML-II Tabular
Cutoff score	Filter the results	0.5
Tabular results	Display top	4 Peptides
PROTEASOME AND IMMUNOPROTEASOME FILTERS		
Proteasome Filter	Standard Proteasome	OFF **Stringency:** 4 %
Immunoproteasome Filter	Inducible Proteasome	OFF **Stringency:** 4 %

Submit sequence Reset

Fig. 2. Sequence submission form of ANNpred.

or local sequence files can be uploaded. All the nonstandard characters such as [*&^%$@#!()_+~=;"'<>?.\|} are ignored from the sequence. A warning is displayed in case input from both or none of the sources is detected.

- Format: The server accepts both the formatted and the unformatted raw sequences. One has to choose whether the sequence uploaded or pasted is plain or formatted before running prediction. The results of the prediction would be erroneous if the chosen format is inappropriate.
- Allele selection: Single or many alleles may be selected. Multiple alleles can be selected through Alt or Meta key, thus checking the promiscuity of the MHC binders and T-cell epitopes. The error message is flashed in case no selection is made.
- Cutoff score: The user, from a list box provided in the form, can choose the value of the cutoff score. Peptides having a score greater than the cutoff score are predicted as MHC binders, otherwise as nonbinders.
- Threshold: Threshold from 1 to 10% can be selected in order to vary the stringency level. A threshold of 1% predicts 1% of the best scoring natural peptides in a protein sequence. The threshold correlates to the peptide score and therefore with HLA–ligand interaction. More importantly, threshold is an indicator of the likelihood that predicted peptide is capable of binding to a given HLA molecule. Lower the threshold (equal to high stringency), lower is the false-positive rate and the higher is the false-negative rate. In contrast, higher the threshold (equal to low stringency), higher is the false-positive rate and lower is the false-negative rate. In short, from the

same protein sequence input, a threshold setting of 1% will predict a lower number of peptide sequences for a lower number of HLA-II alleles, compared with 2% or higher thresholds; however, this will also ensure a higher likelihood of positive downstream experimental results. Normally, at least for a first round of screening, threshold values higher than 3% are not desirable, because the rate of false positives can increase the size of the predicted repertoire to an amount unacceptable for later experimental testing.

- Proteasome and immunoproteasome filters: In higher eukaryotes, the proteasome complex performs the function of generating a pool of peptides for loading onto MHC class I molecules ***(12)***. So, proteasomes have a critical role to play in deciding which MHC binder acts as CTL epitope. The testing of peptides binding to MHC is indispensable. But there still remains a possibility that these peptides may not be generated in vivo by the proteasome. So, it is possible to narrow down the number of CTL epitopes by including the proteasomal preferences in MHC prediction. The user can vary the stringency of locating the proteasomal cleavage site by varying the threshold. The higher the value of threshold, nonstringent will be the prediction. The lower the value of threshold more stringent will be the prediction.
- Output Format: Clicking the submit button in the submission form (*see* **Note 2**) returns the result in user-friendly text formats. Each result display provides comprehensive information about the length of input sequence, nonamers generated, threshold, and the number of selected alleles .

  – HTML I: In this format, all the predicted binders (overlapping or nonoverlapping) appear in separate lines. So, this form of display is much beneficial to locate the overlapping binders as well as their exact position in the sequence (Fig. 3).

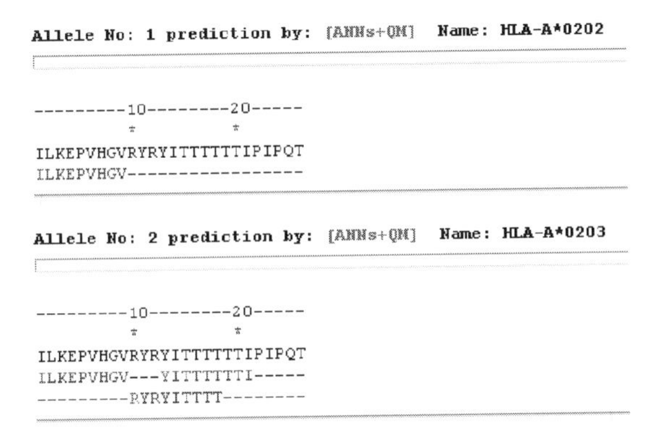

Fig. 3. Prediction of major histocompatibility complex (MHC) binders using ComPred.

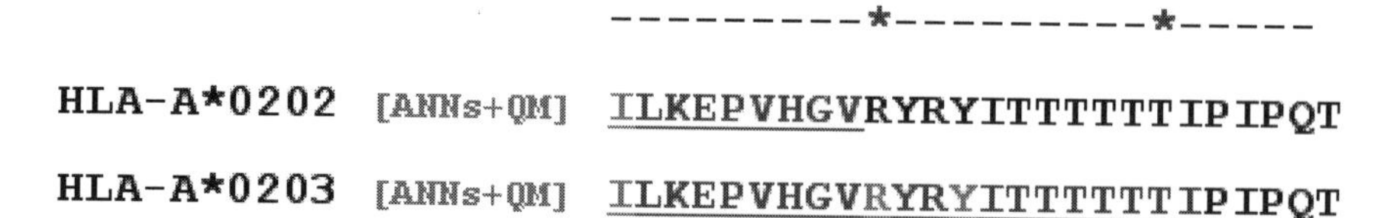

Fig. 4. Mapping of binders of major histocompatibility complex (MHC) class I alleles, useful in detecting promiscuous binding regions.

- HTML II: This format displays the binders of any allele in a single line just by coloring the input sequence. First residue of each predicted binder is shown in gray color and the rest of the residues in black (Fig. 4). This option is very useful in locating the promiscuous MHC binders in the sequence.
- Tabular: In this format, predicted binders of each allele are displayed in separate tables in the descending order of their score (Fig. 5). User can customize the number of top scoring peptides to be displayed in each table.

3.1.2. ComPred

This is a comprehensive platform for prediction of MHC binders from an antigenic sequence for 67 (Table 2) different MHC alleles. The prediction for 30 alleles is based on the hybrid approach of ANN and quantitative matrices. The prediction for the rest 37 alleles is based on the quantitative matrices only. The matrices for 17 MHC alleles out of these 37 alleles have been generated in this study, and rest of the matrices are obtained from BIMAS server. The predicted MHC binders are refined to potential T-cell epitopes by locating the proteasomal cleavage sites.

ALLELE: HLA-A*0205				
Threshold for 4 % with score: 5.950				
Prediction method	Rank	Sequence	Residue No.	Peptide Score
QM	1	ILKEPVHGV	1	9.320
QM	2	TTTIPIPQT	18	4.060
QM	3	YITTTTTTI	13	2.030
QM	4	PVHGVRYRY	5	1.430

Fig. 5. Major histocompatibility complex (MHC) binders in sorting order of binding affinity.

Table 2
Major histocompatibility complex (MHC) class I alleles for which ComPred allows to predict binders

HLA-A1	HLA-A2	HLA-A*0201	HLA-A*0202	HLA-A*0203
HLA-A*0205	HLA-A*0206	HLA-A2.1	HLA-A3	HLA-A*0301
HLA-A11	HLA-A*1101	HLA-A20	HLA-A24	HLA-A*2402
HLA-A31	HLA-A*3101	HLA-A*3302	HLA-A68.1	HLA-A*6801
HLA-A*6802	HLA-B7	HLA-B*0702	HLA-B8	HLA-B14
HLA-B27	HLA-B*2702	HLA-B*2705	HLA-B*3501	HLA-B*3701
HLA-B*3801	HLA-B*3901	HLA-B*3902	HLA-B40	HLA-B*4403
HLA-B*51	HLA-B*5101	HLA-B*5102	HLA-B*5103	HLA-B*5201
HLA-B*5301	HLA-B*5401	HLA-B*5801	HLA-B60	HLA-B61
HLA-B62	HLA-Cw*0301	HLA-Cw*0401	HLA-Cw*0602	HLA-Cw*0702
H2-Db	H2-Dd	H2-Ld	H2-Kb	H2-Kd
H2-Kk	HLA-G	H2-Qa	HLA-B*2706	HLA-B35
Mamu-A*01	HLA-A*0204	HLA-B*2703	HLA-B*2704	HLA-B*2902
HLA-A*3301	HLA-B44			

Instructions for prediction using ComPred

On appearance of the home page of the server nHLAPred, the user should click on Submit Sequence (left bottom) in ComPred column or "ComPred" in menu bar, in order to use ComPred server (Fig. 1). The server will display sequence submission form (Fig. 2).

- Sequence: The server accepts, as input, a protein or peptide sequence (at least of nine amino acid residues) in single-letter amino acid code, in any of the standard formats. The sequence can be pasted in the text box or local sequence files can be uploaded either. All the nonstandard characters such as [*&^%$@#!()_+~=;"'<>?.\|} are ignored from the sequence. A warning is displayed in case input from both or none of the sources is detected.
- Format: The server accepts both the formatted and the unformatted raw sequences. One has to choose whether the sequence uploaded or pasted is plain or formatted before running prediction. The results of the prediction would be erroneous if the chosen format is inappropriate.
- Allele selection: Single or many alleles may be selected. Multiple alleles can be selected through Alt or Meta key, thus checking the promiscuity of the MHC binders and T-cell epitopes. The error message is flashed in case no selection is made.
- Cutoff score: The user, from a list box provided in the form, can choose the value of cutoff score. Peptides having a score greater than the cutoff score are predicted as MHC binders, otherwise as nonbinders. In ComPred, the peptides predicted binders

at a particular cutoff score are filtered by quantitative matrices at nonstringent threshold, for example, at 10%, whereas the predicted nonbinders are filtered by quantitative matrices in stringent conditions, for example, at 1% threshold. This filtration results in reduction of false predictions.

- Other options: ComPred has number of other options such as, threshold, proteasome and immunoproteasome filters, and output format, which are similar to nHLAPred options.

3.2. MHC2Pred

Accessing the above web address leads to the home page of the server MHC2Pred (Fig. 6). Certain menu options appear on the menu bar that are further linked to relevant sections. These options are:

- Home: This link directs to the home page of the server. The home page itself is designed as the submission form of this server (Fig. 7).
- Help: Clicking at this option would lead to a section of the server that provides complete information about the method and stepwise guidance to use it in order to predict MHC class II binders.
- Information: One would be directed to the page containing description of the step-by-step algorithm followed in the development of the server on clicking this option.
- Team: This link is connected to the page containing information about the people involved in the development of the server.

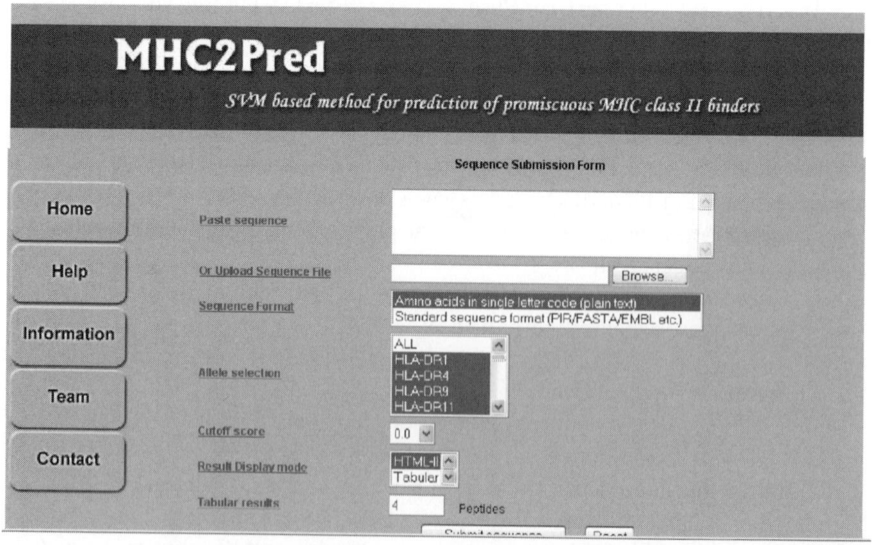

Fig. 6. Snapshot of home page of MHC2Pred server.

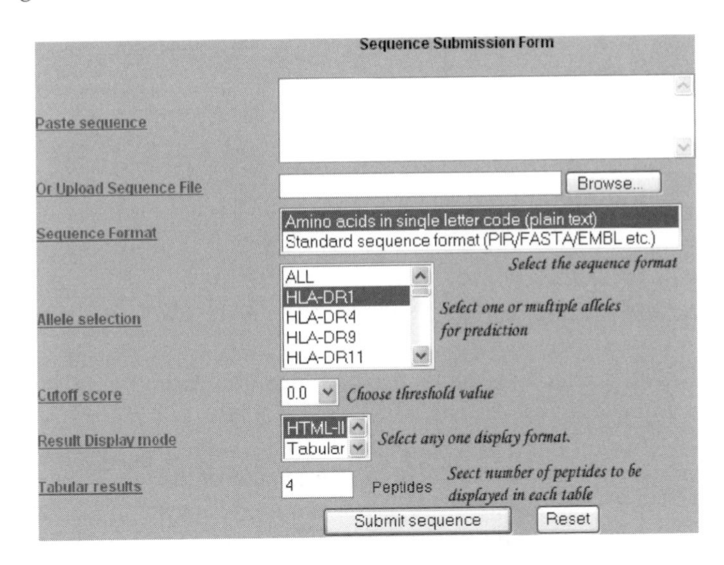

Fig. 7. Sequence submission form of MHC2Pred server.

- Contact: This option is linked to the page having the address and e-mail ID of concerned person to be contacted in case of any query.

Instructions for using the server:

- Antigenic sequence: The query sequence (in single-letter amino acid codes) can be given as input in any of the standard formats. Users can paste plain sequence in the provided text box, or can upload the local sequence files, in single-letter amino acid codes. All the nonstandard characters such as [*&^%$@#!()_+~=;"' <>?.\|} are ignored from the sequence. The query sequence should be at least of nine amino acid residues, otherwise a warning message will be flashed. The warning is also displayed if the input from both or none of the sources is detected.
- Format of sequence: The server allows both the formatted and the unformatted raw antigenic sequences to be given as input. The user should choose whether the sequence uploaded or pasted is plain or formatted before running prediction. Choice of inappropriate format will return incorrect prediction results.
- Selection of allele: The user can vary the number of the alleles before running prediction. The server allows the selection of multiple alleles through Alt or Meta key. The selection of the multiple alleles is crucial for the prediction of promiscuous MHC binders or T-cell epitopes. The error message is displayed in case no MHC allele is selected.
- Threshold: A threshold score between −1.5 and 1.5 can be selected in order to discriminate the binders from nonbinders. The peptides possessing score greater than the cutoff score are predicted as binders or else they are predicted as nonbinders. In case no selection is made, a default threshold of prediction methods will be

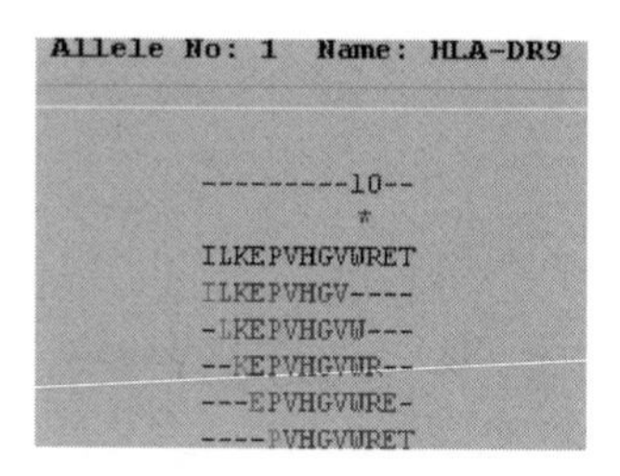

Fig. 8. Example output of MHC2Pred server.

considered. The default threshold is that at which the sensitivity and specificity of prediction methods are nearly same.

- Submit: Clicking on the submit button in the submission form (*see* **Note 2**) sends the request to the server and results in the appearance of the results page.
- Prediction results: The results of the prediction are displayed in user-friendly text formats. Each of the result display format firstly provides a comprehensive account of length of input sequence, prediction approach, nonamers generated, and threshold.
  - HTML I: This format displays all overlapping predicted MHC binders in separate lines. This is very useful in detecting the overlapping MHC binders. The display provides a clear indication about the exact position of predicted binder. An example is shown below (Fig. 8)
  - HTML II: This format displays all predicted binders for specific MHC allele in the single line just by coloring the predicted binders. The starting residue of each predicted binder is shown in gray and rest of the residues in black color. The option is very useful in detecting the promiscuous MHC binder in the sequence (Fig. 9).
  - Tabular: This one is the most common display format used by most prediction methods. The peptides are displayed in table in descending order of their score. The predicted binders of each MHC allele are shown in separate tables. The server also provides the facility to customize the number of the top-scorer peptides to be displayed in each table (Fig. 10).

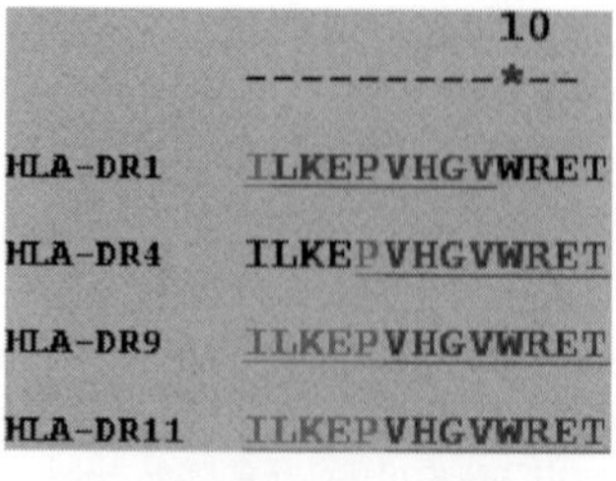

Fig. 9. User-friendly output suitable for identifying promiscuous binders.

ALLELE: HLA-DR9				
Threshold 0.0 as cutoff score				
Prediction method	Rank	Sequence	Residue No.	Peptide Score
SVM	1	KEPVHGWWR	3	1.328
SVM	2	LKEPVHGWW	2	1.054
SVM	3	ILKEPVHGV	1	0.439
SVM	4	PVHGWWRET	5	0.414

Fig. 10. Predicted major histocompatibility complex (MHC) binders in sorting order of affinity.

4. Conclusion

The above-discussed methods have large potential for further improvement of prediction accuracy, especially in view of further extension or growth in data of MHC binders and nonbinders and improvement in methods of binding affinity determination. The methods will assist immunologists in identifying potential vaccine candidates. The method will also find application in cellular immunology, transplantation, vaccine design, immunodiagnostics, immunotherapeutics, and molecular understanding of autoimmune susceptibility.

Acknowledgments

We acknowledge the financial support from the Council of Scientific and Industrial Research (CSIR) and Department of Biotechnology (DBT), Government of India.

Notes

1. The quantitative-based methods (Fig. 11) consider the contribution of each residue at each position in peptide instead of anchor positions/residues. Score of each residue in the sequence at its corresponding position in the matrix is added up to give the final score of the peptide. Quantitative matrices provide a linear model with easy-to-implement capabilities. Another advantage of using the matrix approach is that it covers a wider range of peptides with binding potential, and it gives a quantitative score to each peptide. Their predictive accuracies are considerable. The quantitative matrices used for prediction are available.
2. To avoid misuse of the site, the services are available for the registered users only. Users who are interested to use these servers are required to register themselves at http://www.imtech.res.in/errors/noauth.html. They need to fill up a registration form if they agree to the terms and conditions stated in the form. The user name and password is then sent by e-mail.

Allele HLA-A2									
Amino acid/Position	P1	P2	P3	P4	P5	P6	P7	P8	P9
A	0.52	-0.67	-0.25	-0.29	-0.35	-0.55	-0.10	-0.34	-0.05
C	0.00	-2.00	-0.40	0.29	1.00	1.67	1.33	0.67	1.00
D	-1.60	-2.00	0.08	0.34	-0.75	-0.86	-0.82	-0.40	-1.69
E	-1.41	-1.64	-1.48	-0.05	-0.43	-0.92	-1.08	-0.04	-2.00
F	0.00	-1.08	1.05	-0.40	1.28	0.27	1.39	-0.53	-2.00
G	0.91	-1.82	-0.47	1.18	0.30	-0.40	-0.11	0.13	-1.82
H	0.22	-2.00	0.22	0.22	-0.29	-0.50	0.93	-0.22	-2.00
I	-0.27	0.89	-0.62	-1.09	-0.62	0.00	-0.27	-0.07	0.00
K	0.25	-1.47	-1.14	-0.75	-0.77	-1.56	-1.20	-0.63	-1.43
L	0.51	1.62	1.24	-0.29	0.19	0.44	0.38	0.22	1.31
M	-0.67	1.47	0.29	1.43	1.33	1.67	0.00	0.40	1.00
N	-0.22	-2.00	0.29	-1.00	-1.11	-0.82	-0.22	-0.44	-2.00
P	-0.50	-2.00	-0.50	0.59	0.62	0.88	0.17	0.11	-2.00
Q	-0.75	-1.14	-1.64	0.26	-0.82	-0.35	-0.22	0.33	-1.33
R	0.17	-0.86	-0.29	0.32	-0.11	-1.11	-0.80	-0.15	-1.20
S	0.76	-2.00	0.40	0.50	0.00	0.11	-0.53	0.10	-1.08
T	-0.88	-0.75	-0.81	-0.92	-0.50	-0.67	-0.24	0.92	-0.71
V	-0.81	-0.88	0.22	-0.83	0.00	1.23	0.44	-0.50	1.38
W	-1.38	-1.60	-0.10	-1.64	-0.11	-1.47	-0.86	-1.00	-2.00
X	2.00	2.00	0.00	0.00	0.00	2.00	2.00	0.00	0.00
Y	-0.12	-2.00	0.09	-2.00	0.43	-0.12	-0.25	0.00	-1.43

THRESHOLD (%)	1%	2%	3%	4%	5%	6%	7%	8%	9%	10%
NUMERICAL VALUE	3.45	2.68	2.19	1.82	1.52	1.27	1.04	0.85	0.66	0.50

Fig. 11. Example quantitative matrix.

References

1. Adams, H. P. and Koziol, J. A. (1995) Prediction of binding to MHC class I molecules. *J. Immunol. Methods* **185**,181–90.
2. Brusic, V., Rudy, G., and Harrison, L. (1994) Prediction of MHC binding peptide by using artificial neural networks. In Stonier, R.J., Yu, X.S. (Eds.). *Complex Systems: Mechanism of Adaptation*, Amsterdam, IOS Press, pp. 253–60.
3. Parker, K. C., Bednarek, M. A., and Coligan, J. E. (1994) Scheme for ranking potential HLA-A2 binding peptides based on independent binding of individual peptide side chains. *J. Immunol.* **152**,163–75.
4. Rammensee, H. G., Friede, T., and Stevanovic, S. (1995) MHC ligands and peptide motifs: first listing. *Immunogenetics* **41**,178–228.
5. Buus, S. (1999) Description and prediction of peptide-MHC binding: the "human MHC project". *Curr. Opin. Immunol.* **11**,209–13.
6. Zell, A. and Mamier, G. (1997) Stuttgart Neural Network Simulator version 4.2, University of Stuttgart.

7. Chicz, R. M., Urban, R. G., Gorga, J. C., Vignali, D. A., Lane, W.S., and Strominger, J. L. (1993) Specificity and promiscuity among naturally processed peptides bound to HLA-DR alleles. *J. Exp. Med.* **178**, 27–7.
8. Jardetzky, T. S., Brown, J. H., Gorga, J. C., Stern, L. J., Urban, R. G., Strominger, J. L., and Wiley, D. C. (1996) Crystallographic analysis of endogenous peptides associated with HLA-DR1 suggests a common, polyproline II-like conformation for bound peptides. *Proc. Natl. Acad. Sci. U.S.A.* **93**,734–8.
9. Bhasin, M. and Raghava, G. P. S. (2003) MHCBN: a comprehensive database of MHC binding and non-binding peptides. *Bioinformatics* **19**,665–6.
10. Blythe, M. J., Doytchinova, I. A., and Flower, D.R. (2002) JenPep: a database of quantitative functional peptide data for immunology. *Bioinformatics* **18**,434–9.
11. Joachims, T. (1999) Making large-Scale SVM learning practical. In Scholkopf, B., Burges, C., Smola, A. (Eds.). *Advances in Kernel methods—support vector learning*, Cambridge, MA, London, England MIIT Press, pp. 169–84.
12. Rock, K. L. and Goldberg, A. L. (1999) Degradation of cell proteins and the generation of MHC class I-presented peptides. *Annu. Rev. Immunol.* **17**,739–79.

15

Artificial Intelligence Methods for Predicting T-Cell Epitopes

Yingdong Zhao, Myong-Hee Sung, and Richard Simon

Summary

Identifying epitopes that elicit a major histocompatibility complex (MHC)-restricted T-cell response is critical for designing vaccines for infectious diseases and cancers. We have applied two artificial intelligence approaches to build models for predicting T-cell epitopes. We developed a support vector machine to predict T-cell epitopes for an MHC class I-restricted T-cell clone (TCC) using synthesized peptide data. For predicting T-cell epitopes for an MHC class II-restricted TCC, we built a shift model that integrated MHC-binding data and data from T-cell proliferation assay against a combinatorial library of peptide mixtures

Key Words: T-cell receptors; MHC binding; epitope prediction; support vector machine; artificial intelligence; combinatorial peptide library; vaccine design

1. Introduction

T-cell epitopes are the peptides that bind to the T-cell receptor (TCR) in conjunction with a major histocompatibility complex (MHC) molecule to activate T cells. Peptides degraded from foreign or self-proteins bind to MHC molecules. The MHC–peptide complex can be recognized by TCRs and trigger an immune response. Identifying characteristic patterns of immunogenic peptide epitopes can provide fundamental information for understanding disease pathogenesis and etiology and for the development of preventative and therapeutic vaccines.

We have recently developed two computational approaches to improve the accuracy in T-cell epitope prediction. Support vector machines (SVMs) are particularly appealing for T-cell epitope prediction because of the ability of

From: *Methods in Molecular Biology, vol. 409: Immunoinformatics: Predicting Immunogenicity In Silico*
Edited by: D. R. Flower © Humana Press Inc., Totowa, NJ

SVMs to build effective predictive models when the dimensionality of the data is high and the number of observation is limited. SVMs are based on a strong theoretical foundation for avoiding over-fitting training data, and they do not have the problem of the numerous local minima that limit artificial neural network models *(1)*. We developed a SVM to predict T-cell epitopes with an MHC class I-restricted T-cell clone (TCC) using the synthesized peptide data *(2)*.

For predicting T-cell epitopes with MHC class II restriction, we developed a refined position-specific scoring matrix approach to permit shifts in the binding groove. The model utilizes both quantified peptide–MHC binding information and data from T-cell proliferation assay against a combinatorial peptide library *(3)*. This approach therefore makes use of information that captures all the protein components of the tri-molecular interaction. In previous studies, we analyzed data from positional scanning synthetic combinatorial libraries (PS-SCL) *(4)* utilizing a biometrical approach to predict the spectrum of stimulatory ligands for individual TCCs *(5)*. For TCCs with MHC class II restriction, misalignment of the peptides by a single position in the open-ended binding groove of class II MHC molecules would influence the way they are recognized by the TCR. The library assay results needed to be interpreted with caution due to the shifted alignments of the peptide mixtures. This phenomenon motivated us to develop the current model.

2. Data

2.1. Required Data for SVM

A total of 203 synthetic peptides were tested against the LAU203-1.5 TCC using a chromium release antigen recognition assay *(6)*. LAU203-1.5 is an A*0201-restricted TCC from tumor-infiltrated lymph node cells of a melanoma patient. A peptide with percentage-specific lysis higher than 10% is considered positive.

2.2. Required Data for the Shift Model

2.2.1. Data from Screening Assays Using a Combinatorial Peptide Library

A combinatorial peptide library is screened against a TCC of interest and the appropriate MHC allele. The library of n-mers consists of $20 \times n$ complex mixtures of peptides. Each mixture has n-mers with a specified amino acid in a defined position. The other positions are randomized to contain 19 (cysteine excluded) amino acids in equal proportion. For a 10-mer library, there are

20 amino acids that can be at each of the 10 possible defined positions, and hence, there are 200 combinatorial mixtures in all. The proliferation response pattern of a TCC against a combinatorial peptide library provides a fingerprint of the specificity of the TCC in peptide recognition.

Combinatorial peptide library assay data used to illustrate the shift model here was the proliferation response of the TCC TL3A6. This clone recognizes a peptide of the myelin basic protein (MBP) in complex with the HLA molecule DRB5*0101 ***(5)***. The data set is from 200 library mixtures with two replicate count per minute (cpm) measurements per library mixture.

2.2.2. Peptide–MHC Binding Data

The peptide–MHC binding data for DRB5*0101 are taken from the works of Hammer et al. and Sturniolo et al. ***(7–9)***. The data were presented in a position-specific format with a binding affinity for each amino acid at each position of a peptide. The data represent the relative affinities of peptides with single amino acid substitution from the reference peptide in terms of IC_{50}. Inhibitory concentration IC_{50} is the peptide concentration required to inhibit 50% of binding of the indicator peptide. Thus, its inverse is a measure of binding affinity.

3. Methods

3.1. Support Vector Machine

3.1.1. Theory

SVM classification of a sample with a vector x of predictors is based on:

$$f(x) = sign\left[\sum_i y_i \alpha_i k(x_i, x) + b\right] \quad (1)$$

where the kernel function k measures the similarity of its two vector arguments. For a linear SVM, the inner product kernel function is used. If $f(x)$ is positive, then the sample is predicted to be in class +1, otherwise class −1. The summation is over the set of "support vectors" that define the boundary between the classes. Support vector x_i is associated with a class label y_i that is either +1 or −1. The $\{\alpha_i\}$ and b coefficients are determined by "learning" the data.

An SVM attempts to minimize the generalization error for the independent data rather than minimizing the mean square error for the training set; therefore, it is an approximate implementation of the structural risk minimization induction principle. For two-group classification, the SVM separates the classes with a surface that maximizes the margin between them ***(1)***.

3.1.2. Training and Test Data Sets Preparation

Due to the imbalance of two classes in the data set (36 stimulatory peptides and 167 nonstimulatory peptides), we first divide the data into positive and negative groups. Then in each group random sampling is used to select 80% of the total peptides for training and 20% as a test set. Finally, the positive and negative groups are combined separately in the training and test sets. This procedure is repeated independently ten times.

3.1.3. Dimension Reduction

We encode each amino acid in a peptide by ten factors. These orthogonal factors are obtained from 188 physical properties of 20 amino acids through multivariate statistical analyses by Kidera et al. ***(10)***. They account for 86% of the variance of the 188 physical properties. These factors include alpha-helix or bend-structure preference, bulk, beta-structure preference, hydrophobicity, normalized frequency of double bend, normalized frequency of alpha-region, and pK-C. This encoding reduces the dimension of predictors by half while enabling structural and biophysical properties to be better represented compared to using amino acid indicator variables. Because our peptides are all 10 mers, the total number of input variables is 100.

3.1.4. Model Training

The SVM training is performed using *SVMlight* (version 4.0) ***(11)***. There are 100 input variables, which represent the 10 positions in the peptide. The class values are set to 1 for positive peptides and −1 for negative peptides. The threshold to predict positive or negative peptide is set to 0 by default (*see* **Note 1**).

3.1.5. Cross-Validation

For each training set consisting of 80% of the observations, a fully specified linear SVM is developed. This SVM model is then applied to the 20% test set. Training and testing are repeated ten times for randomly determined training/test set partitions. The final indices are averaged over the ten replicates (*see* **Note 2**).

3.1.6. Predictive Indices

Because identifying stimulatory (positive) peptides is of the greatest concern, sensitivity and positive predictive value (PPV) are used to evaluate the models. Sensitivity is the portion of all stimulating peptides that are correctly identified. PPV is the probability that a peptide predicted to be positive actually does stimulate the TCC. Sensitivity indicates the ability of the classifier to detect real epitopes, whereas PPV reflects the efficiency of the method. A classifier with low PPV will result in the generation of numerous nonstimulatory peptides for the next rounds of testing.

3.1.7. Comparison with other Artificial Intelligence Methods

In order to compare the performance with other prediction methods such as artificial neural networks, relative operating characteristic (ROC) analysis is also used ***(12)***. The area under the ROC curve is independent of the score threshold used for distinguishing positive from negative predictions (*see* **Note 3** and Fig. 1).

3.2. The Shift Model

In principle, screening data from a combinatorial peptide library assay can be a basis of the prediction of novel epitopes specific for the given TCC. In many applications involving MHC class II restriction, however, individual library mixtures for the same amino acid fixed at consecutive positions give indistinguishable T-cell responses. This artifact is thought to be the result of peptides binding to the MHC groove in a "shifted" alignment as shown in Fig. 2.

The main idea of the shift model is that a proliferation response is the sum of the contribution from every possible peptide–MHC binding configuration, rather than the result of the perfect alignment of the peptide mixture with the binding groove. We assume that the average MHC-binding preference of a mixture is dominated by its common amino acid in the fixed position.

3.2.1. Obtaining the Affinity Matrix B^0 from the Available Peptide–MHC Binding Data

The binding score matrix is defined as the log ratio of binding affinity, which is the reciprocal of IC_{50}. The reference peptide has alanine (Ala) in all scanned positions. Hence, the ratio represents the relative affinity of amino acid i compared to Ala for a fixed MHC position j, and the log ratio of binding affinity can be expressed as follows:

$$b_{ij} = -\log \frac{(\mathrm{IC}_{50} \text{ of the substituted peptide : amino acid } i \text{ in position } j)}{(\mathrm{IC}_{50} \text{ of the reference peptide : Ala in position } j)} \quad (2)$$

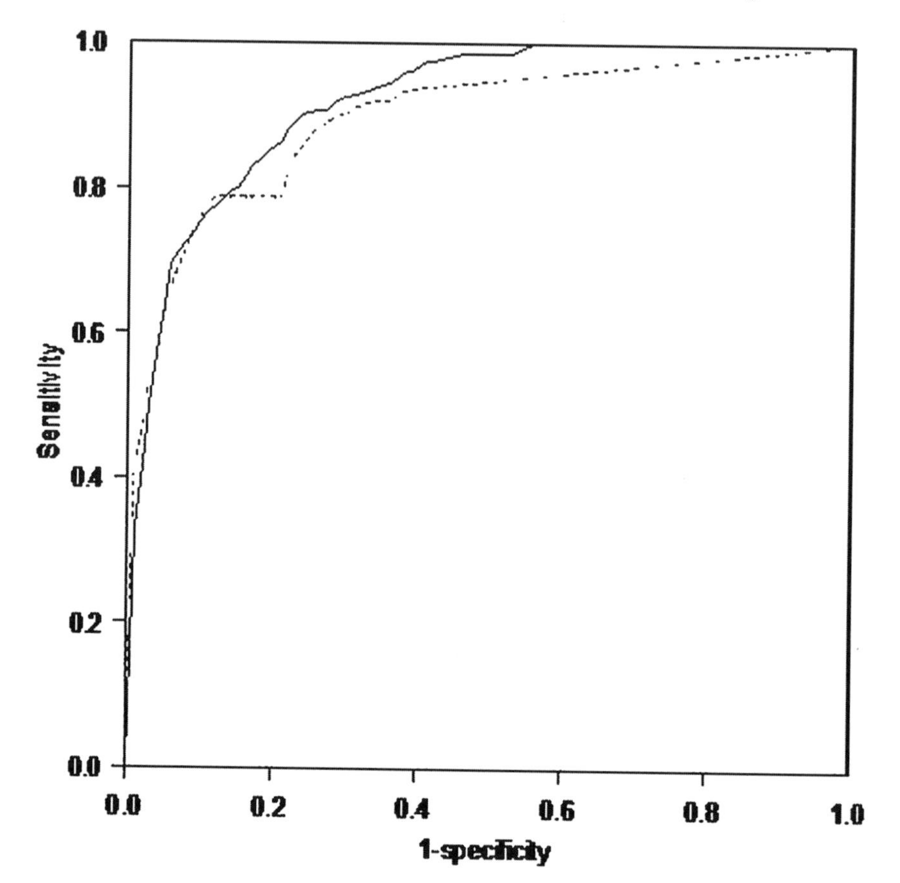

Fig. 1. Relative operating characteristic (ROC) curves for predictions on 203 peptides for melanoma T-cell clone (TCC) LAU203-1.5. Solid line represents averaged predictions using support vector machine (SVM) applied on ten different test sets, whereas dashed line represents averaged predictions using artificial neural network (ANN) applied on ten different test sets.

3.2.2. Estimation of TCR Stimulation Score Matrix S^{MHC} for Peptide–MHC Complexes

The MHC-binding information for a given MHC allele can be expressed as a 20 × 10 matrix B^0_{ij}, where its (i, j)-th entry is the affinity of amino acid i for MHC position j relative to other positions. We model the observed T-cell response S^{pep}_{ij} to the mixture with amino acid i in peptide position j as the

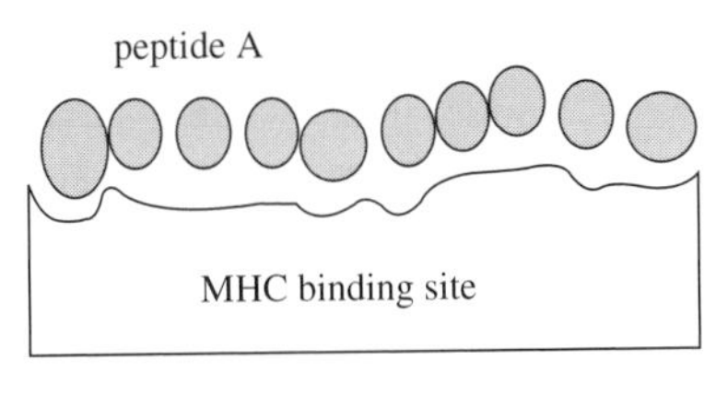

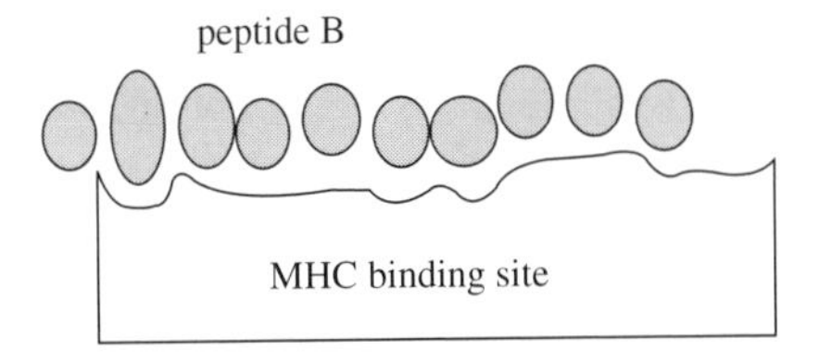

Fig. 2. Peptide–major histocompatibility complexs (MHCs) from the aligned binding (left) and from the shifted binding (right). Because MHC class II molecules have open-ended binding groove, a shifted binding configuration can happen according to the affinity pattern of amino acids for each MHC position.

sum of responses S_{ik}^{MHC} for all allowed MHC position k, each weighted by its affinity B_{ik}^{0}. More precisely, the model can be formulated as

$$S_{ij}^{\mathrm{pep}} = \sum_{k=1}^{10} B_{ijk} S_{ik}^{\mathrm{MHC}} + \varepsilon_{ij} \tag{3}$$

where $B_{ijk} = B_{ik}^{0}$ for $|j-k| \leq 4$ (*see* **Note 4**) and 0 elsewhere.

Ridge regression is used to estimate the S^{MHC} matrix in Equation 3 where we control its scale parameter to stay within a comparable range as S^{pep}. Ridge regression minimizes $\| \varepsilon \|^2 + t \| S^{\mathrm{MHC}} \|^2$, where t is the scale parameter.

3.2.3. Prediction of Novel T-Cell Epitopes

The prediction score for a 10-mer peptide is calculated from the score matrix S^{MHC} by adding the position-specific stimulatory value of each amino acid residue (*see* **Note 5**). A computer program is executed to scan and score all the overlapping 10 mers in a database such as GenPept or a customized set of protein sequences. The 10-mer sequences with high scores are identified as potential T-cell epitopes. These candidate peptides can be synthesized and tested for T-cell stimulation.

3.2.4. Model Evaluation

To test the model, we use S^{MHC} to score synthesized peptides for their ability to stimulate TCC TL3A6 *(5)* and compare the predictions with the experimentally measured responses. This prediction is then compared with the prediction method made from a model without the shift effect. ROC analysis is performed and area under the curve is calculated.

Notes

1. During training an SVM, leave-one-out cross-validation is employed on the training set to automatically optimize the relative misclassification costs for the two classes and to optimize the tuning parameter that reflects the trade-off between the training error and class separation.
2. To ensure that the peptides are sufficiently dissimilar for the cross-validation to be valid, we exclude predictive indices from test sets containing a peptide with Pearson correlation greater than 0.65 with a peptide in the corresponding training set.
3. The average cross-validated sensitivity and PPV of SVMs are 76.3 and 71.6%, respectively, for the ten test sets. Artificial neural network models are optimized by modifying the learning rate and momentum. The optimized models give an average sensitivity of 55.0% and PPV of 81.7% on the ten test sets. Averaged ROC curves for each method applied to the ten different test sets are generated as shown in Fig. 1.
4. We impose a constraint that the minimum peptide–MHC contact is six positions.
5. The score matrix based on S^{MHC} can be used to calculate TCR stimulation scores for peptide–MHC complexes, rather than for (free) peptides. Therefore, it is necessary to first identify the optimal peptide–MHC binding alignment for a given peptide. For the peptide–MHC complex with the highest binding score, the corresponding S^{MHC} components are summed over the ten MHC positions, where an unoccupied position is assigned the position-specific minimum value.

References

1. Vapnik, V. N. (ed.) (1995) *The Nature of Statistical Learning Theory.* Springer-Verlag, New York.
2. Zhao, Y., Pinilla, C., Valmori, D., Martin, R., and Simon, R. (2003) Application of support vector machines in T-cell epitope prediction. *Bioinformatics* **19**, 1978–1984.
3. Sung, M.H., Zhao, Y., Martin, R., and Simon, R. (2002) T-cell epitope prediction with combinatorial peptide libraries. *J. Comput. Biol.* **9**, 527–539.
4. Houghten, R.A. (1985) General method for the rapid solid-phase synthesis of large numbers of peptides: specificity of antigen-antibody interaction at the level of individual amino acids. *Proc. Natl. Acad. Sci. U.S.A.* **82**, 5131–5135.
5. Zhao, Y., Gran, B., Pinilla, C., Markovic-Plese, S., Hemmer, B., Tzou, A., Whitney, L. W., Biddison, W. E., Martin, R., and Simon, R. (2001) Combinatorial peptide libraries and biometric score matrices permit the quantitative analysis of specific and degenerate interactions between clonotypic T-cell receptors and MHC-peptide ligands. *J. Immunol.* **167**, 2130–2141.
6. Rubio-Godoy, V., Dutoit, V., Zhao, Y., Simon, R., Guillaume, P., Houghten, R.A., Romero, P., Cerottini, J.C., Pinilla, C., and Valmori, D. (2002) Positional scanning-synthetic peptide library-based analysis of self- and pathogen-derived peptide

cross-reactivity with tumor-reactive Melan-A-specific CTL. *J. Immunol.* **169**, 5696–5707.

7. Hammer, J., Belunis, C., Bolin, D., Papadopoulos, J., Walsky, R., Higelin, J., Danho, W., Sinigaglia, F., and Nagy, Z.A. (1994) High-affinity binding of short peptides to major histocompatibility complex class II molecules by anchor combinations. *Proc. Natl. Acad. Sci. U.S.A.* **91**, 4456–4460.
8. Hammer, J., Bono, E., Gallazzi, F., Belunis, C., Nagy, Z., and Sinigaglia, F. (1994) Precise prediction of major histocompatibility complex class II-peptide interaction based on peptide side chain scanning. *J. Exp. Med.* **180**, 2353–2358.
9. Sturniolo, T., Bono, E., Ding, J., Raddrizzani, L., Tuereci, O., Sahin, U., Braxenthaler, M., Gallazzi, F., Protti, M.P., Sinigaglia, F., and Hammer, J. (1999) Generation of tissue-specific and promiscuous HLA ligand databases using DNA microarrays and virtual HLA class II matrices. *Nat. Biotechnol.* **17**, 555–561.
10. Kidera, A., Konishi, Y., Oka, M., Ooi, T., and Scheraga, H.A. (1985) Statistical analysis of the physical properties of the 20 naturally occuring amino acids. *J. Protein Chem.* **4**, 23–55.
11. Joachim, T. (1999) Making large scale SVM learning practical, In *Advances in Kernel Methods-Support Vector Learning*. MIT press, Cambridge.
12. Swets, J. (1988) Measuring the accuracy of diagnostic systems. *Science* **240**, 1285–1293.

16

Toward the Prediction of Class I and II Mouse Major Histocompatibility Complex–Peptide-Binding Affinity

In Silico Bioinformatic Step-by-Step Guide Using Quantitative Structure–Activity Relationships

Channa K. Hattotuwagama, Irini A. Doytchinova, and Darren R. Flower

Summary

Quantitative structure-activity relationship (QSAR) analysis is a cornerstone of modern informatics. Predictive computational models of peptide–major histocompatibility complex (MHC)-binding affinity based on QSAR technology have now become important components of modern computational immunovaccinology. Historically, such approaches have been built around semiqualitative, classification methods, but these are now giving way to quantitative regression methods. We review three methods—a 2D-QSAR additive-partial least squares (PLS) and a 3D-QSAR comparative molecular similarity index analysis (CoMSIA) method—which can identify the sequence dependence of peptide-binding specificity for various class I MHC alleles from the reported binding affinities (IC_{50}) of peptide sets. The third method is an iterative self-consistent (ISC) PLS-based additive method, which is a recently developed extension to the additive method for the affinity prediction of class II peptides. The QSAR methods presented here have established themselves as immunoinformatic techniques complementary to existing methodology, useful in the quantitative prediction of binding affinity: current methods for the in silico identification of T-cell epitopes (which form the basis of many vaccines, diagnostics, and reagents) rely on the accurate computational prediction of peptide–MHC affinity.

We have reviewed various human and mouse class I and class II allele models. Studied alleles comprise HLA-A*0101, HLA-A*0201, HLA-A*0202, HLA-A*0203, HLA-A*0206, HLA-A*0301, HLA-A*1101, HLA-A*3101, HLA-A*6801, HLA-A*6802, HLA-B*3501, H2-K^k, H2-K^b, H2-D^b HLA-DRB1*0101, HLA-DRB1*0401, HLA-DRB1*0701, I-A^b, I-A^d, I-A^k, I-A^s, I-E^d, and I-E^k.

In this chapter we show a step-by-step guide into predicting the reliability and the resulting models to represent an advance on existing methods. The peptides used in this study are available

From: *Methods in Molecular Biology, vol. 409: Immunoinformatics: Predicting Immunogenicity In Silico*
Edited by: D. R. Flower © Humana Press Inc., Totowa, NJ

from the AntiJen database (http://www.jenner.ac.uk/AntiJen). The PLS method is available commercially in the SYBYL molecular modeling software package. The resulting models, which can be used for accurate T-cell epitope prediction, will be made are freely available online at the URL http://www.jenner.ac.uk/MHCPred.

Key Words: Major histocompatibility complex; peptides/epitopes; QSAR; additive method; CoMSIA

1. Introduction

Quantitative structure-activity relationship (QSAR) analysis, as a predictive tool of wide applicability, is one of the main cornerstones of modern cheminformatics and, increasingly, bioinformatics. QSARs are used to identify the structural and physical properties underlying the biological activity of a series of peptides. Immunoinformatics, a newly emergent subdiscipline of bioinformatics, which addresses informatic problems within immunology, uses QSAR technology to tackle the crucial issue of epitope prediction. As high-throughput biology reveals the genomic sequences of pathogenic bacteria, viruses, and parasites, such predictions will become increasingly important in the post-genomic discovery of novel vaccines, reagents, and diagnostics. In order to better understand the sequence dependence of peptide–major histocompatibility complex (MHC) binding of the mouse MHC, we have now used our approach to explore the amino acid preferences of various human and mouse alleles.

We have recently developed an immunoinformatic technique for the prediction of peptide–MHC affinities, known as the additive method, a 2D-QSAR technique which is based on the Free-Wilson principle *(1)*, whereby the presence or absence of groups is correlated with biological activity. For a peptide, the binding affinity is thus represented as the sum of amino acid contributions at each position. We have extended the classical Free-Wilson model with terms that account for interactions between side chains of amino acids. An iterative self-consistent (ISC) partial least squares (PLS)-based extension *(2)* of the additive method *(3,4)* has also been developed for prediction of class II peptide-binding affinity and applied to human class II alleles. We now address binding to class II human and mouse alleles for peptides of up to 25 amino acids in length. The ISC additive method assumes that the binding affinity of a large peptide is principally derived from the interaction, with an MHC molecule, of a continuous subsequence of amino acids within it. The ISC is able to factor out the contribution of individual amino acids within the subsequence, which is initially identified in an iterative manner. Using literature data, we have applied the additive method to peptides binding to several human class I *(3–5)* and class II alleles *(2)*.

Three-dimensional QSARs are a technique of significant value in identifying correlations between ligand structures and binding affinity. This value is often enhanced greatly when analyzed in the context of high-resolution ligand-receptor structures. In such cases, enthalpic changes—van der Waals and electrostatic interactions—and entropic changes—conformational and solvent mediated interactions—in ligand binding can be compared with structural changes in both ligand and macromolecule, providing insight into the binding mechanism ***(6,7)***. Although there are many molecular descriptors that account for free-energy changes, 3D-QSAR techniques, which use multivariate statistics to relate molecular descriptors in the space around ligands to binding affinities, have become preeminent because of their robustness and interpretability ***(8)***. In the case of comparative molecular similarity index analysis (CoMSIA), a Gaussian-type functional form is used so that no arbitrary definition of cut-off threshold is required, and interactions can be calculated at all grid points. The obtained values are evaluated using PLS analysis ***(9)***. CoMSIA allows each physicochemical descriptor to be visualized in 3D using a map that denotes binding positions that are either "favored" or "disfavored."

Recently, CoMSIA has been used to produce predictive models for peptide binding to human MHCs: HLA-A*0201 ***(10)*** and the HLA-A2 and HLA-A3 supertypes ***(11,12)***. We show how CoMSIA has been applied to certain class I MHC alleles. These models were used to both evaluate physicochemical requirements for binding and explore and define preferred amino acids within each pocket. The explanatory power of such a 3D-QSAR method is considerable, not only in its direct prediction accuracy but also in its ability to map advantageous and disadvantageous interaction potentials onto the structures of the peptides being studied. The data are highly complementary to the detailed information obtained from crystal structures of individual peptide–MHC complexes.

2. Method Theory

2.1. Additive Method—Class I and Class II Alleles

A program was developed and implemented into the QSAR module of SYBYL to transform the nine amino acid peptide sequences into a matrix of 1 and 0. A term is equal to 1 when a certain amino acid at a certain position or a certain interaction between two side chains exists and 0 when they are absent. For example, 180 columns account for the amino acids contributions (20aa $\times$ 9 positions); 3,200 columns account for the adjacent side chains or 1–2 interactions (20 $\times$ 20 $\times$ 8); and 2,800 columns account for every second side chain or 1–3 interactions (20 $\times$ 20 $\times$ 7). As these two models were roughly equivalent in terms of statistical quality, we applied the principle of Occam's razor and sought the simplest explanation, choosing the amino acids only model, which will be discussed in this study.

The matrix was assessed using PLS ***(13)***, an extension of multiple linear regression (MLR). The method works by producing an equation or QSAR, which relates one or more dependent variables to the values of descriptors and uses them as predictors of the dependent variables (or biological activity) ***(13)***. The IC_{50} values (the dependent variable y) were represented as negative logarithms (pIC_{50}). The predictive ability of the model was validated using "Leave-One-Out" Cross-Validation (LOO-CV) method.

2.1.1. Cross-validation Using the "LOO-CV" Method

Cross-validation (CV) is a reliable technique for testing the predictivity of models. With QSAR analysis in general and PLS methods in particular, CV is a standard approach to validation. CV works by dividing the data set into a set of groups, developing several parallel models from the reduced data with one or more of the groups excluded, and then predicting the activities of the excluded peptides. When the number of excluded groups is the same as the number in the set, the technique is called LOO-CV. The predictive power of the model is assessed using the following parameters: cross-validated coefficient (q^2) and the standard error of prediction (SEP), which are defined in Equations 1 and 2.

$$q^2 = 1.0 - \frac{\sum_{i=1} \left(p\mathrm{IC}_{50(\mathrm{exp})} - p\mathrm{IC}_{50(\mathrm{pred})}\right)^2}{\sum_{i=1} \left(p\mathrm{IC}_{50(\mathrm{exp})} - p\mathrm{IC}_{50(\mathrm{mean})}\right)^2} \text{ or simplified to } q^2 = 1.0 - \frac{\mathrm{PRESS}}{\mathrm{SSQ}} \qquad (1)$$

where $pIC_{50(pred)}$ is a predicted value and $pIC_{50(exp)}$ is an actual or experimental value. The summations are over the same set of pIC_{50} values. PRESS is the predictive error sum of squares, and SSQ is the sum of squares of $pIC_{50(exp)}$ corrected for the mean.

$$\mathrm{SEP} = \sqrt{\frac{\mathrm{PRESS}}{p-1}} \qquad (2)$$

where p is the number of the peptides omitted from the data set. The optimal number of components (NC) resulting from the LOO-CV is then used in the noncross-validated model which was assessed using standard MLR validation terms, explained by variance r^2 and standard error of estimate (SEE) which are defined in Equations 3 and 4.

$$r^2 = 1.0 - \frac{\sum_{i=1}^{n} \left(pIC_{50(exp)} - pIC_{50(calc)}\right)}{\sum_{i=1}^{n} \left(pIC_{50(exp)} - pIC_{50(mean)}\right)} = 1 - \frac{ESS}{SSQ} \tag{3}$$

$$SEE = \sqrt{\frac{ESS}{n - c - 1}} \tag{4}$$

where $pIC_{50(calc)}$ is the pIC_{50} value calculated by the non cross-validated model and ESS is the estimated error sum of square where n is the number of peptides and c is the number of components (NCs). In the present case, a component in PLS is an independent trend relating measured biological activity to the underlying pattern of amino acids within a set of peptide sequences. Increasing the NCs improves the fit between target and explanatory properties; the optimal NCs corresponds to the best q^2. Both SEP and SEE are standard errors of prediction and estimate the distribution of errors between observed and predicted calculated values in the regression models.

2.2. ISC Algorithm—Class II Alleles

An ISC PLS-based additive method was applied to the set of class II alleles. The ISC PLS-based algorithm *(2)* works by generating a set of nonameric subsequences extracted from the parent peptide. Values for pIC_{50} corresponding to this set of peptides were predicted using PLS and compared to the experimental pIC_{50} value for each parent peptide. The best predicted nonamer was selected for each peptide, that is, those with the lowest residual between the experimental and predicted pIC_{50}. LOO-CV was then employed to extract the optimal NCs, which was then used to generate the noncross-validated model. Each new model is built from the chosen set of optimally scored nonamers. The method works by comparing the new set of peptide sequences with the old set, and if the new set is different, the next iteration is begun. The process is repeated until the set of extracted nonameric peptide sequences identified by the procedure have converged. The resulting coefficients of the final noncross-validated model describe the quantitative contributions of each amino acid at each of the nine positions.

2.3. CoMSIA

2.3.1. Molecular Modeling

Wherever possible, an X-ray crystallographic structure for the nonameric/octameric peptide binding to the various class I alleles was chosen as a starting conformation. Using the crystallographic peptide as a template, all the studied peptides were built and then subjected to an initial geometry optimization using the Tripos molecular force field and charges derived using the MOPAC AM1 Hamiltonian semiempirical method ***(14)***. Molecular alignment was based on the backbone atoms of the peptides, which was defined as an aggregate during optimization.

2.3.2. CoMSIA Method

Five physicochemical descriptors (steric, electrostatic, hydrophobic, and hydrogen-bond donor and hydrogen-bond acceptor) were evaluated using a probe atom placed within a 3D grid. The atom had a radius of 1 Å and charge, hydrophobic interaction, and hydrogen-bond donor and acceptor properties all equal to +1. The grid was extended beyond the molecular dimensions by 4.0 Å in the *X*, *Y*, and *Z* directions. The spacing between probe points within the grid was set at 2.0 Å and was increased in steps of 0.5 Å. CoMSIA analysis for each allele was carried out using PLS ***(15)***, and models were then validated through the LOO-CV method, as previously described.

2.3.3. CoMSIA Maps

The results of the noncross-validated CoMSIA models were displayed as contour maps, with each physicochemical descriptor highlighted in different colors, reflecting favorable or disfavorable changes in the peptide structure and its influence on MHC binding. These maps were created using the standard deviation coefficient option based on actual values. The CoMSIA steric bulk map is shown using green (more bulk is favored) and yellow (less bulk is disfavored) contours. The electrostatic potential map is shown with blue (negative potential is disfavored) and red (negative potential is favored) contours. CoMSIA hydrophobic interaction fields are colored yellow (where hydrophobic interaction enhances affinity) and white (where hydrophilic interactions enhance affinity). The hydrogen-bond donor map is shown using cyan (donors on the ligand are preferred) and purple (donors are disfavored) contours. Finally, in the hydrogen-bond acceptor map, favored areas are shown in magenta and disfavored in yellow.

3. Methodology

3.1. Peptide Database

The information and data based on the peptide sequences and their binding affinities were obtained from the AntiJen database, a development of JenPep ***(16,17)*** (URL: http://www.jenner.ac.uk/AntiJen). These include human class I (HLA-A*0101, HLA-A*0201, HLA-A*0202, HLA-A*0203, HLA-A*0206,

HLA-A*0301, HLA-A*1101, HLA-A*3101, HLA-A*6801, HLA-A*6802, and HLA-B*3501), mouse class I (H2-K^k, H2-K^b, and H2-D^b), human class II (HLA-DRB1*0101, HLA-DRB1*0401, and HLA-DRB1*0701), and mouse class II (I-A^b, I-A^d, I-A^k, I-A^s, I-E^d, and I-E^k). Compilations of quantitative affinity measures for peptides binding to class I and class II MHCs were carried out with known binding affinities (IC_{50}). The binding affinities were originally assessed by a competition assay based on the inhibition of binding of the radiolabeled standard peptide to detergent-solubilized MHC molecule ***(18,19)***. Predicted pIC_{50} can be related to changes in the free energy of binding: $\Delta G^o_{bind} = -RT \ln IC_{50}$. The values were predicted from a combination of the contributions (p) of individual amino acids at each position of the peptide.

Several QSAR methodologies have been applied to both the class I and class II alleles, and their procedures are described as follows. For the purposes of this study, we shall focus on results from the mouse class I H2-D^b and mouse class II I-A^b allele.

3.2. Computer Software

All QSAR and molecular modeling calculations were carried out on a Silicon Graphics octane workstation using the SYBYL 6.9 molecular modeling package ***(20)*** and Microsoft Excel 2000.

3.3. Additive Method—Class I and Class II Alleles

This section is a step-by-step guide to the additive PLS method used in this study. Please note that some modules discussed in the following sections are explicit for our methods.

3.3.1. Build Initial Additive Model

1. Extract peptides and their IC_{50} values from AntiJen database and import into Excel.
2. Extracted IC_{50} values were first converted to $\log[1/IC_{50}]$ values (or $-\log_{10}[IC_{50}]$ or pIC_{50}) to be used as the dependent variables in a QSAR regression.
3. Convert the list of peptides and pIC_{50} values from ".xls" file into a ".txt" file and import into SGI workstation.
4. Open SYBYL (version 6.9).
5. The first step is to convert the '.txt' file into a '.csv' file via the implemented script written for the additive method (see Section 2.1):

    a. Select '*Jenner*' followed by 'Class I Additive CSV' from the tool bar
    b. Enter 'name of file containing peptides and IC_{50}': original excel '.txt' file containing list of peptides
        'length of peptides' = 9 (in this case)
        'name of CSV file' = '.csv'

c. Select *'File'* followed by 'Molecular Spreadsheet'
d. Then in the following order select: *'open'* – *'format'* – 'ASCII *file*' – *'.csv'* – *'open'* – *'merge'* – *'OK'* (Fig. 1).
e. You are now ready to move on to the QSAR module for the PLS part of the calculation.

3.3.2. PLS Calculation

1. Select *'QSAR'* from the toolbar followed by 'PLS' option.
2. From the PLS module, the following options were selected in order to perform the cross-validation method:

   a. *'Leave-1-Out'*
   b. *'Dependent Columns'* = IC_{50}
   c. *'Components'* = 6 = 6
   d. *'Scaling'* = none
   e. *'Column Filtering'* = no
   f. *'Use SAMPLS'* = no
   g. *DO PLS*

3. After calculation, save PLS analysis (.pls).
4. Re-select *'QSAR'* followed by *'Report QSAR'* and save file name (.lis).
5. Go to *'File'* and read in the new '.lis' file.
6. Open the '.lis' file, find the residual values list, and identify the peptide that has a residual value of ±2.000 and remove it from the molecular spreadsheet (.csv). Re-save the table under a new filename.

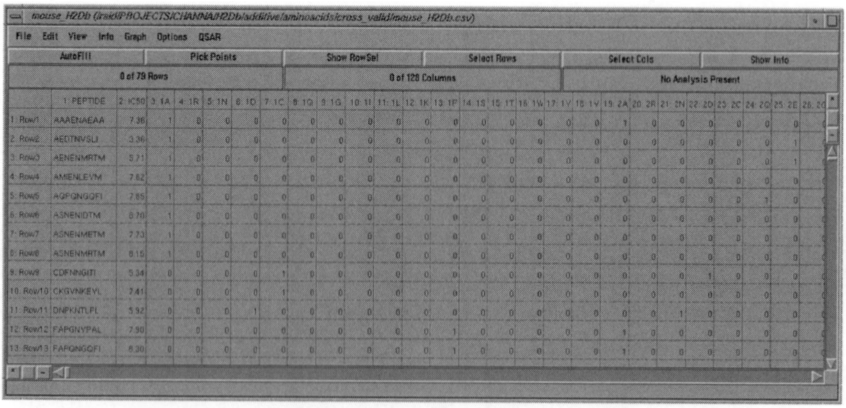

mouse_H2Db (/raid/PROJECTS/CHANNA/H2Db/additive/aminoacids/cross_valid/mouse_H2Db.csv)

File Edit View Info Graph Options QSAR

AutoFill | Pick Points | Show RowSel | Select Rows | Select Cols | Show Info

0 of 79 Rows | 0 of 128 Columns | No Analysis Present

	1: PEPTIDE	2: IC50	3: 1A	4: 1R	5: 1N	6: 1D	7: 1C	8: 1Q	9: 1G	10: 1I
1: Row1	AAAENAEAA	7.36	1	0	0	0	0	0	0	0
2: Row2	AEDTNVSLI	3.36	1	0	0	0	0	0	0	0
3: Row3	AENENMRTM	5.71	1	0	0	0	0	0	0	0
4: Row4	AMIENLEVM	7.62	1	0	0	0	0	0	0	0
5: Row5	AGPQNGQFI	7.85	1	0	0	0	0	0	0	0
6: Row6	ASNENIDTM	8.70	1	0	0	0	0	0	0	0
7: Row7	ASNENMETM	7.73	1	0	0	0	0	0	0	0
8: Row8	ASNENMRTM	8.15	1	0	0	0	0	0	0	0
9: Row9	CDFNNGITI	5.34	0	0	0	0	1	0	0	0
10: Row10	CKGVNKEYL	7.41	0	0	0	0	1	0	0	0
11: Row11	DNPKNTLPL	5.92	0	0	0	1	0	0	0	0
12: Row12	FAPGNYPAL	7.90	0	0	0	0	0	0	0	0
13: Row13	FAPQNGQFI	8.30	0	0	0	0	0	0	0	0

	11: 1L	12: 1K	13: 1F	14: 1S	15: 1T	16: 1W	17: 1Y	18: 1V	19: 2A	20: 2R	21: 2N	22: 2D	23: 2C	24: 2Q	25: 2E	26: 2G
1: Row1	0	0	0	0	0	0	0	0	1	0	0	0	0	0	0	[illegible]
2: Row2	0	0	0	0	0	0	0	0	0	0	0	0	0	0	1	[illegible]
3: Row3	0	0	0	0	0	0	0	0	0	0	0	0	0	0	1	[illegible]
4: Row4	0	0	0	0	0	0	0	0	0	0	0	0	0	0	0	[illegible]
5: Row5	0	0	0	0	0	0	0	0	0	0	0	0	0	1	0	[illegible]
6: Row6	0	0	0	0	0	0	0	0	0	0	0	0	0	0	0	[illegible]
7: Row7	0	0	0	0	0	0	0	0	0	0	0	0	0	0	0	[illegible]
8: Row8	0	0	0	0	0	0	0	0	0	0	0	0	0	0	0	[illegible]
9: Row9	0	0	0	0	0	0	0	0	0	0	0	1	0	0	0	[illegible]
10: Row10	0	0	0	0	0	0	0	0	0	0	0	0	0	0	0	[illegible]
11: Row11	0	0	0	0	0	0	0	0	0	0	1	0	0	0	0	[illegible]
12: Row12	0	0	1	0	0	0	0	0	1	0	0	0	0	0	0	[illegible]
13: Row13	0	0	1	0	0	0	0	0	1	0	0	0	0	0	0	[illegible]

Fig. 1. Additive table.

7. Repeat PLS method again, following steps 1–7.
8. The cross-validation method stops until at a cut-off point where the residual value is < 2.000 and/or $q^2 > 0.4$.

The optimal NCs leading to the highest $q^2_{(LOO)}$ and the lowest SEP were used to derive the noncross-validated model, by carrying out the following steps.

9. Select *'QSAR'* from the toolbar, followed by *'PLS'* option.
10. From the PLS module the following options were selected:
    a. *'No Validation'*
    b. *'Dependent Columns'* = IC_{50}
    c. *'Components'* = optimal number from cross-validation result
    d. *'Scaling'* = none
    e. *'Column Filtering'* = no
    f. *'Use SAMPLS'* = no
    g. *DO PLS*
11. After calculation, save PLS analysis (.pls).
12. Re-select *'QSAR'* followed by *'Report QSAR'* and save new output file name (.lis).
13. The final step involves the analysis of the noncross-validation '.lis' file. This is done by exporting the file into excel and looking at the regression equation in order to examine the positive and negative binding interactions of amino acids at each of the 9 positions (Fig. 2).

3.4. ISC Algorithm—class II Alleles

In order to carry out the ISC method on class II alleles, each peptide of varying lengths 10–25 must be broken down into peptide lengths of 9. This is done by taking fragments of nonamers from positions 1–9, 2–10, 3–11, 4–12 … 17–25 (assuming that 25 is the longest peptide). This can be carried out in excel or by writing a Perl script. Each nonamer gets the parent IC_{50} value.

Once you have your list of class II nonamers and their respective pIC_{50} values, follow the same procedure as described in Sections 3.3.1 and 3.3.2. Once the PLS calculation has been carried out and the noncross-validation (.lis) file is saved, this must be edited using 'nedit', 'vi,' or 'jot' commands in the SGI window. Within the file the following must be carried out:

1. Keep only the IC_{50} regression equation.
2. Make a note of the 'constant' value.
3. Add five spaces down (return key) at the end of the file.
4. Save as a .txt file.

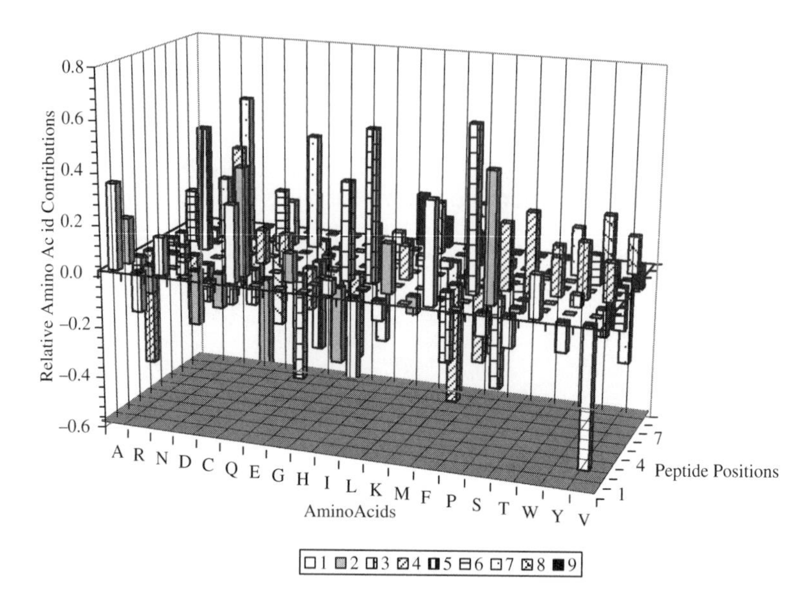

Fig. 2. Relative contributions of position-wise amino acids at each binding positions 1–9 for the H2-D^b allele. The contribution made by different individual amino acids at each position of the 9-mer binding peptide. The contribution is equivalent to a position-wise amino acid regression coefficient obtained by partial least squares (PLS) regression (as described in the text).

The ISC algorithm is an in-house development implemented within SYBYL 6.9; the following steps demonstrate that have been taken for class II prediction.

1. Select *'Jenner'* from the toolbar menu.
2. Select *'Peptide to Fasta'* from the drop-down menu:
   a. *'file of peptide sequences'* = the original exported '.txt' file containing list of peptides
   b. *'fasta file'* = fasta_file
3. Go back to *'Jenner'* and select *'Class I Additive Regression'*:
   a. *'name of text file containing regression equation'* = enter the edited '.txt' non-cross validated file
   b. *'name of additive model output file'* = '.out'
4. Open and edit the newly created *'.out'* file and edit the following (Fig. 3):
   a. Add the noted *'constant'* value.
   b. Change any values of –1.000 to 0.000 (this indicates where any amino acids are absent).

	P1	P2	P3	P4	P5	P6	P7	P8	P9
A	–0.016	–0.008	0.265	–0.115	0.066	–0.442	0.050	0.447	–0.034
C	**0.000**	0.083	0.037	–0.051	0.090	0.050	0.216	0.079	–0.139
D	–0.065	**0.000**	–0.067	**0.000**	**0.000**	0.107	–0.077	–0.041	–0.203
E	–0.028	–0.129	**0.000**	**0.000**	**0.000**	**0.000**	**0.000**	–0.048	**0.000**
F	**0.000**	**0.000**	**0.000**	–0.283	**0.000**	**0.000**	**0.000**	**0.000**	**0.000**
G	–0.286	–0.039	0.050	–0.011	**0.000**	**0.000**	–0.003	**0.000**	–0.067
H	–0.003	–0.013	**0.000**	**0.000**	**0.000**	**0.000**	**0.000**	0.213	**0.000**
I	–0.043	0.090	–0.364	–0.090	**0.000**	–0.244	–0.351	**0.000**	–0.069
K	0.094	**0.000**	**0.000**	**0.000**	–0.069	**0.000**	**0.000**	**0.000**	**0.000**
L	**0.000**	–0.215	–0.110	0.094	–0.162	**0.000**	–0.003	–0.242	0.066
M	0.008	–0.067	**0.000**	0.258	0.223	0.154	0.017	–0.027	0.082
N	**0.000**	0.298	**0.000**	0.042	–0.003	–0.069	0.064	–0.097	–0.455
P	0.100	**0.000**	0.032	0.090	0.030	0.201	0.080	**0.000**	0.280
Q	–0.013	**0.000**	–0.235	**0.000**	**0.000**	**0.000**	**0.000**	–0.067	–0.051
R	0.164	–0.286	0.066	0.122	–0.233	0.120	0.213	–0.229	0.216
S	–0.051	0.090	0.161	0.036	–0.078	0.041	–0.125	**0.000**	0.213
T	0.054	0.151	0.079	–0.060	0.233	**0.000**	–0.079	0.012	0.161
V	–0.069	–0.048	**0.000**	0.064	**0.000**	**0.000**	**0.000**	**0.000**	**0.000**
W	**0.000**	**0.000**	–0.029	**0.000**	–0.097	–0.003	**0.000**	**0.000**	**0.000**
Y	0.155	0.092	0.116	–0.097	**0.000**	0.085	**0.000**	**0.000**	**0.000**

Fig. 3. Additive model for the binding affinity prediction to the I-A^b allele. *constant = 6.044 (The constant accounts, at least nominally, for the peptide backbone contribution). **0.000 represents position where amino acids are absent within matrix.

   c. Re-save as a *'.txt'* file.

5. Go back to *'Jenner'* and select *'Run Class II Additive Method'* and enter the following details:

   a. *'name of fasta file containing sequences to be predicted'* = fasta_file
   b. *'name of the additive model data file'* = edited '.out' file
   c. *'minimum* pIC_{50} *to be output'* = -
   d. *'name of* output *file'* = '.out'-
   e. *'name of output file'* = '.out'
   f. *'name of output file of best nonamers'* = '.txt'

6. Take the final output '.txt' file and repeat the whole process again following the steps described in Sections 3.3.1, 3.3.2, and 3.4.
7. The procedure is repeated until the peptides in the final output file have converged.

3.5. CoMSIA—Class I Alleles

The following section describes the steps needed to build a CoMSIA model. If no X-ray data are available for the particular peptide molecule complex you are studying, then the closest crystallographic structure for the peptide binding to that particular allele should be downloaded and used as the starting conformation. This can be done by searching the Protein Data Bank (PDB) (http://www.rcsb.org/pdb/). For example, the X-ray structure of the nonameric peptide FAPGVFPYM[50] bound to the H2-D^b allele was used for this study.

3.5.1. Steps for Building a CoMSIA Model

Once an X-ray structure has been searched and saved, the initial procedure for collecting and preparing the data is the same as in Section 3.3.1, see steps 1–5. The following are the necessary steps for building the CoMSIA model:

1. Open SYBYL.
2. From the main toolbar, go to *'Read'* and open the downloaded '.pdb' file.
3. A pop-up request will ask you to 'center the molecule' – answer *'yes.'*
4. Select *'Build/Edit'* from the main toolbar, followed by:
   a. *'delete'*
   b. *'substructure'*
   c. Select the 9 nonamers from the bottom of the list
   d. Select *'OK'* then *'invert'* and finally *'OK'* again.

The next step is to add hydrogens to the '.pdb' peptide backbone:

5. Select *'Biopolymer'* from the main toolbar menu, followed by:
   a. *'add hydrogens'*
   b. *'all'*
   c. *'OK'* – the pop-up request will ask you to select *'essential'* or *'all'* – select *'all'*
   d. Save as a new '.pdb' file.

The final step in the initial steps of building a CoMSIA model is introducing all the peptide sequences you wish to study by following this simple procedure:

6. Select *'Jenner'* from the main menu.
7. Then select 'Build CoMSIA' and enter the following details:
   a. *'file of peptide sequences'* = enter the original '.txt' file list of peptides.
   b. *'OK'*

3.5.2. Steps for Aligning CoMSIA Model

The following section describes the procedure to align the desired peptides from the database to the backbone of the X-ray structure peptide:

1. Select *'File'* from the main toolbar, followed by:
   a. *'database'*
   b. *'open'* – '.mdb' file (the newly created database file from Section 3.5.1—step 7).

c. *'open'* – *'update'*
d. *'OK'*

2. Select 'File' from the main toolbar again, followed by:

a. *'database'*
b *'get molecule'* and highlight the first peptide sequence in the list.
c. *'OK'*

3. Delete hydrogens and side chains, keeping only the backbone chain (Fig. 4):

a. Save as new '.mol' file to use as template.

4. Finally, to align the peptides go to *'File'*:

a. *'align database'* – and enter the following details:
'database to align' = '.mdb' file
'template molecule' = '.mol' file
b. *'GO'* (Fig. 5)

You should now see all the peptides in the database aligning with the backbone of the X-ray structure.

3.5.3. Steps for Creating the CoMSIA Grid

Once the CoMSIA model is created, the next part of the method is to calculate the descriptors to be used as interaction points between probe points within a 3D grid and the atoms on the aligned peptides. This is carried out as follows:

1. Select *'File'* from the main toolbar, followed by *'Molecular Spreadsheet.'*
2. From the 'molecular spreadsheet,' select *'new'* – *'database'* and open the '.mdb' file.
3. At this stage you must manually input the pIC_{50} values into *'column 1.'*

The next step is to calculate the five descriptors (steric, electrostatic, hydrophobic, and hydrogen-bond donor and hydrogen-bond acceptor) found

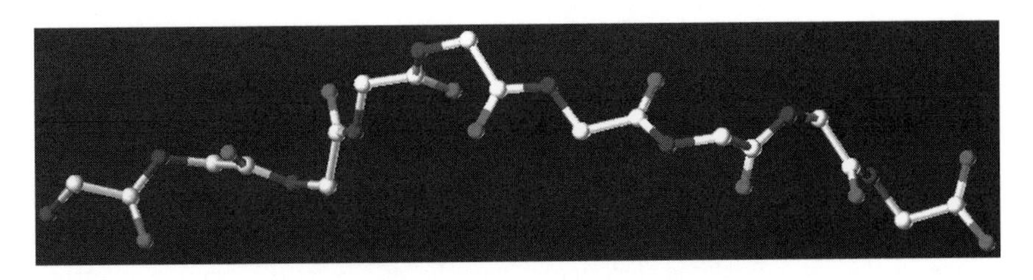

Fig. 4. Backbone chain of the X-ray structure peptide.

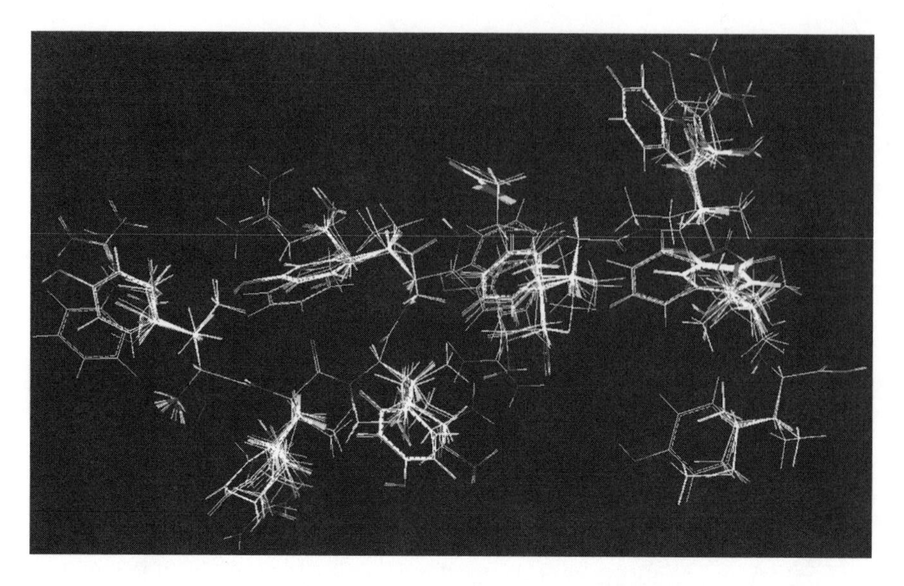

Fig. 5. Superimposed alignment of peptide molecules for the H2-D[b] alleles.

within the QSAR module of SYBYL and to create and add a 3D grid around the aligned peptides. This is carried out as follows:

4. Highlight *'column 2'*
   a. Select *'autofill'*
   b. Followed by *'CoMSIA'* from the list and *'open'*
   c. Select a field type (a descriptor)
   d. Keep the *'attenuation factor'* as the default setting of 0.3
   e. Finally, select *'use existing region'* and *'define'*

When the 3D grid appears, extend in the *X*, *Y*, and *Z* directions so that it encompasses the peptides but allow for an extra 4 Å in each direction (Fig. 6).

5. After creating the grid save as '.rgn' file
6. Highlight the next column on the molecular spreadsheet and repeat steps 4 and 5. The final step in the CoMSIA method is the PLS calculation in order to reach a model using the calculated descriptors and the effect on the peptide interactions

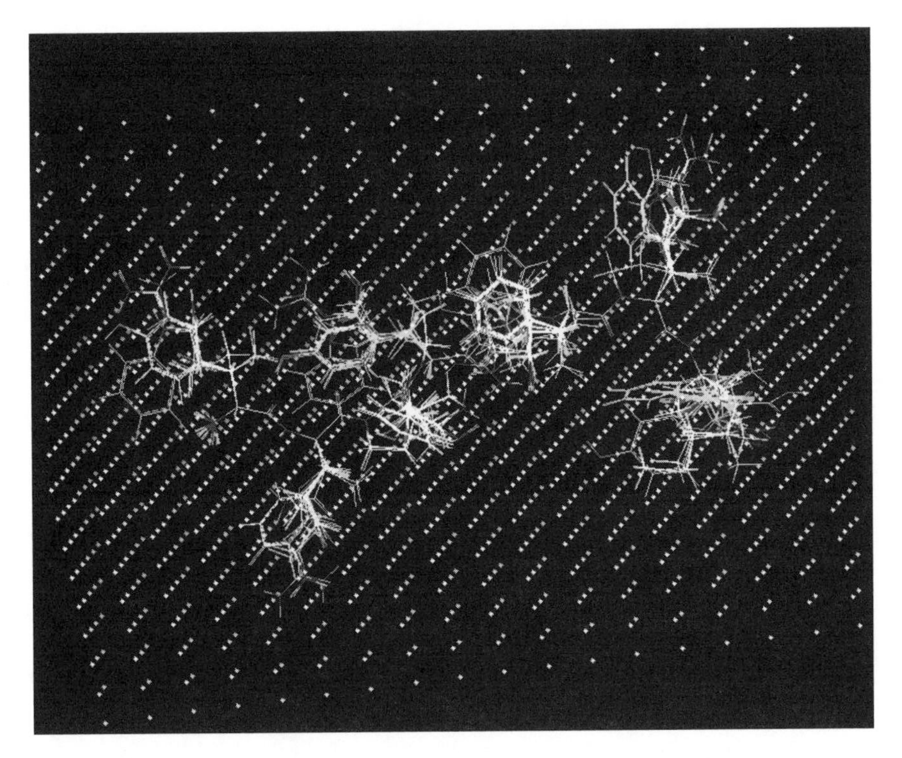

Fig. 6. Superimposed H2-D^b peptide molecules placed within 3D grid lattice.

within the grid. This procedure is carried out in the same way as described in Section 3.3.2. The only significant difference on the PLS menu is:

a. *'scaling'* = CoMFA Standard
b. *'column filtering'* = 1.0

3.5.4. Analyzing the CoMSIA Model

Once the PLS calculation has been carried out and a statistical model is achieved, we can now examine the CoMSIA maps to see where the interactions lie between each descriptor and the peptides. This is carried out as follows:

1. Select *'File'* from the main toolbar.
2. Select *'Read'* and open the *'.mol'* file.

3. Go back to *'File.'* select *'Align Database'* and enter the following details:

   a. *'database to align'* = '.mdb' file
   b. *'template molecule'* = pick the first peptide from list

4. Select *'GO'* – —a pop-up caution message appears, just click *'OK.'*
5. Select *'file'* from the toolbar and open the final *'non-cross validation model'* (.csv) file from the molecular spreadsheet.
6. From the molecular spreadsheet, select *'QSAR,'* followed by 'PLS' and open the *'non-cross validation model'* (.lis) file.
7. Go back to *'QSAR'*:

   a. select *'view QSAR'*
   b. followed by *'CoMSIA'*

8. A pop-up message now appears asking *'do you wish to remove it?'* – answer *'NO.'*

The CoMSIA map module now appears. From this module, select "contribution (%)" and unselect the "examine predicted." It is now possible to examine the CoMSIA maps with respect to the interactions of descriptors and peptides (Fig. 7A–E).

3.6. Designing New Predicted Class I epitopes

The main aim of using QSARs in this type of model building is using the results to design new peptides and/or superbinders. By looking at the relative positive contributions of amino acids at positions 1–9 (Fig. 2), it is possible to take the best contributing amino acids at each position and create a list of all

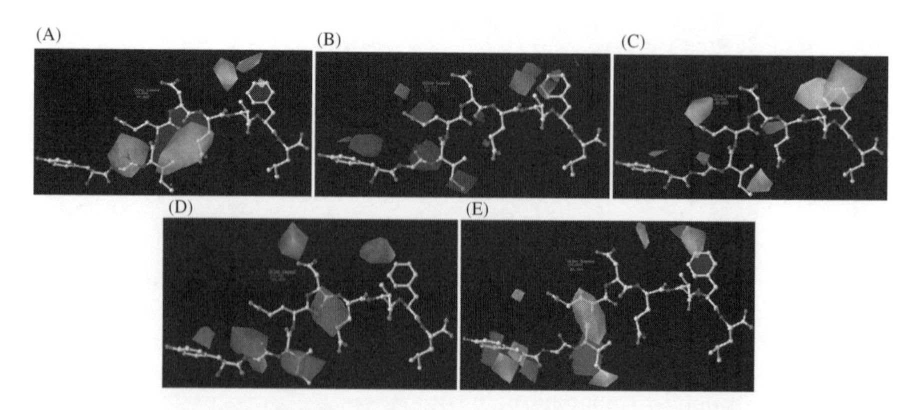

Fig. 7. H2-D^b steric (**A**), electrostatic (**B**), hydrophobic (**C**), hydrogen-bond acceptor (**D**), and hydrogen-bond donor (**E**) potential maps.

possible sequences. A further "in-house" method has been written, developed, and implemented with SYBYL to create these new binders. The following are the necessary steps for designing new peptides:

1. Import the file containing the new predicted list of peptides (.txt) to the SGI workstation.
2. Open SYBYL
3. Select *'Jenner'* from the main toolbar followed by *'Create Peptide List Combinatorially'* and enter the following details:
   a. *'File with position dependent residue preference'* = '.txt'
   b. *'Name of output peptide file'* = '.dat'
4. Go back to *'Jenner'* and select *'Peptide to Fasta'* and enter the following details:
   a. *'File of peptide sequences'* = prediction.dat
   b. *'fasta_file'* = fasta_file
5. Open and edit the final noncross-validation model file *'.lis'* from the ADDITIVE/PLS method and carry out the following steps:
   a. keep IC_{50} equation only and add five spaces below the regression equation.
   b. make a note of the constant value (e.g., 5.519).
   c. re-save *'.txt'* file.
6. Go back to *'Jenner'* and select *'Class I Additive Regression'* and enter the following details:
   a. *'Name of text file containing regression equation'* = original predicted '.txt' file
   b. *'Name of additive model out out file'* = .out (change .data to .out)
7. Now open and edit the new '.out' file and carry out the following steps:
   a. add the noted constant value (e.g., 5.519)
   b. if '-1.000' occurs in the matrix, it means that an amino acid is absent at that position—so change to 0.000.
   c. re-save '.out' file
8. Go back to *'Jenner'* and select *'Class I Additive Method'* and enter the following details:
   a. *'Name of fasta file containing sequences to be predicted'* = fasta_file
   b. *'Name of additive model data file'* = '.out'
   c. *'Minimum IC_{50} to be output'* = -
   d. *'Name of output file'* = new '.out' file

When the prediction calculation has finished, export the new predicted '.out' file to excel and examine the newly predicted binding affinities of the peptides.

References

1. Kubinyi, H., and Kehrhahn, O.H., 1976, Quantitative structure-activity relationships. 3.1 A comparison of different Free-Wilson models. *J. Med. Chem.* 19: 1040–1049.
2. Doytchinova, I.A., and Flower, D.R., 2003, Towards the in silico identification of class II restricted T-cell epitopes: a partial least squares iterative self-consistent algorithm for affinity prediction. *Bioinformatics* 19: 2263–2270.
3. Doytchinova, I.A., Blythe, M.J., and Flower, D.R., 2002, Additive method for the prediction of protein-peptide binding affinity. Application to the MHC class I molecule HLA-A*0201. *J. Proteome Res.* 1: 263–272.
4. Guan, P., Doytchinova, I.A., and Flower, D.R., 2003, HLA-A3 supermotif defined by quantitative structure-activity relationship analysis. *Protein Eng.* 16: 11–18.
5. Hattotuwagama, C.K., Guan, P., Doytchinova, I.A., Zygouri, C., and Flower, D.R., 2004, Quantitative online prediction of peptide binding to the major histocompatibility comlex. *J. Mol. Graph. Model.* 22: 195–207.
6. Klebe, G., Abraham, U., and Mietzner, T., 1994, Molecular similarity indices in a comparative analysis (CoMSIA) of drug molecules to correlate and predict their biological activity. *J. Med. Chem.* 37: 4130–4146.
7. Klebe, G., and Abraham, U., 1999, Comparative molecular similarity index analysis (CoMSIA) to study hydrogen-bonding properties and to score combinatorial libraries. *J. Comput. Aided Mol. Des.* 13: 1–10.
8. Bohm, M., Sturzebecher, J., and Klebe, G., 1999, Three-dimensional quantitative structure-activity relationship analyses using comparative molecular field analysis and comparative molecular similarity indices analysis to elucidate selectivity differences of inhibitors binding to trypsin, thrombin, and factor Xa. *J. Med. Chem.* 42: 458–477.
9. Stahle, L., and Wold, S., 1988, Multivariate data analysis and experimental design in biomedical research. *Prog. Med. Chem.* 25: 291–338.
10. Doytchinova, I.A., and Flower, D.R., 2002, Physicochemical explanation of peptide binding to HLA-A*0201 major histocompatibility complex: a three-dimensional quantitative structure-activity relationship study. *Proteins* 48: 505–518.
11. Doytchinova, I.A., and Flower, D.R., 2002, A comparative molecular similarity index analysis (CoMSIA) study identifies an HLA-A2 binding supermotif. *J. Comput. Aided Mol. Des.* 16: 535–544.
12. Guan, P., Doytchinova, I.A., and Flower, D.R., 2003, A comparative molecular similarity indices (CoMSIA) study of peptide binding to the HLA-A3 superfamily. *Bioorg. Med. Chem.* 11: 2307–2311.
13. Wold, S., 1995, PLS for multivariate linear modelling. *Chemometric Methods In Molecular Design* (H. van de Waterbeemd, ed.), VCH, Weinheim, pp. 195–218.
14. Dewar, M.J.S., Zoebisch, E.G., Healy, E.F., and Stewart, J.J.P., 1985, AM1: a new general purpose quantum mechanical molecular model *J. Am. Chem. Soc.* 107: 3902–3909.

15. Young, D., 2001, *Computational Chemistry: A Practical Guide for Applying Techniques to Real World Problems*. Wiley Inter-Science, New York, p. 243.
16. Blythe, M., Doytchiniva, I.A., and Flower, D.R. 2002, JenPep: a database of quantitative functional peptide data for immunology. *Bioinformatics* 18: 434–439.
17. McSparron, H., Blythe, M.J., Zygouri, C., Doytchinova, I.A., and Flower, D.R., 2003, JenPep: a novel computational information resource for immunology and vaccinology. *J. Chem. Inf. Comput. Sci.* 43: 1276–1287.
18. Ruppert, J., Sidney, J., Celis, E., Kubo, R.T., Grey, H.M., and Sette, A., 1993, Prominent role of secondary anchor residues in peptide binding to HLA-A*0201 molecules. *Cell* 74: 929–937.
19. Sette, A., Sidney, J., del Guercio, M.-F., Southwood, S., Ruppert, J., Dalberg, C., Grey, H.M., and Kubo, R.T., 1994, Peptide binding to the most frequent HLA-A class I alleles measured by quantitative molecular binding assays. *Mol. Immunol.* 31: 813–822.
20. Sybyl 6.9, Tripos Inc., 1699. Hanley Road, St. Louis, MO 63144.

17

Predicting the MHC–Peptide Affinity Using Some Interactive-Type Molecular Descriptors and QSAR Models

Thy-Hou Lin

Summary

The ligand–receptor interaction between some peptidomimetic inhibitors and a class II major histocompatibility complex (MHC)–peptide presenting molecule, the HLA-DR4 receptor, can be modeled using some 3D quantitative structure-activity relationship (QSAR) methods such as the comparative molecular field analysis (CoMFA) and some molecular descriptors using the *Cerius2* program. The structures of these peptidomimetic inhibitors can be generated theoretically, and the conformations used in the 3D QSAR studies can be defined by aligning them against the known structure of HLA-DR4 receptor through a least-square fitting procedure. The best CoMFA models can be constructed using the aligned structures of the best fitting result. The principal components analysis (PCA) module of the *Cerius2* program can be used to trim outliers of the CoMFA columns generated. Procedures for a direct QSAR analysis using the *Cerius2* descriptors and regression analysis by the genetic function module are also presented

Key Words: 3D QSAR; PLS; PCA; CoMFA; structure alignment

1. Introduction

The major histocompatibility complex (MHC) class II molecules are cell-surface proteins that perform an essential function in immunological detection using T-helper cells. They are encoded by the genes *HLA-DR, HLA-DQ*, and *HLA-DP*. Each MHC molecule consists of an α-chain and β-chain. In the case of the DR molecule, the two chains are encoded by the genes *HLA-DRA* and *HLA-DRB1*, and only DRB1 is polymorph, that is, only the gene has a number of different alleles existing in the population *(1)*. In addition, each individual possesses two DRB1 alleles, one from each parent.

From: *Methods in Molecular Biology, vol. 409: Immunoinformatics: Predicting Immunogenicity In Silico*
Edited by: D. R. Flower © Humana Press Inc., Totowa, NJ

The serological typification of the DR alleles leads to the differentiation between ten different classes, HLA-DR1-DR10 *(2)*. Molecular genetic typification shows that these classes can be further split, for example, DR2 has been divided into DR15 and DR16 *(3)*. Within these classes it is possible to distinguish between a number of subtypes. Up until now, 33 subtypes of DR4 have been described and are termed HLA-DRB1*0401-*0433 *(3–5)*. Rheumatoid arthritis (RA) or chronic polyarthritis is an intermittent systemic autoimmune disease that occurs in ~1% of the population *(6–9)*. The etiology of the disease is unknown. It has been shown that there is a genetic disposition for RA caused by several alleles of the HLA-DRB1 region *(7,8)*. RA is associated with the HLA-DRB1*04 subtypes DRB1*0401, *0404, *0405, and *0408 and also in some different ethnic groups with the subtypes DRB1*0101, *0102, and DRB1*1001 *(8,9)*.

Recently, the general features of the molecular recognition between antigenic peptide and the binding site on several MHC class II molecules have been elucidated through crystallization of several MHC molecular complexes *(10–14)*. Both the MHC α-chain and β-chain contribute to the peptide-binding site, which is made up of a β-sheet floor topped by two roughly parallel α-helical regions *(15–17)*. The peptide-binding motifs for some heptapeptides binding to the DR alleles have also been determined through phage display libraries and synthetic peptides *(18–21)*. Peptides bind in an extended conformation in the groove between the two helices, with about ten residues able to interact with the MHC protein while the peptide termini extending from the binding site *(20,21)*. The conformation places 4–6 of the peptide side chains into pockets within the overall groove. The residues lining these pockets vary between allelic variants, providing different peptide sequence-binding specificity. The interaction buries about 70% of the peptide surface area in the central region of a bound peptide, leaving the remainder available for interaction with the antigen receptors on T cells *(21)*.

The binding of peptides to human and mouse MHC class II molecules is characterized by several conserved side chain-binding pockets, namely, p1–p9 within the overall peptide-binding groove *(18,19,22)*. The pockets are numbered along the peptide relative to a large usually hydrophobic pocket near the peptide-binding site. The importance of residues at p1, p2, p4, p6, and p7 on binding has been addressed by panning M13 phage-expressed random peptide libraries *(22)*. An immunodominant peptide epitope of hemagglutinin (HA) (HA306-318) from influenza A virus H3N2 has been found to bind with different DR alleles of the MHC class II molecules *(23)*. The α/β T-cell receptor (TCR) HA1.7

specific for the HA antigen peptide is HLA-DR1 restricted but cross-reactive with the HA peptide presented by the MHC class II molecule HLA-DR4 *(23)*. The overall structures of the HA1.7/DR4/HA and HA1.7/DR1/HA complexes are found to be very similar, though there is a difference in the amino acid sequence of DR1 and DR4 located deeply inside the peptide-binding groove and out of reach by direct contact by the TCR *(23)*. The binding of peptides to HLA-DR1 has been strengthened by incorporating an *N*-methyl substitution at p7 of the peptide *(24)*. The *N*-methyl group oriented in the p6/p7 pocket is shown to displace one of the waters usually bound in this pocket, and the corresponding MHC–peptide complexes generated are able to activate the antigen-specific T cells *(24)*. The binding between MHC class II molecule HLA-DR4 and its peptide ligands can be predicted using some interactive-type molecular descriptors such as comparative molecular field analysis (CoMFA) and quantitative structure-activity relationship (QSAR) techniques described in the following sections.

2. Materials

Construction of structures for the peptide ligands should be based on the X-ray structure of an MHC–peptide ligand complex obtained from the Protein Data Bank (PDB). The construction of the ligand structure can be done within the MHC active site by replacing side chains of the template with other groups by using the SYBYL 7.1 program *(25)* (Build/Edit >> Sketch Molecule >> Draw). The hydrogen atoms should be added for each structure. Each of these structures should be rotated into the coordinate frame of the X-ray ligand before being merged with the ligand-depleted MHC receptor. Then, each structure of the ligand–receptor complex should be subjected to a brief energy minimization using the SYBYL 7.1 program (Compute >> Minimize) and the Tripos60 Force Field Engine. The Gasteiger–Hückel *(26)* and KOLL_ALL *(27)* charges can be deployed, respectively, for ligands and receptor, and a nonbonding cutoff of 8 Å can be used for each structure complex. Each ligand structure thus constructed is extracted from each structure complex using the SYBYL 7.1 program (Build/Edit >> Extract >> Substructures). The biological activity of each ligand is expressed in pIC_{50}. The entire compound set is divided into two sets, namely, a training and test set according to the following rules *(28)*: (1) the entire set should contain at least 16 compounds to assure statistical significance of the pharmacophore model, (2) the activity range of the compounds should span at least 4 orders of magnitude, (3) each order of magnitude should be represented by at least three compounds, (4) the most active and inactive compounds should be included, and (5) two compounds with similar structure must differ in activity by an order of magnitude to be included.

3. Methods

3.1. The CoMFA Descriptors Using the SYBYL 7.1 Program

1. Clear all the molecules in the display area [Build/Edit >> Zap (Delete) Molecule].
2. Make a database for the structures of peptide ligands built (File >> Database >> New >> Put Molecule) (*see* **Notes 1–3** for dividing data set into a training set and a test set).
3. Perform the structural alignment for the database by using a common structure template (File >> Align Database >>> Database to Align: >>> Template Molecule: >>> Location of Substructure: >>> Put Molecules Into: >>> Align:) (*see* **Notes 4–6** for aligning structures in a set).
4. Make a molecular spreadsheet for all the ligand structures in the aligned database (File >> Molecular Spreadsheet >> New >>> Database >>> Open). All the ligand structures in the database are read in and entered as rows in the spreadsheet.
5. The binding activity of each ligand expressed in pKi [log(1/Ki)] should be typed into the first column of the spreadsheet. One can also use MSS (panel on the molecular spreadsheet): File >>> Import to import the binding activity if it has been saved into a file. Label the column as the ACTIVITY column.
6. The process of adding CoMFA fields involves scanning all the aligned molecules to establish an encompassing region and computing somewhat more than 33,000 energies. On the MSS panel, select "empty column 2" and press Autofill. Select COMFA as the new column type and press OK. (Use the default selections in the Add New CoMFA Column dialog box. Use the Tripos Standard CoMFA Field class. The other options are available only with an Advanced CoMFA license. *See* **Note 7** for choosing a grid space for computing the CoMFA column.)
7. Perform the partial least-square (PLS) analysis on the CoMFA column created by pressing MSS:QSAR >>> Partial Least Squares, the PLS analysis dialog box appears.
8. Input COMFA2 as the *Column* to use and then input the ACTIVITY column as the Dependent Column whose values will be predicted from the resulting PLS analysis.
9. Perform a PLS analysis with Leave-One-Out validation where the number of groups is equal to the number of rows selected. Toggle the SAMPLS box off (*see* **Notes 8** and **9** on this action). One can speed up the computation process

by using the SAMPLS option if the data set has been cross-validated beforehand. Set Components as 5, Scaling as CoMFA Standard, Column Filtering on 2.0 (to omit those columns whose energy variance is less than 2.0 kcal/mol). Type one as the Analysis Name and press Do PLS. Select End when prompted to save the analysis. Meanwhile, look into the analysis details shown in the text window and pay attention to the cross-validated r^2 value and the optimal number of components obtained. A meaningful model can be established only when the cross-validated r^2 computed is greater than 0.5.

10. Derive the best predictive model for use in prediction and in graphic presentation by using the best cross-validated result. Set the options as follows while the PLS analysis dialog box is still on: Validation: No Validation, perform a PLS analysis without any validation, this is typically done at the end of PLS analysis; Components: 5; Scaling: CoMFA Standard; Column Filtering: on. Then, type two as the Analysis Name, press Do PLS and then OK when asked to save the analysis. The text window will show the r^2 measure of fit, the contribution of electrostatic and steric fields all in percentage. Press End to close the PLS dialog box.
11. Examine the CoMFA results from the MSS panel by pressing QSAR >>> View CoMFA. The view CoMFA dialog box is displayed. In the Display option menu, select a mode appropriate for your terminal (*see* **Note 10** on this action.). Press Show and Quit to exit the view CoMFA dialog box. Read the information in the text window. Close the spreadsheet by pressing Close in the MSS panel. Answer Yes or No, depending on whether you want to save the spreadsheet in a table file. If yes, type in a name for the table file. By default, the table uses the same base name as the database and adds the extension .tbl.

3.2. The QSAR Prediction Using the Cerius2 Program

1. Start a new session of *Cerius2* (*29*) by typing Cerius2 at the Unix prompt and the *Visualizer and Cerius2 Models* windows will appear.
2. Go to the Build/3D-Sketcher panel and select the desired structural template buttons to draw the structure.
3. Clean the structure by clicking Preferences button in the 3D_Sketcher card and a Cleaner Controls card will appear. Select One Short Clean and Watch one-short progress buttons. Close the Cleaner Controls card using the >< buttons and click CLEAN in the 3D-Sketcher.
4. Minimize the drawn structure by going to OFF SETUP card and select Load Force Field? and select *cvff950_1.0.1* as the force field to use.
5. Start the RUN by clicking the radio button on the Energy Minimization card.

6. Load molecules into the QSAR study table by going to QSAR/Show Study table. Click the Add all button under Molecules pull-down. Type the activity data into the column labeled as Activity.
7. Select the default options corresponding to the QSAR application by pressing the Preferences/Default Set/QSAR menu item in the study table menu bar.
8. Add a set of default descriptors to the study table by selecting the Descriptors/Add Default menu item in the study table menu bar.
9. Select the column labeled as Activity in the study table by clicking the column heading. Mark this column as dependent variables (Y) by selecting the Variables/Set Y menu item in the study table menu bar.
10. Select all the descriptor columns in the study table by using the < Shift >-clicking the column headings. Mark these columns as independent variables by selecting Variables/Set X menu item in the study table menu bar.
11. Use the genetic function approximation (GFA) method to generate a QSAR equation by setting the Methods pop-up to GFA. Go to Preferences in the menu bar of the study table and click Statistical Method. Click Configure GFA and then verify that the linear term is selected. Click the RUN button to start the GFA calculation.
12. Select the first (best) equation for validation using the cross-validation method. The equations generated are downloaded into the Equation Viewer control panel and sorted by the lack-of-fit (LOF) parameter. The QSAR equation is automatically inserted as a new column labeled as *GFA Predicted Activity* in the study table along with a column showing the residuals (observed—predicted activity values) and labeled as *GFA Residuals Activity*. The cross-validation results of the QSAR equation are shown in the text window.
13. Select the Tools/Equation Viewer menu in the study table menu bar to view the terms, coefficients, and statistics of the equation. Click the More button in the QSAR Equation section of the Equation Viewer control panel to open the preferences control panel for QSAR equations. Then press the Auto update 2D Plot button. Select the QSAR equation number 1 in the Equation Viewer control panel and click the Plot Equation action button to view the 2D plot of predicted vs. observed activity.
14. Click the Save Equations button in the Equation Viewer control panel to save the QSAR equation. Set the pop-up to Current Equations Set in the Save QSAR equations control panel open.
15. Predict the activity of a new molecule using the QSAR equation by adding the molecule into the study table. Go to the File/Load Model menu item in the *Cerius2* Visualizer panel and then load the molecule into *Cerius*2. Add the molecule to the study table by selecting the Molecules/Add Current menu item in the study table menu bar. (The new molecule is added at the bottom of the study table and the predicted activity is automatically shown in the column *GFA Predicted Activity*.)

3.3. The Principal Components Analysis Using the Cerius2 Program

1.

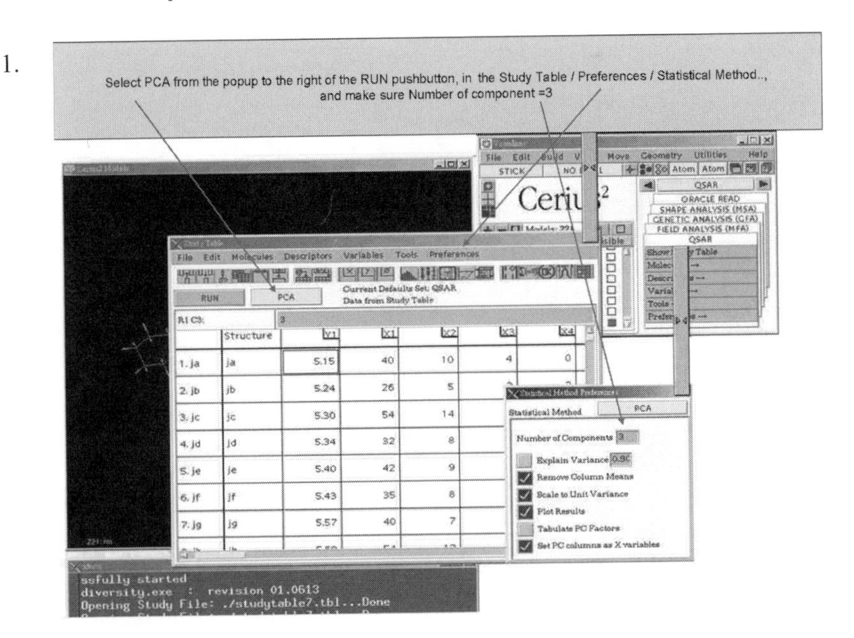

Select PCA from the popup to the right of the RUN pushbutton, in the Study Table / Preferences / Statistical Method..,
and make sure Number of component =3

2.

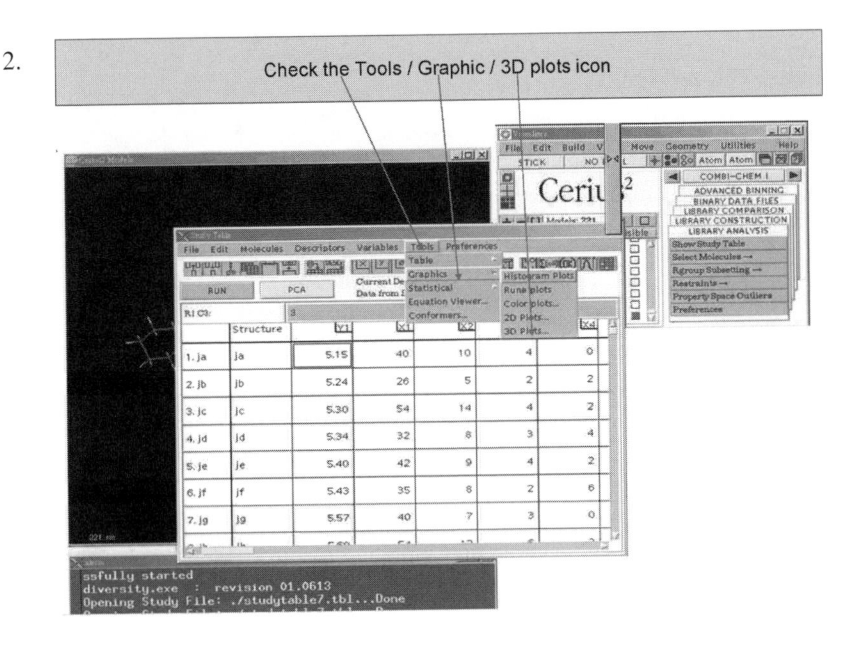

Check the Tools / Graphic / 3D plots icon

3.

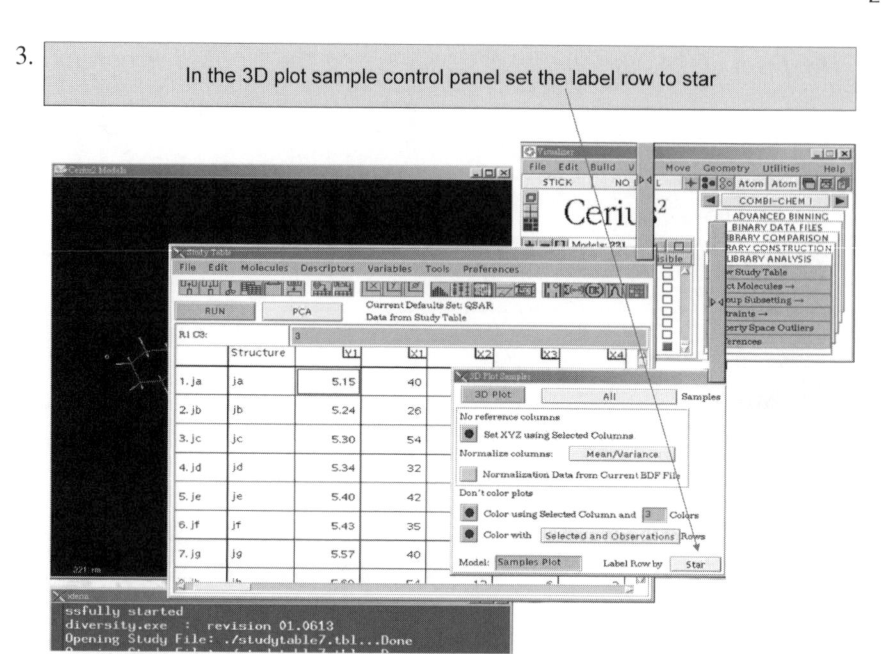
In the 3D plot sample control panel set the label row to star
Study Table
File Edit Molecules Descriptors Variables Tools Preferences
RUN
PCA
Current Defaults Set: QSAR
Data from Study Table
3D Plot
All
Samples
No reference columns
Set XYZ using Selected Columns
Normalize columns:
Mean/Variance
Don't color plots
Color with
Selected and Observations
Model:
Samples Plot
Label Row by
Star
COMBI-CHEM I
ADVANCED BINNING
BINARY DATA FILES
ssfully started
diversity.exe : revision 01.0613
Opening Study File: ./studytable7.tbl...Done

4.

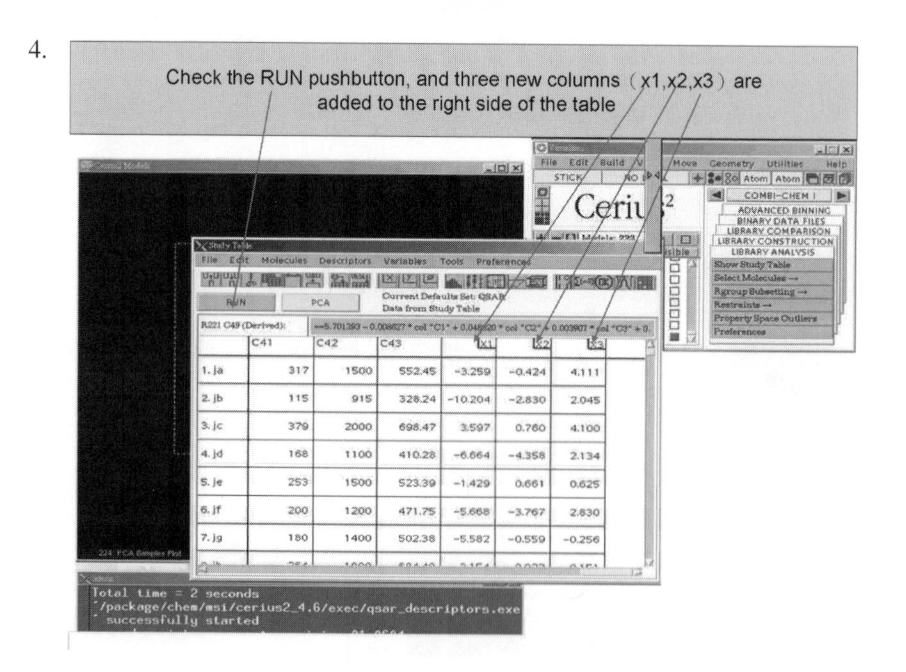
Check the RUN pushbutton, and three new columns（x1,x2,x3）are
added to the right side of the table
Study Table
File Edit Molecules Descriptors Variables Tools Preferences
RUN
PCA
Current Defaults Set: QSAR
Data from Study Table
COMBI-CHEM I
ADVANCED BINNING
BINARY DATA FILES
LIBRARY COMPARISON
LIBRARY CONSTRUCTION
LIBRARY ANALYSIS
Show Study Table
Property Space Outliers
Preferences
Total time = 2 seconds
"/package/chem/msi/cerius2_4.6/exec/qsar_descriptors.exe
" successfully started

5.

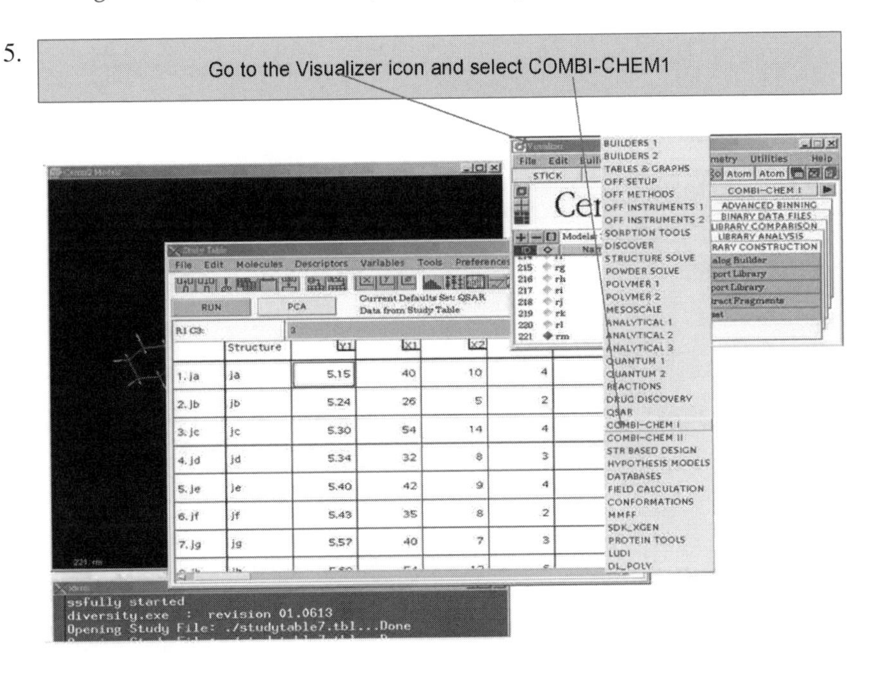
Go to the Visualizer icon and select COMBI-CHEM1

6.

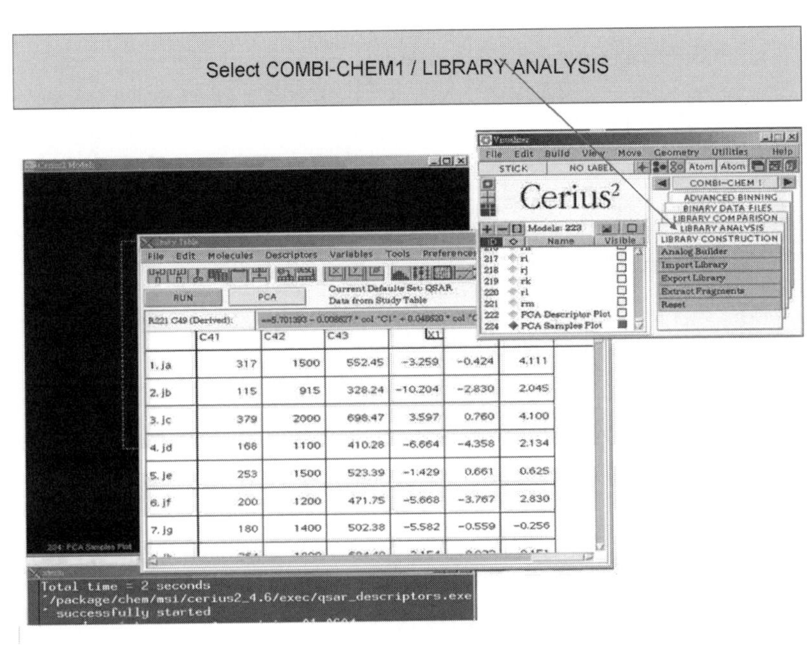
Select COMBI-CHEM1 / LIBRARY ANALYSIS

7.

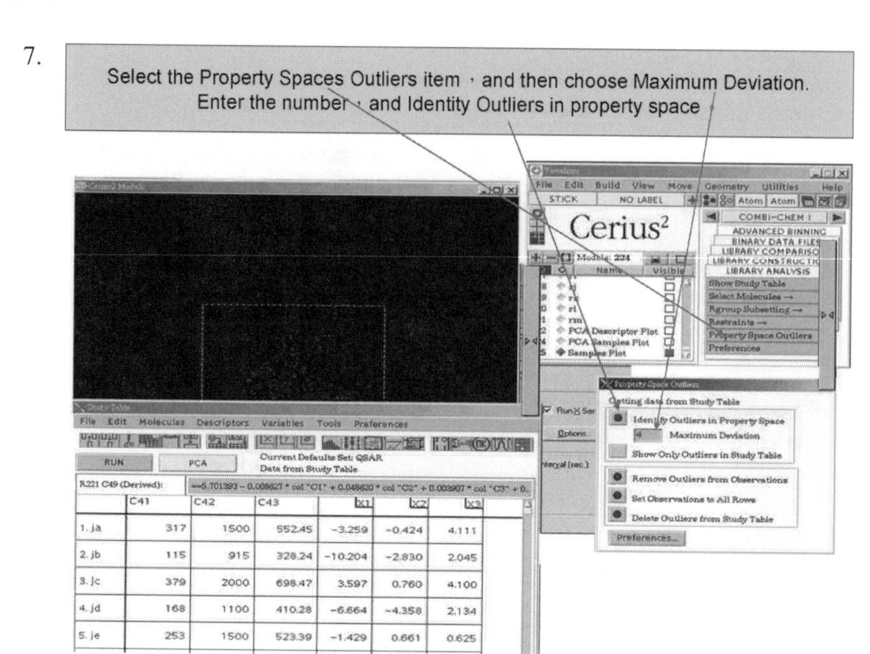
Select the Property Spaces Outliers item , and then choose Maximum Deviation.
Enter the number , and Identity Outliers in property space
Cerius²
COMBI-CHEM I
LIBRARY ANALYSIS
Show Study Table
Property Space Outliers
Preferences
Identify Outliers in Property Space
Maximum Deviation
Show Only Outliers in Study Table
Remove Outliers from Observations
Set Observations to All Rows
Delete Outliers from Study Table
RUN
PCA
Current Defaults Set: QSAR
Data from Study Table

8.

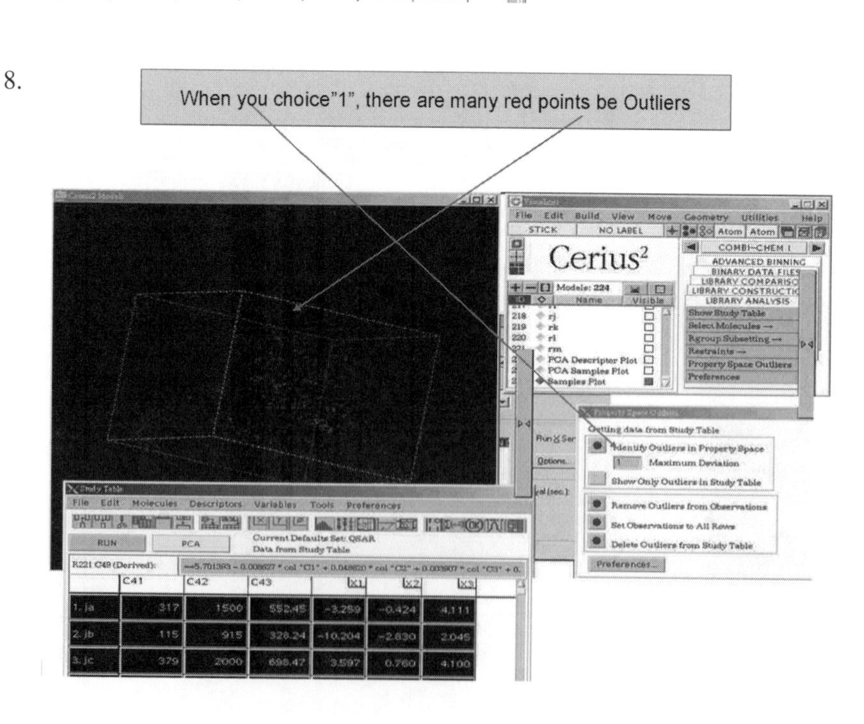
When you choice"1", there are many red points be Outliers
Cerius²
COMBI-CHEM I
LIBRARY ANALYSIS
Property Space Outliers
Identify Outliers in Property Space
Maximum Deviation
Show Only Outliers in Study Table
Remove Outliers from Observations
Set Observations to All Rows
Delete Outliers from Study Table
RUN
PCA
Current Defaults Set: QSAR
Data from Study Table

9. 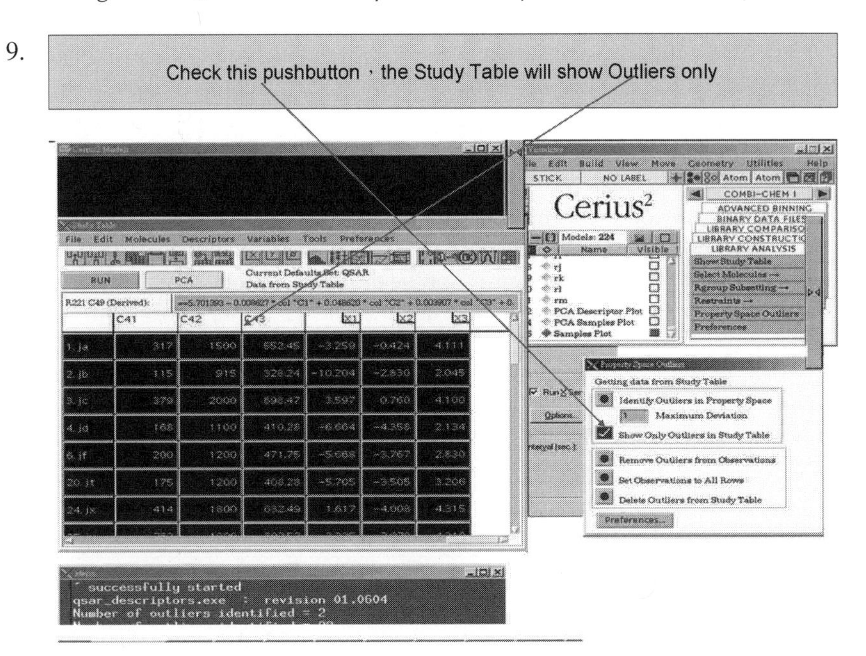

10. The PCA can be used to relieve redundancy among possibly correlated variables. The analysis allows one to visualize most of the variance of the data set by visualizing the first three principal components.

Acknowledgment

This work is supported in part from a grant (NSC94-2313-B007-001) of National Science Council, Taiwan.

Notes

1. The training set compounds should span a parameter space in which all data points are more or less equally distributed.
2. To cover the property space with the smallest possible number of objects, one should remove redundancy. However, in the case of poor test set prediction, some redundancies may be included in the training set to improve the statistics.
3. A broad variety of structural features should be included in the training set in order to allow reliable predictions for the test set compounds.
4. The structures of both training and test sets are required to be aligned by hand or by an appropriate field fit in performing CoMFA.
5. The most popular method for molecular aligning molecular structures is least-square fitting method, which gives the best matching of positions of atoms that have

been decided to correspond in advance. However, aligning dissimilar molecules should be done by methods other than least-squares matching of atom positions.

6. Obtaining biologically relevant conformations when receptor-bound crystal structures are missing is often achieved by the active-analog approach developed by Marshall et al. ***(30)***. The active conformations of flexible compounds are determined by systematic conformational searches using geometrical constraints of a rigid template analog. However, it must be assumed that neither the receptor structure nor the binding mode varies for the different molecules that are examined.
7. A smaller grid space other than the default one (2 Å) may be used only for enclosing the activity region identified by the initial CoMFA run. Decreasing the grid spacing will increase grid points, and the noise in data is also increased, thereby the overall statistics is deteriorated.
8. A linear regression analysis cannot be directly applied to the CoMFA descriptors due to the enormous number of x variables generated. Both PCA and PLS can be used to obtain a linear equation for the CoMFA descriptors. SAMPLS is a modification of PLS analysis implemented in the SYBYL package. Owing to a much smaller number of arithmetic operations, SAMPLS operates a few to several orders of magnitude faster in cross-validation runs than ordinary PLS analysis.
9. In the most common leave-one-out cross-validation, one object is omitted from the training set, and a PLS model is derived from the residual compounds. This model is used to predict the biological activity value of the compound, which was not included in the model. To yield more stable PLS model for larger data sets, several objects are eliminated from the data set at a time, randomly or in a systematic manner, and the excluded objects are predicted by the corresponding model.
10. The CoMFA results are usually presented as a set of contour maps. These contour maps show favorable and unfavorable steric regions or electropositive or electronegative substituents in certain positions. Predictions for the test set and for other compounds can be made either by a qualitative inspection of these contour maps or in a quantitative manner, by calculating the fields of these molecules and by inserting the grid values into the PLS model.

References

1. Watts, C. Capture and processing of exogenous antigens for presentation on MHC molecules. *Annu. Rev. Immunol.* **1997**, *15*, 821–850.
2. Rudensky, A.; Prestoa-Hurlburt, P.; Hong, S. C.; Barlow, A.; Janeway, C. A. Jr. Sequence analysis of peptides bound to MHC class II molecules. *Nature* **1991**, *353*, 622–627.
3. Chicz, R. M.; Urban, R. G.; Lone, W. S.; Gorga, J. C.; Stern, L. J.; Vignali, D. A.; Strominger, J. L. Predominant naturally processed peptides bound to HLA-DR1 are derived from MHC-related molecules and are heterogeneous in size. *Nature* **1992**, *358*, 764–768.

4. Tiwari, J.; Terasaki, P. HLA and disease association. Springer-Verlag, New York, **1985**.
5. Rowley, M. J.; Stockman, A.; Bond, C. A.; Tait, B. D.; Rowley, G. L.; Sherritt, M. A.; Mackay, I. R.; Muirden, K. D.; Bernard, C. C. The effect of HLA-DRB1 disease susceptibility markers on the expression of RA. *Scand. J. Rheumatol.* **1997**, *26*, 448–455.
6. Weyand, C. M.; Goronzy, J. J. Inherited and noninherited risk factors in rheumatoid arthritis. *Curr. Opin. Rheumatol.* **1995**, *7*, 206–213.
7. Nepom, G. T.; Gersuk, V.; Nepom, B. S. Prognostic implications of HLA genotyping in the early assessment of patients with rheumatoid arthritis. *J. Rheumatol. Suppl.* **1996**, *44*, 5–9.
8. Wagner, U.; Kaltenhauser, S.; Sauer, H.; Arnold, S.; Seidel, W.; Hantzschel, H.; Kalden, J. R.; Wassmuth, R. HLA markers and prediction of clinical course and outcome in rheumatoid arthritis. *Arthritis Rheum.* **1997**, *40*, 341-351.
9. Perdriger, A.; Chales, G.; Semana, G.; Guggenbuhl, P.; Meyer, O.; Quillivic, F.; Pawlotsky, Y. Role of HLA-DR-DR and DR-DQ association in the expression of extraarticular manifestations and rheumatoid factor in rheumatoid arthritis. *J. Rheumatol.* **1997**, *24*, 1272–1276.
10. Stern, L. J.; Brown, J. H.; Jardetzky, T. S.; Gorga, J. C.; Urban, R. G.; Strominger, J. L.; Wiley, D. C. Crystal structure of the human class II MHC protein HLA-DR1 complexed with an influenza virus peptide. *Nature* **1994**, *368*, 215–221.
11. Jardetzky, T. S.; Brown, J. H.; Stern, L. J.; Urban, R. G.; Chi, Y. I.; Stauffacher C.; Strominger, J. L.; Wiley, D. C. Three dimensional structure of a human class II histocompatibility molecule complexed with superantigen. *Nature* **1994**, *368*, 711–718.
12. Ghosh, P.; Amaya, M.; Mellins, E.; Wiley, D. C. The structure of an intermediate in class II MHC maturation: CLIP bound to HLA-DR3. *Nature* **1995**, *378*, 457–462.
13. Brown, J. H.; Jardetzky, T. S.; Gorga, J. C.; Stern, L. J.; Urban, R. G.; Strominger, J. L.; Wiley, D. C. Three-dimensional structure of the human class II histocompatibility antigen HLA-DR1. *Nature* **1993**, *364*, 33–39.
14. Dessen, A.; Lawrence, C. M.; Cupo, S.; Zaller, D. M.; Wiley, D. C. X-ray crystal structure of HLA-DR4 (DRA0101, DRB10401) complexed with a peptide from human collagen II. *Immunity* **1997**, *7*, 473–481.
15. Garboczi, D. N.; Ghosh, P.; Utz, F.; Oing, R.; Biddison, W. E.; Wiley, D. C. Structure of the complex between human T-cell receptor, viral peptide and HLA-A2. *Nature* **1996**, *384*, 134–141.
16. Garcia, K. C.; Degano, M.; Stanfield, R. L.; Brunmark, A.; Jackson, M. R.; Peterson, P. A.; Teyton, L.; Wilson, I. A. An αβ T cell receptor structure at 2.5Å and its orientation in the TCR-MHC complex. *Science* **1996**, *274*, 209–219.
17. Reinherz, E. L.; Tan, K.; Tang, L.; Kern, P.; Liu, J.; Xiong, Y.; Hussey, E.; Smolyar, A.; Hare, B.; Zhong, R.; Joachimiak, A.; Chang, H.; Wagner, G.; Wang, J. The crystal structure of a T-cell receptor in complex with peptide and MHC class II. *Science* **1999**, *286*, 1913–1921.

18. Hammer, J.; Takacs, B.; Sinigaglia, F.; Identification of a motif for HLA-DR1 binding peptides using M13 display libraries. *J. Exp. Med.* **1993**, *176*, 1007–1013.
19. Hammer, J.; Valsasnini, P.; Tolba, K.; Bolin, D.; Higelin, J.; Takacs, B.; Sinigaglia, F. Promiscuous and allele-specific anchors in HLA-DR binding peptides. *Cell* **1994**, *74*, 197–203.
20. Hammer, J.; Bono, E.; Gallazzi, F.; Belunis, C.; Nagy, Z. A.; Sinigaglia, F. Precise prediction of major histocompatibility complex class II-peptide interaction based on peptide side chain scanning. *J. Exp. Med.* **1995**, *180*, 2353–2358.
21. Hammer, J.; Callazzi, F.; Bono, E.; Karr, R. W.; Guenot, J.; Valsasin, P.; Nagy, Z. A.; Sinigaglia, F. Peptide binding specificity of HLA-DR4 molecules: correlation with rheumatoid arthritis association. *J. Exp. Med.* **1995**, *181*, 1847–1855.
22. Bolin, D. R.; Swain, A. L.; Ramakanth, S.; Berthel, S. J.; Gillespie, P.; Huby, N. J. S.; Makofske, R.; Orzechowski, L.; Perrotta, A.; Toth, K.; Cooper, J. P.; Jiang, N.; Falcion, F.; Campbell, R.; Cox, D.; Gaizband, D.; Belunis, C. J.; Vidovic, D.; Ito, K.; Crowther, R.; Kammlott, U.; Zhang, X.; Palermo, R.; Weber, D.; Guenot, J.; Nagy, Z.; Olson, G. L. Peptide and peptide mimetic inhibitors of antigen presentation by HLA-DR class II MHC molecules. Design, structure-activity relationship, and X-ray crystal structure. *J. Med. Chem.* **2000**, *43*, 2135–2148.
23. Hennecke, J.; Wiley, D. C. Structure of a complex of the human α/β T cell receptor (TCR) HA 1.7, influenza hemagglutinin peptide and major histocompatibility complex class II molecule, HLA-DR4 (DRA*0101 and DRB1*0401): insight into TCR cross-restriction and alloreactivity. *J. Exp. Med.* **2002**, *195*, 571–581.
24. Zavala-Ruize, Z.; Sundberg, E. J.; Stone, J. D.; DeOliveira, D. B.; Chan, I. C.; Svendsent, J.; Mariuzza, R. A.; Stern, L. J. Exploration of the p6/p7 region of the peptide-binding site of the human class II major histocompatibility complex protein HLA-DR1. *J. Biol. Chem.* **2003**, *278*, 44904–44912.
25. SYBYL 7.1; The Tripos Associates; 1699. Hanley Road, St. Louis, MO, USA.
26. Gasteiger, J.; Marsili, M. Iterative partial equalization of orbital electronegativity - a rapid access to atomic charges, *Tetrahedron* **1980**, *36*, 3219–3228.
27. Weiner, S. J.; Kollman, P. A.; Case, D. A.; Singh, U. C.; Ghio, C.; Alagona, G.; Profeta, S. Jr.; Weiner, P. A new force field for molecular mechanical simulation of nucleic acids and proteins. *J. Am. Chem. Soc.* **1984**, *106*, 765–784.
28. Golbraikh, A.; Tropsha, A. Beware of q2! *J. Mol. Graph. Model.* **2002**, *20*, 269–276.
29. Accelrys Inc.; *Cerius2* Modeling Environment, Release 4.0, San Diego, Accelrys Inc., 2002.
30. Marshall, G. R.; Barry, C. D.; Bosshard, H. E.; Dammkoehler, R. A.; Dunn, D. A. The conformational parameter in drug design: the active analog approach. In Olson, E.C. and Christoffersen, R.E. (Eds.) *Computer-Assisted Drug Design, ACS Symp. Series*, Vol 112. American Chemical Society, Washington, DC. **1979**, pp. 206–226.

18

Implementing the Modular MHC Model for Predicting Peptide Binding

David S. DeLuca and Rainer Blasczyk

Summary

The challenge of predicting which peptide sequences bind to which major histocompatibility complex (MHC) molecules has been met with various computational techniques. Scoring matrices, hidden Markov models, and artificial neural networks are examples of algorithms that have been successful in MHC–peptide-binding prediction. Because these algorithms are based on a limited amount of experimental peptide-binding data, prediction is only possible for a small fraction of the thousands of known MHC proteins. In the primary field of application for such algorithms—vaccine design—the ability to make predictions for the most frequent MHC alleles may be sufficient. However, emerging applications of leukemia-specific T cells require a patient-specific MHC–peptide-binding prediction. The modular model of MHC presented here is an attempt to maximize the number of predictable MHC alleles, based on a limited pool of experimentally determined peptide-binding data.

Key Words: Modules; pockets; HLA; MHC; class I; class II; peptide; binding; prediction

1. Introduction

The major histocompatibility complex (MHC) is a highly polymorphic collection of genes encoding membrane surface proteins, which plays an important role in the immune system. MHC binds short peptide sequences and presents them on the cell surface for inspection by T cells *(1)*. In humans, MHC is known as human leukocyte antigen (HLA).

Because of MHC's role in recognizing pathogenic and cancerous peptides, these genes are under high environmental pressure to be very polymorphic.

From: *Methods in Molecular Biology, vol. 409: Immunoinformatics: Predicting Immunogenicity In Silico*
Edited by: D. R. Flower © Humana Press Inc., Totowa, NJ

Presently, 2,088 HLA alleles have been identified *(2)*. Predicting which peptide sequences will bind to specific MHC alleles is dependent on the amount of experimentally determined peptide-binding data available for each allele. Such data are only available for a small fraction of all the alleles. The goal of the modular concept is to take advantage of similarities among alleles by utilizing existing peptide-binding data to make predictions for alleles, for which no peptides are available.

Although MHC polymorphism can be caused by point mutation, it is mainly a result of gene conversion and recombination *(3)*. Therefore, although a specific MHC is unique, it may be identical to a second MHC in one region and identical to a third MHC in another region. Such similarities can be exploited by breaking down MHC into modules and correlating these modules with the available peptide-binding data *(4,5)*. In this way, peptide-binding data specific for a small number of MHC variants can be applied to an expanded number of variants.

The part of the MHC–peptide-binding groove that interacts with a specific position in the bound peptide is known as a pocket. Originally these pockets were designated A–F *(6)*. Further analysis of crystallographic data in class I HLA has provided more complete definitions of which positions in HLA are responsible for binding certain positions in the peptide *(7,8)*. Because of the side chain orientation in the protein's three-dimensional structure, the positions responsible for peptide binding are not sequential. For example, the particular residues in HLA class I that interact with the N-terminal amino acid (P1 = peptide position 1) in the peptide are at positions 5, 7, 33, 59, 62, 63, 66, 99, 159, 163, 167, and 171 *(7)*. These positions are used to define a module. A module is the sequence of amino acids found at these positions in a specific MHC allele. For a 9-mer peptide, a given allele will have nine modules (P1, P2, . . . P9). Because of similarities among MHC alleles, different MHCs can share modules when they posses the same amino acids at the defined positions (Tables 1 and 2).

The result of this modular concept is an expanded number of MHC alleles, for which peptide binding can be predicted.

2. Implementation

The modular prediction algorithm available via the PeptideCheck (http://www.peptidecheck.org) website was written in Java and runs on a Tomcat application server, utilizing servlets, java server pages, and a MySQL database.

Table 1
Modules for A*0101 and A*7401 at P1

A*0101												
Position	5	7	33	59	62	63	66	99	159	163	167	171
Amino acid	M	Y	F	Y	Q	E	N	Y	Y	R	G	Y
Other alleles with this module:	A*0102, A*0103, A*0106, A*0107, A*0110											
A*7401												
Position	5	7	33	59	62	63	66	99	159	163	167	171
Amino acid	M	Y	F	Y	Q	E	N	Y	Y	T	W	Y
Other alleles with this module:	A*0256, A*0301-14, A*1104, A*3001–6, 8, 9, 11, 12, A*3101, 3, 4, 6, 9, A*3201–4, 6–8, A*3601–3, A*7402, 3, 5–10											

The positions listed here are positions in the HLA protein, which are likely to affect the binding of amino acids at P1 in the peptide. The amino acids listed are those amino acids which occur at the given positions in A*0101 and A*7401, respectively. These lists of nonsequential amino acids are the modules at P1. The alleles listed under "Other alleles with this module" possess the same amino acids at these positions and therefore possess the same P1 modules.

Table 2
Number of modules for each peptide position

Peptide positions	1	2	3	4	5	6	7	8	9
Number of modules	176	365	424	72	298	458	282	82	405

The total number of modules for each peptide position is less than the number of HLA proteins, because related alleles share certain modules. These numbers are based on all class I HLA-A, HLA-B, and HLA-C proteins from the IMGT/HLA database version 2.10.0, which contains 1,098 class I proteins.

2.1. HLA Sequence Data

HLA protein sequences are available in the IMGT/HLA database and are regularly updated *(2)*. Sequences can be downloaded directly from the file transfer protocol (FTP) server under *ftp://ftp.ebi.ac.uk/pub/databases/imgt/mhc/hla/*. Nucleotide and protein sequences are available in various formats. Sequence alignments for all HLA genes are available as zip files. Because many of the HLA sequences are incomplete (e.g., only certain exons have been determined), sequence alignments are necessary. Programmers may either download the individual sequences, and align them locally, or download the alignment files, and extract the sequence information.

2.2. Peptides

The module-based peptide-binding prediction requires collections of peptide, which have been experimentally proven to bind MHC. Databases such as SYFPEITHY, MHCBN, and AntiJen are good sources of peptide-binding data ***(9–11)***. Although some databases provide binding affinities, the algorithms described here require only that a distinction is made between binders and nonbinders. Nonbinders are often a limiting factor. Alternatively, random sequences of peptides can be generated and assumed to be nonbinders. This assumption will be true for the vast majority of sequences because less than 1% of possible peptide sequences are thought to bind HLA class I ***(12)***. The use of random nonbinders has several precedents ***(13,14)***. In this implementation, random nonamers were generated by randomly choosing human proteins from the Entrez protein database. Segments of nine amino acids were then randomly chosen.

2.3. Modules

At the heart of the modular concept lies the pocket definition. For our purposes, a pocket is the list of positions in HLA, which is responsible for binding a particular amino acid position in the peptide. In this study, the pockets were defined as per Chelvanayagam's analysis of crystallographic HLA data ***(7)***. Alternative definitions have been provided by Saper and Reche ***(6,8)***.

A module is the sequence of amino acids found at the pocket positions for a given allele. Modules are generated by combining the pocket definitions provided by Chelvanayagam or others with the HLA protein sequences (Table 1). Although many related alleles produce the same module sequences, only unique sequences should be stored in the database. A second database table can be used to correlate the module sequences with the alleles that posses them.

2.4. Matrices and Prediction

The simplest implementation of modular peptide-binding prediction is using a scoring matrix. When predicting binding to nonamers, the matrices are 9×20 and contain values for each amino acid at each position peptide (Table 3). The following pseudocode demonstrates how to generate the matrix:

Modular Matrix

For each module
  Retrieve all alleles that have this module
  For all alleles with this module
    Retrieve all binders
    For each binder
      Count the amino acid at the position corresponding to this module
  Divide the scores by the number of binders found for this module

A score for a peptide's binding ability is generated by multiplying the nine corresponding values from the matrix. This score is indicative of the likelihood that this peptide is a binder and can be compared to a threshold to predict binding. In the modular matrix, the values are based on the frequencies of the amino acids, among binding peptides, specific to a particular module (*see* **Note 1**). Because different alleles can have certain modules in common, the module-specific values are based on peptides that bind to all the alleles which have that module.

2.5. Evaluating Predictive Performance

Predictive performance can be calculated using the area under the receiver operating characteristic curve (A_{ROC}). The ROC curve is based on the prediction's sensitivity

$$SE = {}^{TP}/_{(TP+FN)}$$

and specificity

$$SP = {}^{TN}/_{(TN+FP)}$$

where TP = true positives—correctly predicted binders; FN = false negatives—binders incorrectly predicted to be nonbinders; TN = true negatives—correctly predicted nonbinders; and FP = false positives—nonbinders incorrectly predicted to bind. The ROC curve is a plot of SE versus 1 SP over a range of thresholds (Fig. 1).

Using the same peptides for training as well as testing is for obvious reasons taboo. Peptides used in testing should be excluded from the matrix scores. This can be done by splitting the peptide data into separate training and testing pools (e.g., two-thirds for training and one-third for testing). A method that

Table 3
Modular matrix for A*0201

Amino acids	Positions in peptide								
	P1	P2	P3	P4	P5	P6	P7	P8	P9
A	15	3	10	5	10	10	15	10	6
C	1	0	1	2	0	2	1	1	1
D	0	0	4	5	3	2	1	1	0
E	1	0	2	12	3	2	3	6	0
F	7	0	5	1	6	4	9	5	0
G	7	0	7	12	10	4	2	8	0
H	2	0	1	1	1	1	3	3	0
I	6	10	4	3	5	7	6	3	11
K	12	0	2	8	2	2	0	5	0
L	8	62	12	5	9	12	12	10	30
M	2	8	3	0	1	2	1	1	1
N	1	0	6	2	3	2	3	3	0
P	1	0	3	11	8	10	5	5	0
Q	1	1	3	5	5	3	2	4	0
R	4	0	1	4	3	1	2	4	0
S	8	0	6	6	3	5	4	7	0
T	2	5	2	3	4	4	5	8	3
V	5	5	5	4	9	14	9	3	41
W	1	0	4	0	3	0	1	2	0
Y	7	0	6	0	3	1	2	2	0

Contributors	Num.	P1	P2	P3	P4	P5	P6	P7	P8	P9
A*0201	735	+	+	+	+	+	+	+	+	+
A*0202	75	+	+	−	−	−	−	−	+	−
A*0203	65	+	+	−	−	−	−	−	+	+
A*0204	38	+	+	−	+	−	−	−	−	−
A*0205	23	+	−	−	−	−	−	−	+	−
A*0206	81	+	−	−	+	+	−	+	+	+
A*0209	5	+	+	+	+	+	+	+	+	+
A*0211	4	+	+	+	+	−	−	−	−	−
A*0214	8	+	−	−	+	+	−	+	+	−
A*0207	19	−	−	−	+	+	−	+	+	+
A*0210	3	−	−	−	+	+	−	+	+	+
A*0217	1	−	−	−	+	−	−	−	−	−
A*6901	3	−	−	−	−	−	−	+	+	−
A*2603	2	−	−	−	−	−	−	−	+	−
A*6601	10	−	−	−	−	−	−	−	+	−
A*6802	40	−	−	−	−	−	−	−	+	−

Each column represents a position in the peptide. The rows are given with the one-letter code for the amino acids. The values represent the frequencies of those amino acids at those positions, based on the peptides that are available for each module. The lower portion of the table shows which alleles contributed to the scores above and the number of peptides (Num.) used. The "+" symbol indicates that this allele shares a module with A*0201 at the given peptide position.

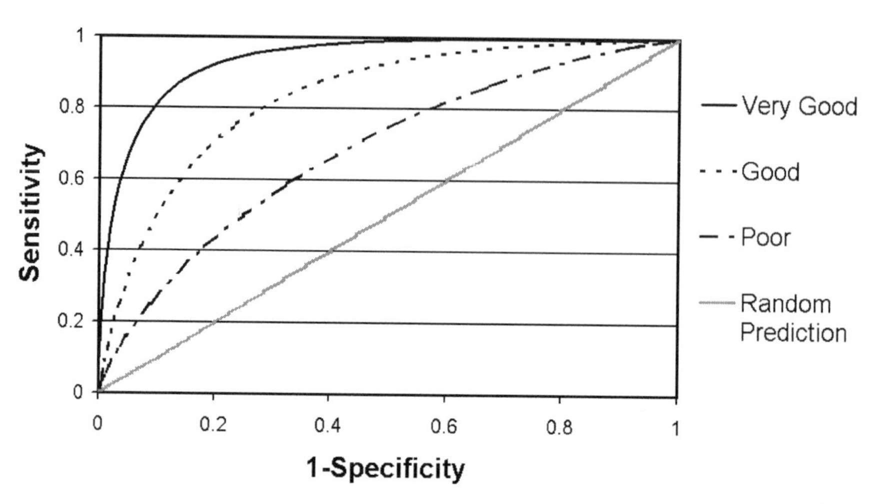

Fig. 1. Receiver operating characteristic (ROC) curves. The ROC curve is a function of specificity as well as sensitivity. The area under the ROC curve (A_{ROC}) is the standard measure of accuracy for major histocompatibility complex (MHC)–peptide-binding prediction. Random prediction refers to the expected results when randomly guessing whether the peptide is a binder or nonbinder.

delivers better result, especially when few peptides are available, but is more computationally intensive is the "jackknife" technique. Before performing the prediction for a given peptide, the peptide and all peptides with only one amino acid difference are removed from the training data, and the matrices were calculated without these peptides.

A goal of the modular concept is to make prediction possible for alleles, for which no peptide data are available. To test the modular concept, a "no-self" evaluation is necessary. In this implementation, the values in the modular matrix were generated and tested for a given allele, without using peptide-binding data for that allele. For example, predictions were made for A*0201 using binding data from other alleles (A*0202–0206, 0209, 0211, 0214, 0207, 2603, 6601, 6802, and 6901) but excluding peptides proven to bind A*0201.

3. Application

The module-based HLA–peptide-binding prediction is available as part of the PeptideCheck website (http://www.peptidecheck.org).

3.1. Predicting HLA–peptide Binding

In the simplest case, the user can enter a peptide sequence and choose an HLA allele. The result is a score representing the probability that the given peptide is bound by the given allele. Alternatively, the user may enter a protein sequence, and all possible resulting peptides are scored. Conveniently, more than one HLA allele can be chosen at a time.

The prediction algorithm generates a score. To determine whether this score is indicative of binding or nonbinding, it must be compared to a threshold. Choosing a threshold is dependent on experimental context. For example, if the user is intent on finding peptides that will have the highest chance of binding in the laboratory, a very high threshold is recommended. If the question is whether a peptide is or is not a minor histocompatibility antigen (peptide derived from a variant region of a non-HLA protein) then a balanced threshold is necessary. The threshold suggested in the PeptideCheck website is the point at which the sensitivity and specificity curves cross. Unfortunately, it is not possible to suggest thresholds for all predictable alleles. One can only generate sensitivity and specificity curves when peptide-binding data are available. However, modular peptide-binding prediction allows for prediction when no data are available (*see* **Note 2**). In this case, no threshold can be suggested, and it is recommended that the user compares scores to find peptides that represent the most likely binders.

3.2. Predicting Peptide Presentation Profile/Individual's Peptide-binding characteristics

In the area of leukemia-specific T-cell therapy, it is important to compare the peptide-binding profile of the patient. Peptide-binding profiles can be created by entering the patient's HLA genotype. In the case of a full heterozygosity, this includes two alleles from each of the HLA-A, HLA-B, and HLA-C loci. The user can either provide a peptide, one or more protein sequences, or a single-nucleotide polymorphism (SNP) profile for analysis. The resulting table displays the best binders, the proteins that they stem from, the binding score, and to which alleles they bind.

3.3. Exploring Modular Relations Between HLA Alleles

To understand the relations between various HLA alleles, it can be useful to compare them at the modular level. This is particularly useful when choosing which HLA alleles to study when determining peptide-binding motifs. After selecting an allele, the user is presented with the list of modules that this allele possesses. Clicking on a module brings up the list of alleles that possess this module. If binding motifs are available, they are also displayed. In this way, the user can choose an allele and find information about its binding motif based on the binding data for other alleles. Conversely, the user may determine which other alleles would benefit from the binding data of the target allele, if its peptides were to be purified and sequenced. In this way, researchers can choose those alleles for study, which are the most informative on a modular level. Prioritizing alleles in this way will ensure that peptide-binding data be found most efficiently to maximize modular peptide prediction.

Notes

1. Although the modular concept of HLA has been shown to be successful in expanding the number of predictable HLA alleles, the implementation described here has several drawbacks. The matrix scores are based on the assumption that there is a correlation between the rate of occurrence of particular amino acids at particular positions in the peptides and the importance of those amino acids in peptide binding. Although this may be true for pool sequences, many of the peptides in the peptide databases are of synthetic origin. The synthetic peptides are based on known binders but contain specific amino acid substitutions, with the goal of uncovering the roles of certain positions in the peptide. These synthetic peptides invalidate the assumption mentioned above. Drawing a correlation between peptide sequences and binding affinity is certainly a solution to this problem.
2. The modular concept will be expanded in the future to make prediction possible for more alleles, through the clustering of modules. There are module sequences that differ only slightly from each other, and which bind the same amino acids, despite small differences. Such modules will be clustered together in future implementations to maximize the usability of the provided peptide-binding data. Module-based supertypes are also an interesting consequence of such an analysis.

References

1. Marsh, S. G., Parham, P. & Barber, L. D. (2000). *The HLA FactsBook.* Academic Press, London.
2. Robinson, J., Waller, M. J., Parham, P., de Groot, N., Bontrop, R., Kennedy, L. J., Stoehr, P. & Marsh, S. G. (2003). IMGT/HLA and IMGT/MHC: sequence databases for the study of the major histocompatibility complex. *Nucleic Acids Res* 31, 311–4.
3. Kotsch, K. & Blasczyk, R. (2000). The noncoding regions of HLA-DRB uncover interlineage recombinations as a mechanism of HLA diversification. *J Immunol* 165, 5664–70.

4. Bade-Doeding, C., Eiz-Vesper, B., Figueiredo, C., Seltsam, A., Elsner, H. A. & Blasczyk, R. (2005). Peptide-binding motif of HLA-A*6603. *Immunogenetics* 56, 769–72.
5. DeLuca, D. S., Khattab, B. & Blasczyk, R. (2007). A modular concept of HLA for comprehensive peptide binding prediction. *Immunogenetics* 59, 25–35.
6. Saper, M. A., Bjorkman, P. J. & Wiley, D. C. (1991). Refined structure of the human histocompatibility antigen HLA-A2 at 2.6 A resolution. *J Mol Biol* 219, 277–319.
7. Chelvanayagam, G. (1996). A roadmap for HLA-A, HLA-B, and HLA-C peptide binding specificities. *Immunogenetics* 45, 15–26.
8. Reche, P. A. & Reinherz, E. L. (2003). Sequence variability analysis of human class I and class II MHC molecules: functional and structural correlates of amino acid polymorphisms. *J Mol Biol* 331, 623–41.
9. Rammensee, H., Bachmann, J., Emmerich, N. P., Bachor, O. A. & Stevanovic, S. (1999). SYFPEITHI: database for MHC ligands and peptide motifs. *Immunogenetics* 50, 213–9.
10. Bhasin, M., Singh, H. & Raghava, G. P. (2003). MHCBN: a comprehensive database of MHC binding and non-binding peptides. *Bioinformatics* 19, 665–6.
11. Blythe, I. A. D., & Flower, D. R. (2001). JenPep: a database of quantitative functional peptide data for immunology. *Bioinformatics* 18, 434–9.
12. Yewdell, J. W. & Bennink, J. R. (1999). Immunodominance in major histocompatibility complex class I-restricted T lymphocyte responses. *Annu Rev Immunol* 17, 51–88.
13. Reche, P. A., Glutting, J. P., Zhang, H. & Reinherz, E. L. (2004). Enhancement to the RANKPEP resource for the prediction of peptide binding to MHC molecules using profiles. *Immunogenetics* 56, 405–19.
14. Donnes, P. & Elofsson, A. (2002). Prediction of MHC class I binding peptides, using SVMHC. *BMC Bioinformatics* 3, 25.

19

Support Vector Machine-Based Prediction of MHC-Binding Peptides

Pierre Dönnes

Summary

The use of major histocompatibility complex (MHC) class I binding peptides for immunotherapeutic purposes has shown promising results in recent years. The identification of such peptides mostly starts with predicting MHC-binding peptides, given a protein of interest. An accurate prediction method can reduce the number of peptides that needs to be tested experimentally. This protocol describes in this describes how support vector machines (SVMs) can be used for predicting MHC class I binding peptides. Focus is given on data representation, the concept of cross-validation, and how optimal SVM-specific parameters are obtained.

Key Words: Support vector machines (SVMs); MHC binding; immunotherapy

1. Introduction

Major histocompatibility complex (MHC)–peptide binding is a prerequisite for T-cell activation in the immune system. In recent years, MHC-binding peptides have shown promising results for immunotherapeutic purposes and in vaccine development. In silico identification of MHC-binding peptides can reduce the number of peptides that need to be tested experimentally, and many different approaches have been proposed. These include sequence-based methods such as position-specific scoring matrices (PSSMs) *(1,2)* and

Address for correspondence: Bioinformatics, F. Hoffmann-La Roche Ltd., CH-4070 Basel, Switzerland, Email: pierre.doennes@roche.com

From: *Methods in Molecular Biology, vol. 409: Immunoinformatics: Predicting Immunogenicity In Silico*
Edited by: D. R. Flower © Humana Press Inc., Totowa, NJ

neural networks *(3)*. Several structure-based methods such as threading *(4)* and molecular dynamics approaches *(5)* have also been presented. Furthermore, support vector machines (SVMs) have been applied for this prediction task and have shown higher accuracy compared to both the SYFPEITHI *(1)* and BIMAS *(2)* methods for many MHC alleles *(6)*. PSSM methods typically assign a score for each amino acid in every position of the peptide, hence assuming an independent contribution of each amino acid to the overall binding energy, whereas SVMs can model the data in a "nonlinear" fashion.

SVMs is a supervised machine learning method, able to learn the input/output functionality of a given problem. Machine learning methods in general can be seen as a descendant from statistical learning, and the term "learning" is used because early methods were inspired by the learning process of the brain. The field of molecular biology has been described as tailor-made for machine learning approaches *(7)*, where a vast amount of data are available, but the underlying theory is not fully understood. In this case, the input is a peptide and the output is one of the classes, MHC binding or non-MHC binding. SVMs use linear functions to separate data points, and the nonlinearity is given by a "kernel" mapping of the input data. Detailed theory of SVMs can be found in several comprehensive textbooks *(8,9)*. The aim of this protocol is merely to describe how SVMs can be applied for predicting MHC class I binding peptides.

The starting point of this protocol is a set of peptides known to bind a certain MHC allele. Furthermore, a set of nonbinders is needed in order to train the SVMs to discriminate between binders and nonbinders. The peptide information then has to be represented in a format that can be used by the SVM software. Some preprocessing of the data, such as removing sequences containing unknown amino acids or duplicate entries, is also carried out. The next step is to generate and test SVM models using different parameter settings. Because no real theory exists for choosing the kernel and related parameters, a systematic search of the parameter space is carried out. To find the best parameter setting, a measure of prediction performance is needed. Here we will use fivefold cross-validation together with Matthews correlation coefficient (MCC) *(10)* to obtain a measure of the prediction accuracy. The optimal SVM model finally obtained can be used for predicting new peptides likely to bind the MHC molecule of interest.

2. Materials

The material needed to develop the prediction method is a data set of MHC-binding and nonbinding peptides. Furthermore, an implementation of an SVM learning algorithm is needed. Basic knowledge in a scripting language (e.g., Perl, php, or python) and a functioning UNIX/Linux environment is also required.

2.1. Peptide Data

2.1.1. MHC-binding Peptides

Major histocompatibility complex-binding peptides can be extracted from existing databases, such as SYFPEITHI *(1)* and MHCPEP *(11)*, or obtained from own experiments. On what allele resolution the data is taken might vary. In some cases all HLA-A*02 binding peptides might be used, whereas in other cases only verified HLA-A*0204 peptides are used. MHC class I binding peptides typically have a length between eight and ten amino acids. Here the sequence itself is used for SVM training, meaning that different models should be generated for different peptide lengths. Here we will assume that a data set of 100 9-mer peptides is used.

2.1.2. Nonbinding Peptides

Because the aim is to make a classification between MHC-binding and nonbinding peptides, a set of nonbinding peptides is also needed. Most public databases do not contain such information, and one way to obtain such a set is to extract random peptides of the desired length from existing protein databases. Here the SWISSPROT database *(12)* is used for extracting a set of nonbinding peptides. Extracting peptides randomly from a protein database induces a risk of allowing some MHC-binding peptides into the nonbinding data set. However, in most cases, the probability of doing so is relatively small, because very few peptides of a given protein usually bind a certain MHC allele *(13)*.

2.2. SVM Implementation

The SVM implementation used here is SVM^{LIGHT} *(14)*, which can be downloaded from http://svmlight.joachims.org/ (also *see* **Note 1**). SVM^{LIGHT} is implemented in the C programming language, and it has been used for many different bioinformatics classification task. The SVM^{LIGHT} software can be downloaded and installed by the following steps:

1. Download the svm_light.tar.gz file from http://svmlight.joachims.org/
2. Create a new directory: `$mkdir svm_light`
3. Unpack everything: `$gunzip -c svm_light.tar.gz | tar xvf -`
4. `$make`

This will generate two executable files `svm_learn` and `svm_classify`. The `svm_learn` module is used to learn the input/output functionality, given a labeled set of training data. The `svm_classify` module can then use the classification model generated by `svm_learn` for prediction.

3. Methods

In this section, all the important steps for generating an SVM-based prediction method for MHC class I binding peptides are outlined. Focus is put on important concepts, and examples are given for data representation and cross-validation. No example code is given for processing files containing peptide sequence into files in the SVM software format, and so on. This is a trivial task assuming some basic knowledge in any scripting language. It is also suggested that functions using the SVM software by system commands are implemented in the scripting language of choice. By doing this, small scripts can be used for data processing, SVM training, and performance evaluation.

3.1. Peptide Data

The peptide data needs some preprocessing before it can be represented in the data format used for SVM training. The data set of binders should be processed in order to remove duplicate entries and sequences containing unknown amino acids (usually indicated by the letter "X"). Some extra care should also be taken with sequences obtained from alanine scan experiments, and so on (*see* **Note 2**).

A data set of nonbinding peptides can be constructed from proteins in the SWISSPROT database. A local copy of the SWISSPROT database can be obtained from ftp://ftp.ebi.ac.uk/pub/databases/swissprot/. The following steps can be carried out in order to extract a data set of nonbinders.

1. Randomly extract 10,000 proteins from the database.
2. Chop these protein sequences into peptides of the desired length (same as the MHC-binding peptides data set) and store them as keys in a hash map (will remove duplicate entries). Randomly pick the number of peptides desired, for example, the same number as in the binding data set in order to receive a balanced data set. Peptides also found in the data set of binders should not be included in the nonbinder data set.

*3.2. Using SVM^LIGHT^*

3.2.1. Using `svm_learn`

The `svm_learn` module is for reading an input file and to generate a model for prediction. It is used in the following way:

```
$svm_learn [options] training_data model_file
```

The `training_data` file contains the labeled examples for which a functional mapping should be found, and the format of this file is described below. Here, the use of the radial basis function (RBF) kernel is described, which means that the [options] will look like:

```
-t 2 -c C -g G
```

where -t 2 defines that the RBF kernel should be used (also *see* **Note 3**), C defines the trade-off between error and margin, and G defines the kernel-specific parameter gamma. How to find the optimal C and G is described below. More information about the [options] available for svm_learn can be found in the manual or by:

```
$svm_learn -?
```

The SVM model generated by svm_learn is saved in the model_file and can be used by svm_classify for prediction.

3.2.2. Using Svm_classify

The svm_classify module is used in the following way:

```
$svm_classify [options] example_file model_file
output_file
```

where the example_file contains the data to be classified, the model_file is the prediction model generated by svm_learn, and the output_file is where the prediction results will be written. Once again more information about the [options] parameters can be found in the manual or by:

```
$svm_classify -h
```

3.3. Data Representation

The peptide data need to be processed into the format used by the SVM software. SVM^{LIGHT} reads input data in the form:

```
Example 1:class feature1:value1 feature2:value2....
Example 2:class feature1:value1 feature2:value2....
```

This means that each peptide is represented by a separate row in the training_data and example_file files. The class is represented by +1 for binders and −1 for nonbinders in this case. The features and related values depend on the type of data representation used. Here, binary sparse, encoding is used to represent the peptides. Each amino acid is represented as a bit vector of 20 elements, where "1" indicates the type of amino acid, see Fig. 1 (alternative data representations are described in **Note 4**). The total length of a vector representing a nine amino acid–long peptide is $9 \times 20 = 180$ positions. Figure 2 exemplifies how the dipeptides "AD" and "YC" are represented in sparse binary format. Furthermore, Fig. 2 shows how the input files for both `svm_learn` and `svm_classify` can be written in a more compact way, because all features that are not given an explicit value are considered to be "0" (which is the case for most positions of the 180 element–long input vector).

3.4. SVM Training and Evaluation

3.4.1. Creating Data Sets for Cross-validation

Most machine learning methods run the risk of overfitting the prediction model, that is, the model perfectly reproduces the training data but lacks any form of generalization ability on novel data. In order to obtain a fair estimate of the prediction performance, fivefold cross-validation is applied. Here, both the data sets of binders and nonbinders are split into five subsets. Four subsets of binders and nonbinders, respectively, are then combined into a training data set,

A	10000000000000000000
C	01000000000000000000
D	00100000000000000000
E	00010000000000000000
F	00001000000000000000
G	00000100000000000000
H	00000010000000000000
I	00000001000000000000
K	00000000100000000000
L	00000000010000000000
M	00000000001000000000
N	00000000000100000000
P	00000000000010000000
Q	00000000000001000000
R	00000000000000100000
S	00000000000000010000
T	00000000000000001000
V	00000000000000000100
W	00000000000000000010
Y	00000000000000000001

Fig. 1. In sparse binary representation, each amino acid is represented as a vector of 20 elements. The vector contains 19 "0" positions and a "1" indicating the type of amino acid.

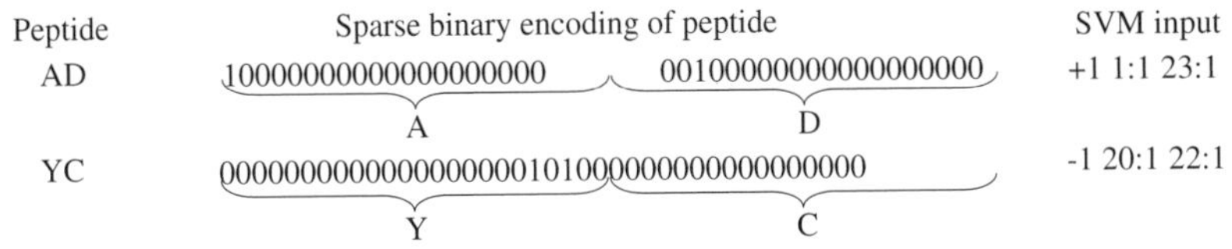

Peptide	Sparse binary encoding of peptide	SVM input
AD	10000000000000000000 00100000000000000000	+1 1:1 23:1
	A D	
YC	0000000000000000000101000000000000000000	-1 20:1 22:1
	Y C	

Fig. 2. An example of how the two dipeptides "AD" and "YC" are encoded using binary sparse encoding. Furthermore, the format of these in the SVM input file is shown, assuming "AD" is a binder (+1) and "YC" as a nonbinder (−1). Here the features having a "0" value can be left out, giving a more compact representation.

and the two left out are used for testing (also *see* **Note 5**). Assuming a data set of 100 binders and a data set of 100 nonbinders, each training set will contain 160 sequences (80 binders and 80 nonbinders), and each test set will contain 40 sequences (20 binders and 20 nonbinders).

3.4.2. SVM Parameter Optimization

Because the optimal parameters for each classification task are not known from the start, it is necessary to test different parameters in order to find the optimal ones. This is best done by a systematic sampling of the parameter space by a grid search. Here, all combinations of two parameters are tested, given the start, stop, and step size for each of the parameters.

In the case of an RBF kernel, the two parameters c and g can be optimized. Hsu et al. ***(15)*** suggest that a good strategy is to try exponentially growing parameters. In the RBF case, this could mean $c = 2^{-5}, 2^{-3}, \ldots 2^{15}$ and $g = 2^{-15}, 2^{-13}, \ldots 2^{3}$. For a given parameter setting, five SVM models are generated and used to predict the test data sets. By comparing the known labels with the predicted values, the prediction accuracy can be calculated (positive prediction scores mean that the predicted class is +1 and negative scores mean predicted −1 class). Four variables are defined and used for this purpose: true positives (TP)—the number of binders predicted as such, true negatives (TN)—the number of nonbinders predicted as such, false positives (FP)—the number of predicted binders that actually are nonbinders, and false negatives (FN)—the number of predicted nonbinders that actually are binders. From these values, the MCC can be defined as:

$$\mathrm{MCC} = \frac{(\mathrm{TP} \times \mathrm{TN}) - (\mathrm{FP} \times \mathrm{FN})}{\sqrt{(\mathrm{TN}+\mathrm{FN})\,(\mathrm{TN}+\mathrm{FP})\,(\mathrm{TP}+\mathrm{FN})\,(\mathrm{TP}+\mathrm{FP})}}$$

A perfect correlation between predicted and real values would give an MCC of 1, random predictions an MCC of 0, and anti-correlated predictions a value of −1. Furthermore, the specificity (SP) and sensitivity (SE) of the prediction can be defined as:

$$\mathrm{SP} = \frac{\mathrm{TN}}{\mathrm{TN}+\mathrm{FP}}$$

$$\mathrm{SE} = \frac{\mathrm{TP}}{\mathrm{TP}+\mathrm{FN}}$$

A general procedure for the whole SVM optimization is given below, assuming that the data have been split into training and test files as described above.

1. Choose initial kernel parameters `c` and `g`.
2. Train five different SVM models, using the parameters chosen in order to obtain five prediction models.
3. Use the test files and their respective model files for prediction. The results are saved in five result files.
4. Compare the five result files with the labels in the respective test files. Sum up the total number of TP, FP, TN, and FN. Calculate the performance measures MCC, SP, and SE. Save the performance measures and parameters to a file.
5. Update the kernel parameters and go back to Step 2. Alternatively if all parameter combinations have been tested, move on to Step 6.
6. Search the files containing the parameter setting and parameters for the best MCC.
7. Use the optimal parameter setting and all data to train a SVM model that subsequently can be used for prediction.

3.5. Predicting New Data

The optimized model generated above can now be used for prediction of new data. Typically this involves the identification of candidate binders from a given protein. The following steps can be carried out for such a prediction:

1. Use a sliding window to generate all peptides from a given query protein.
2. Write these peptides into an `example_file` that can be used by the `svm_classify` module. (It does not matter if the peptides are labeled +1 or −1)
3. Use `svm_learn` and the optimal SVM model for prediction.
4. Associate all peptides with their corresponding prediction score and sort everything according to the scores. The most likely binders have the highest scores.

Notes

1. A good overview of available SVM implementations can be found at http://www.kernel-machines.org/. A number of tutorials and related information about kernel machines can also be found here.

2. Some databases contain many synthetic peptides with a high degree of similarity. This might lead to a bias of the data, and sometimes it is useful to remove too similar sequences from the training data. For example, the maximal sequence identity between two peptides in the training data set might be six out of nine positions.
3. SVM^{LIGHT} also allows easy usage of the *linear*, *polynomial*, and *tanh* functions. These also have kernel-specific parameters that have to be optimized, and grid search strategies can be applied here as well. More information about these kernels are found in the SVM^{LIGHT} manual.
4. An alternative way to represent the data is to use the amino acid properties found in the AAIndex database ***(16)***. These include, for example, hydrophobicity, size, and charge. A 9-mer peptide where each amino acid is represented by two features will then have a total of 18 features.
5. Cross-validation can generally be conducted at an N-fold level. If very little data are available leave-one-out cross-validation can be used, where one example at the time is used as test data and all other data are used for training.

References

1. Rammensee, H.-G., Bachman, J., Philipp, N., Emmerich, N., Bachor, O. A., and Stevanovic, S. (1997) SYFPEITHI: a database for MHC ligands and peptide motifs. *Immunogenetics* **50**, 213–219.
2. Parker, K. C., Bednarek, M. A., and Coligan, J. E. (1994) Scheme for ranking potential HLA-A2 binding peptides based on independent binding of individual peptide side-chains. *J. Immunol.* **152**, 163–175.
3. Gulukota, K., Sidney, J., Sette, A., and DeLisi, C. (1997) Two complementary methods for predicting peptides binding major histocompatibility complex molecules. *J. Mol. Biol.* **267**, 1258–1267.
4. Altuvia, Y., Schueler, O., and Margalit, H. (1995) Ranking potential binding peptides to MHC molecules by a computational threading approach. *J. Mol. Biol.* **249**, 244–250.
5. Rognan, D., Lauemoller, S., Holm, A., Buus, S., and Tschinke, V. (1999) Predicting binding affinities of protein ligands from three-dimensional models: application to peptide binding to class I major histocompatibility proteins. *J. Med. Chem.* **42**, 4650–4658.
6. Dönnes, P. and Elofsson, A. (2002) Prediction of MHC class I binding peptides, using SVMHC. *BMC Bioinformatics* **3**, 25.
7. Shavlik, J., Hunter, L., and Searls, D. (1995), Introduction. *Mach. Learn.* **21**, 5–9
8. Vapnik, V. N. (1999) *The Nature of Statistical Learning Theory*, Wiley, New York, USA.
9. Cristianini, N. and Shawe-Taylor, J. (2000) *An Introduction to Support Vector Machines and Other Kernel-Based Learning Methods*, Cambridge University Press, The Edinburgh Building, Cambridge, UK.

10. Matthews, B. W. (1975) Comparison of predicted and observed secondary structure of T4 phage lysozyme. *Biochim. Biophys. Acta* **405**, 442–451.
11. Brusic, V., Rudy, G., and Harrsison, L. C. (1998) MHCPEP, a database of MHC-binding peptides: update 1997. *Nucleic Acids Res.* **26**, 368–171.
12. Boeckmann, B., Bairoch, A., Apweiler, R., Blatter, M.-C., Estreicher, A., Gasteiger, E., Martin, M. J., Michoud, K., O'Donovan, C., Phan, I., Pilbout, S., and Schneider, M. (2003) The SWISS-PROT protein knowledgebase and its supplement TrEMBL in 2003. *Nucleic Acid Res.* **31**, 365–370
13. Mamitsuka, H. (1998) Predicting peptides that bind to MHC molecules using supervised learning of hidden Markov models. *Proteins* **33**, 460–474.
14. Joachims, T. (1998) Making large-scale SVM learning practical, In *Advances in Kernel Methods - Support Vector Learning* (Schölkopf, B., Burges, C., and Smola, A. eds.), MIT Press, Cambridge, MA.
15. Hsu, C.-H., Chang, C.-C., and Lin, C.-J. (2003) A Practical Guide to Support Vector Classification, www.csie.ntu.edu.tw/~cjlin/papers/ guide/guide.pdf
16. Kawashima, S., Ogata, H., and Kanehisa, M. (1999) AAindex: Amino Acid Index Database. *Nucleic Acids Res.* **27**, 368–369
17. Baldi, P., Brunak, S., Chauvin, Y., Andersen, C. A., and Nielsen, H. (2000) Assessing the accuracy of prediction algorithms for classification: an overview. *Bioinformatics* **16**, 412–424.

20

In Silico Prediction of Peptide–MHC Binding Affinity Using SVRMHC

Wen Liu, Ji Wan, Xiangshan Meng, Darren R. Flower, and Tongbin Li

Summary

The binding between peptide epitopes and major histocompatibility complex (MHC) proteins is a major event in the cellular immune response. Accurate prediction of the binding between short peptides and class I or class II MHC molecules is an important task in immunoinformatics. SVRMHC which is a novel method to model peptide–MHC binding affinities based on support vector machine regression (SVR) is described in this chapter. SVRMHC is among a small handful of quantitative modeling methods that make predictions about precise binding affinities between a peptide and an MHC molecule. As a kernel-based learning method, SVRMHC has rendered models with demonstrated appealing performance in the practice of modeling peptide–MHC binding.

Key Words: SVR; SVRMHC; epitope binding; modeling

1. Introduction

Major histocompatibility complex (MHC) molecules are polymorphic glycoproteins found on cell membranes. They are capable of binding small peptide fragments derived from pathogen proteins, forming MHC–antigenic peptide complexes. These complexes are then recognized by the T-cell receptors (TCRs) on the T-cell surface, inducing cellular immune responses. There are two major types of MHCs. Class I MHCs are expressed by nearly all nucleated cells in vertebrates, and they are recognized by CD8-expressing T-cytotoxic (T_c) cells. The peptides that bind class I MHCs are often cleavage products from intracellular proteins, and their lengths are usually short

From: *Methods in Molecular Biology, vol. 409: Immunoinformatics: Predicting Immunogenicity In Silico*
Edited by: D. R. Flower © Humana Press Inc., Totowa, NJ

(8–11 amino acids). Class II MHC molecules are expressed only by antigen-presenting cells, and they are recognized by CD4-expressing T-helper (T_h) cells. The peptides recognized by class II MHCs are antigen peptides that are longer and more variable in lengths (between 10–20 amino acids). Because of the central role played by MHCs in the cellular immune responses, in silico prediction of peptide–MHC binding has remained a critical task in immunoinformatics. A wide variety of methodologies have been applied in this field, including motif searching ***(1)***, position-specific scoring matrices (PSSMs) ***(2–5)***, artificial neural networks (ANNs) ***(6–8)***, hidden Markov models (HMMs) ***(9)***, support vector machine (SVM) classification ***(10,11)***, three-dimensional quantitative structure-activity relationship (3D QSAR) ***(12–14)***, and partial least square (PLS)-based modeling ***(15–18)***. Here, we describe a recently developed support vector machine regression (SVR)-based method, named SVRMHC. As a kernel-based learning method, SVRMHC exhibits pleasant predicting performance enjoyed by other SVM methods such as SVMHC ***(10)*** and HLA-DR4Pred ***(11)***. Meanwhile, as a regression or quantitative modeling method, SVRMHC renders models capable of providing precise information about peptide–MHC binding, i.e., binding affinities, a feature shared by only a handful of recent methods such as 3D QSAR ***(12–14)*** and the "additive method" ***(15–18)***.

2. SVRMHC Modeling Overview

SVRMHC was developed with SVMs, a class of supervised learning methods based on the principle of structural risk minimization, rooted in the *Statistical Learning Theory* by Vapnik ***(19)***. A distinguished characteristic of SVMs is the use of nonlinear "kernels" to implicitly map the input space into a very high dimensional feature space, where an optimal separating hyperplane is constructed (in the case of a classification task), or linear regression is conducted with an ε-insensitive loss function (in the case of a regression task, or SVR). A kernel is a function in the form of $K(x, y)$ that satisfies Mercer's condition, i.e.,

$$\iint K(x, y)g(x)g(y)dxdy \geq 0,$$

so that the mapping between the input space and the feature space can be done by a dot product operation. Commonly used kernel functions include the linear kernel, polynomial kernel, and the radial basis function (RBF) kernel.

Kernel name	Kernel function
Linear kernel	$K(x, y) = x \bullet y$
Polynomial kernel	$K(x, y) = (x \bullet y + 1)^d$
RBF kernel	$K(x, y) = e^{-\gamma \|x-y\|^2}$

More information about SVM and SVR can be found in ***(19–21)***.

In SVRMHC, the input peptide sequences can be represented by either the commonly used "sparse encoding" method ***(10,22)*** or by a "11-factor encoding" scheme, which were constructed to include several important general physicochemical properties (polarity, isoelectric point, and accessible surface area) and a number of properties that were identified in 3D QSAR analysis ***(23)*** as key determinants of peptide–MHC interaction (volume, number of hydrogen-bond donors, and hydrophobicity) ***(24)***. Our experience is that the 11-factor encoding scheme often renders more accurate models than sparse encoding. The binding affinity of a peptide is represented in the form of pIC_{50} (the negative logarithm of IC_{50}) or pBL_{50} (the negative logarithm of the half-maximal binding level BL_{50}). A SVRMHC model is constructed by training, with a set of (sequence:pIC_{50}) or (sequence:pBL_{50}) pairs. After the model is constructed, it can make prediction of the binding affinities of untested peptide sequences.

The source code of SVRMHC and executables for the Linux and DOS operating systems can be downloaded at http://svrmhc.umn.edu/SVRMHC/download/.

3. Execution

3.1. Training a Model for a Class I MHC Molecule

The following command will construct a SVRMHC model for a class I MHC molecule with the default setting:

```
svrmhc -train -i input_file -m model_file
```

The `input_file` should be a multiline text file, each line consisting of the peptide sequence and the pIC_{50} of the peptide separated by the tab character "`\t`". The peptide sequences should all be of the same length (8, 9, or 10). The output of this command `model_file` stores the information of the constructed model.

In the default setting, the 11-factor encoding scheme is applied; a five-fold cross-validation scheme is used, that is, the data set is randomly split into five subsets, model parameters are optimized by assessing the predicting performance on each of the five subsets in turn; the RBF kernel is used; and an iterative outlier removal procedure is taken, where a residual threshold of 2 is used for outlier determination. In other words, after a model is constructed, prediction is made on each peptide in the training data set. If at least one of the peptides exhibits a higher residual than the default residual threshold 2, then the peptide with the highest residual is excluded as an outlier, and the model is re-trained. This process is repeated until no more outliers can be identified. A warning message will be generated if an excessive number of peptides (>5% of the training data set) are determined as outliers by this method. In this case, the user should adjust the residual threshold or change the setting for model construction, for example, choose another encoding method or another kernel function.

To change the encoding method, the option `-e` can be used: `-e sparse` denotes the using of the sparse encoding scheme. The option `-k` can be used to change the kernel function, for example, `-k polynomial` denotes the using of the polynomial kernel, and `-k linear` denotes the using of the linear kernel. The residual threshold for outlier determination can be changed by the `-o` option, for example, `-o 2.5` will adjust the residual threshold to 2.5. Or else, the outlier removal can be turned off all together, by the option `-r no_remove`.

The way by which the cross-validation is done can be adjusted by the `-v` option, for example, `-v 7` denotes the using sevenfold cross-validation. Leave-one-out (LOO) cross-validation can be specified by `-v LOO`.

The following example shows how to construct a SVRMHC model with sparse encoding scheme, polynomial kernel function, LOO cross-validation, and an outlier removal procedure with 2.5 as the residual threshold:

```
svrmhc -train -e sparse -k polynomial -v LOO -o 2.5 -i
        input_file -m model_file
```

The achievement of an accurate SVRMHC model requires proper choosing of kernel parameters, namely, γ for the RBF kernel, n for the polynomial kernel, and kernel-independent parameters, namely, ε and C [see, for example, ***(25)*** for explanations about the meaning of these parameters]. By default, an automatic parameter optimization procedure is invoked, in which a grid search is performed in default parameter ranges with equal step size on the logarithm scale. The default ranges for ε and C are between the 1/10 of the recommended values and ten times the recommended values; the recommended values are calculated according to ***(26)***. The default range for γ for the RBF kernel is

[0.001, 1]. For these three parameters, ε, C, and γ, the default number of steps used in the grid search is 5. For the kernel parameter n of the polynomial kernel, three possible values 2, 3, and 4 are searched by default. Our experience is that the automatic parameter selection procedure invoked in the default setting suffices to find the best model in most situations. However, the user can choose to manually adjust the search ranges and the number of steps in grid search. The program documentation provides more details about these options.

When the training of the model initializes, key information about the setting of the training (e.g., encoding method, the way cross-validation is done, kernel function, and residual threshold for outlier determination) is displayed on the screen. As the execution of the program continues, information about the progress of the execution is shown on the screen. At the completion of the program, a summary of the training is displayed that includes the RMS error, cross-validated correlation coefficient r, and cross-validated q^2 of the resulting model. The program documentation provides the definitions of these performance measuring metrics.

3.2. Making predictions with an established class I model

The following command will make predictions about pIC_{50}/pBL_{50} for a set of peptides with an established class I SVRMHC model:

```
svrmhc -predict -i input_file -m model_file -s result_file
```

The `input_file` needs to be a single line or multiline text file, each line containing a peptide sequence. The `result_file` will contain the predicted pIC_{50}/pBL_{50} values listed together with the peptide sequences.

3.3. Training a model for a class II MHC molecule

The extension of the SVRMHC methodology to the case of the class II MHC molecules is done largely in accordance with ***(16)***. The additional considerations about the constructing and using of class II models include (a) aligning the peptide sequences in the training data set: this is accomplished by applying the *iterative self-consistent algorithm* that performs a greedy search of the optimal alignment; (b) limiting the number of possible alignments of input sequences to a controllable size: this is done by specifying a key anchor position and allowing a limited number of residues to occur at this anchor position; and (c) positioning of the sequences in the test data set: this is carried out by providing three options for positioning—*mean*, *max*, and *combi*, as described in ***(16)***.

The following command will construct a SVRMHC model for a class II MHC molecule with default setting:

```
svrmhc -classII -train -a anchor_position -l anchor_residues
        -i input_file -m model_file
```

In this command, `-classII -train` indicates the training of a class II SVRMHC model. The `anchor_position` is an integer specifying the key anchor position for the purpose of limiting the number of possible alignments. The `anchor_residues` is a string specifying a list of possible amino acid residues allowed at the key anchor position. For popular MHC molecules, such information can be obtained from the SYFPEITHI database *(3)*. The `input_file` should conform to the same format requirement as for class I models. Each peptide sequence in this file should be of length ≥ 9. The `model_file` will store the information about the constructed model.

By default, only one iteration of the *iterative self-consistent algorithm* is executed in the searching of optimal alignment for the input sequences. This is because our experience shows that one iteration of the greedy search often produces a good model already. Sometimes, the search does not converge, and the model performance actually deteriorates with increasing number of iterations. The number of iterations can be changed manually with the option `-n`, e.g., `-n` 4 denotes that four iterations of the greedy search will be executed.

Other options, including those for sequence encoding scheme, kernel function, outlier removal procedure, cross-validation, and the selection of kernel parameters and kernel-independent parameters, can be manually adjusted similarly to the case of class I model construction.

3.4. Making predictions with an established class II model

The following command will make predictions for a set of peptides with a class II SVRMHC model:

```
svrmhc -classII -predict -I input_file -m model_file
        -s result_file
```

By default, *combi* is used for the positioning of the sequences in the testing set. This can be adjusted with the option `-p`: `-p mean` and `-p max` will set the positioning method to *mean* and *max*, respectively.

4. Discussion

Based on kernel learning machines, SVRMHC is a method to construct quantitative models about the binding between peptides and MHC molecules. The SVRMHC method not only demonstrated good performance in quantitative modeling, but it was also shown to outperform several prominent qualitative modeling methods in the ability to identify strong binding peptides *(24)*. We have constructed and tested SVRMHC models for more than 40 class I or class II MHC molecules. Most of these models exhibit fairly decent performance. How good a model can be constructed is determined by many factors. First and foremost is the input data quality. The AntiJen database *(27)* has been our major data source, and the data sets it provides are often of satisfactory quality. The number of peptides in the data set is also an important factor. The "rule of thumb" we apply requires the binding data of at least 30 peptides for constructing a model of a class I MHC molecule, and at least 50 peptides are required for constructing a model for a class II MHC molecule. Understandably, the encoding method and kernel function make discernable difference in the performance of the resulting models. The models constructed with the "11-factor encoding × RBF kernel" setting most often offers the best performance; however, for some MHC molecules, other settings ("11-factor encoding × polynomial kernel," "sparse encoding × RBF kernel," and "sparse encoding × polynomial kernel") lead to more accurate models. Therefore, it is sensible that all these settings be tried and the best model obtained be used. The way by which cross-validation is done in theory will not affect the performance of the model constructed. The adjustment of this option is often dictated by running time consideration—if the data set is not big, for example, containing $\leq$100 sequences, we often chose to use LOO cross-validation. Otherwise, five-fold or seven-fold cross-validation is used to avoid excessive execution time. The modeling for class II MHC molecules is, not surprisingly, more problematic than modeling class I molecules. Meticulous tuning of encoding scheme, kernel function, number of iterations, and residual threshold for outlier determination is often needed to obtain a satisfactory model.

References

1. Sette, A., Buus, S., Appella, E., Smith, J.A., Chesnut, R., Miles, C., Colon, S.M. and Grey, H.M. (1989) Prediction of major histocompatibility complex binding regions of protein antigens by sequence pattern analysis. *Proc Natl Acad Sci USA*, **86**, 3296–3300.

2. Nielsen, M., Lundegaard, C., Worning, P., Hvid, C.S., Lamberth, K., Buus, S., Brunak, S. and Lund, O. (2004) Improved prediction of MHC class I and class II epitopes using a novel Gibbs sampling approach. *Bioinformatics*, **20**, 1388–1397.
3. Rammensee, H., Bachmann, J., Emmerich, N.P., Bachor, O.A. and Stevanovic, S. (1999) SYFPEITHI: database for MHC ligands and peptide motifs. *Immunogenetics*, **50**, 213–219.
4. Parker, K.C., Bednarek, M.A. and Coligan, J.E. (1994) Scheme for ranking potential HLA-A2 binding peptides based on independent binding of individual peptide side-chains. *J Immunol*, **152**, 163–175.
5. Reche, P.A., Glutting, J.P. and Reinherz, E.L. (2002) Prediction of MHC class I binding peptides using profile motifs. *Hum Immunol*, **63**, 701–709.
6. Nielsen, M., Lundegaard, C., Worning, P., Lauemoller, S.L., Lamberth, K., Buus, S., Brunak, S. and Lund, O. (2003) Reliable prediction of T-cell epitopes using neural networks with novel sequence representations. *Protein Sci*, **12**, 1007–1017.
7. Brusic, V., Rudy, G., Honeyman, G., Hammer, J. and Harrison, L. (1998) Prediction of MHC class II-binding peptides using an evolutionary algorithm and artificial neural network. *Bioinformatics*, **14**, 121–130.
8. Honeyman, M.C., Brusic, V., Stone, N.L. and Harrison, L.C. (1998) Neural network-based prediction of candidate T-cell epitopes. *Nat Biotechnol*, **16**, 966–969.
9. Mamitsuka, H. (1998) Predicting peptides that bind to MHC molecules using supervised learning of hidden Markov models. *Proteins*, **33**, 460–474.
10. Donnes, P. and Elofsson, A. (2002) Prediction of MHC class I binding peptides, using SVMHC. *BMC Bioinformatics*, **3**, 25.
11. Bhasin, M. and Raghava, G.P. (2004) SVM based method for predicting HLA-DRB1*0401 binding peptides in an antigen sequence. *Bioinformatics*, **20**, 421–423.
12. Doytchinova, I.A. and Flower, D.R. (2002) Quantitative approaches to computational vaccinology. *Immunol Cell Biol*, **80**, 270–279.
13. Doytchinova, I.A. and Flower, D.R. (2002) A comparative molecular similarity index analysis (CoMSIA) study identifies an HLA-A2 binding supermotif. *J Comput Aided Mol Des*, **16**, 535–544.
14. Doytchinova, I.A. and Flower, D.R. (2001) Toward the quantitative prediction of T-cell epitopes: coMFA and coMSIA studies of peptides with affinity for the class I MHC molecule HLA-A*0201. *J Med Chem*, **44**, 3572–3581.
15. Hattotuwagama, C.K., Toseland, C.P., Guan, P., Taylor, D.L., Hemsley, S.L., Doytchinova, I.A. and Flower, D.R. (2005) Class II mouse major histocompatibility complex peptide binding affinity: in silico bioinformatic prediction using robust multivariate statistics. *J Chem Inf Mod*, **46**(3), 1491–502. (2006)
16. Doytchinova, I.A. and Flower, D.R. (2003) Towards the in silico identification of class II restricted T-cell epitopes: a partial least squares iterative self-consistent algorithm for affinity prediction. *Bioinformatics*, **19**, 2263–2270.

17. Doytchinova, I.A., Blythe, M.J. and Flower, D.R. (2002) Additive method for the prediction of protein-peptide binding affinity. Application to the MHC class I molecule HLA-A*0201. *J Proteome Res*, **1**, 263–272.
18. Hattotuwagama, C.K., Guan, P., Doytchinova, I.A. and Flower, D.R. (2004) New horizons in mouse immunoinformatics: reliable in silico prediction of mouse class I histocompatibility major complex peptide binding affinity. *Org Biomol Chem*, **2**, 3274–3283.
19. Vapnik, V. (1998) *Statistical Learning Theory*. John Wiley & Sons, New York.
20. Vapnik, V. (1995) *The Nature of Statistical Learning Theory*. Springer-Verlag, New York.
21. Cristianini, N. and Shawe-Taylor, J. (2000) *An Introduction to Support Vector Machines and Other Kernel-Based Learning Methods*. Cambridge University Press, Cambridge, UK.
22. Baldi, P. and Brunak, S. (2001) *Bioinformatics: The Machine Learning Approach*. The MIT Press, Cambridge, MA.
23. Doytchinova, I.A. and Flower, D.R. (2002) Physicochemical explanation of peptide binding to HLA-A*0201 major histocompatibility complex: a three-dimensional quantitative structure-activity relationship study. *Proteins*, **48**, 505–518.
24. Liu, W., Meng, X., Xu, Q., Flower, D.R. and Li, T. (2006) Quantitative prediction of mouse class I MHC peptide binding affinity using support vector machine regression (SVR) models. *BMC Bioinformatics*, **7**, 182.
25. Chang, C.C. and Lin, C.J. (2004) A practical guide to SVM classification, LibSVM documentation.
26. Cherkassky, V. and Ma, Y. (2004) Practical selection of SVM parameters and noise estimation for SVM regression. *Neural Netw*, **17**, 113–126.
27. Toseland, C.P., Clayton, D.J., McSparron, H., Hemsley, S.L., Blythe, M.J., Paine, K., Doytchinova, I.A., Guan, P., Hattotuwagama, C.K. and Flower, D.R. (2005) AntiJen: a quantitative immunology database integrating functional, thermodynamic, kinetic, biophysical, and cellular data. *Immunol Res*, **1**, 4.

21

HLA–Peptide Binding Prediction Using Structural and Modeling Principles

Pandjassarame Kangueane and Meena Kishore Sakharkar

Summary

Short peptides binding to specific human leukocyte antigen (HLA) alleles elicit immune response. These candidate peptides have potential utility in peptide vaccine design and development. The binding of peptides to allele-specific HLA molecule is estimated using competitive binding assay and biochemical binding constants. Application of this method for proteome-wide screening in parasites, viruses, and virulent bacterial strains is laborious and expensive. However, short listing of candidate peptides using prediction approaches have been realized lately. Prediction of peptide binding to HLA alleles using structural and modeling principles has gained momentum in recent years. Here, we discuss the current status of such prediction.

Key Words: HLA–peptide binding; modeling; dynamics simulation; threading; optimization; free energy; virtual matrix; virtual pockets; QSAR

1. Introduction

The human leukocyte antigen (HLA)–peptide project ***(1,2)*** was explored with great momentum during the last decade by defining anchor residues in peptide-binding motifs ***(3)*** and by the subsequent utilization of those principles in computational tools ***(4–8)***. The difficulty in collecting immunologic or biochemical binding data, after rigorous cloning, in vitro expression and purification of every HLA allele to perform binding assay with a combinatorial library of peptides, sets the limitation to define anchor residues in peptide motifs for a wide array of HLA alleles.

Simultaneous progress in computational structural biology ***(9)*** provided a structure-based methodology for HLA–peptide prediction ***(10,11)***. Assigning

From: *Methods in Molecular Biology, vol. 409: Immunoinformatics: Predicting Immunogenicity In Silico*
Edited by: D. R. Flower © Humana Press Inc., Totowa, NJ

quantitative prediction score for the derived three-dimensional (3D) models and the subsequent grouping *(12)* or ranking *(10,11,13)* of the modeled complexes is of particular interest. Ranking of the modeled HLA–peptide complexes using pair-wise potentials has been proved to preferentially select hydrophobic–hydrophobic interaction *(10)*. Recent attempts were made to exploit the available structural data on HLA–peptide binding to predict the binding affinities for HLA–peptide ligands *(14)*. However, the analysis was restricted to very few alleles, and future cross-validation requires more biophysical information on diversified MHC–peptide complexes to recalibrate interaction functions.

Addressing the fundamental question of what these complexes do and, most importantly, how they do it by taking into account the subtle polymorphic changes in the HLA binding groove is of interest. This will establish an understanding of the consequences of multiple interactions involved in the cascade of events during cell-mediated immune response. Both class I and class II HLA molecules possess a peptide-binding functional groove characterized by a similar structural fold *(15)*. Subsequently, models were developed using the definition of virtual binding pockets, mapping of virtual pockets (VPs) to position-specific peptide residue anchors, and estimation of peptide residue—virtual binding pocket compatibility *(16)*. VPs are defined using information gleaned from eight unique HLA alleles, and the mapping of VP to position-specific residue anchors is done using the 29 HLA–peptide structures analyzed in this study. Here, we discuss the progress made in HLA–peptide-binding prediction using molecular modeling principles.

2. Prediction Methods

2.1. Molecular Dynamics Simulation

Rognan et al. *(17)* showed the use of molecular dynamics simulation (MDS) to discriminate peptide binders from nonbinders to HLA molecules. The free-energy change was calculated in AMBER force field for six peptides with HLA-B*2705. Structural and dynamical properties of the solvated protein–peptide complexes (atomic fluctuations, solvent-accessible surface areas, and hydrogen-bonding pattern) were found to be in qualitative agreement with the available binding data in these cases. This method is not suitable for high-throughput prediction due to the high computing requirement simulation.

2.2. Self-Consistent Ensemble Optimization and Threading

Kangueane and colleagues *(12)* modeled peptides in the HLA binding groove using self-consistent ensemble optimization (SCEO) discriminated binders from nonbinders using van der Waals clashes. Altuvia, Schueler, and Margalit threaded ***(9,18,19)*** peptides into the HLA binding groove ***(10,11,20,21)***. This approach is dependent on the (1) availability of appropriate peptide structural template for threading and (2) the choice of a pair-wise potential table. The binding affinity is estimated, and peptides are ranked using a suitable pair-wise potential table ***(22–24)***.

2.3. Free Energy Scoring Function

A number of free energy scoring functions (FRESNOs) have been developed for different purposes, and these functions are used for peptide binding to HLA molecules. Rognan and colleagues ***(14,25,26)*** developed FRESNO for predicting peptide binding to HLA molecules. However, this approach requires appropriate structural templates for model building. An extension to this work is EpiDock ***(26)*** applied to predict potential T-cell epitopes from viral proteomes.

2.4. Virtual Matrix

Virtual matrices (VMs), like quantitative matrices, provide a detailed model in which binding of each peptide residue with HLA pockets is quantified using pocket profiles as shown by Hammer and colleagues ***(27)***. VMs are derived by assigning and combining pocket-specific binding properties using structural features or homology principles from known HLA structures and extrapolating to other alleles, whereas quantitative matrices are obtained using peptide data with known allele-specific binding data. The advantage over quantitative matrices is that the method is generic and can be applied to any given allele. One implementation of the algorithm is the software package TEPITOPE ***(27)***. The model is demonstrated for 11 HLA-DR alleles. Furthermore, they have been successfully applied to predict T-cell epitopes in oncology, allergy, and autoimmune diseases ***(17,28–31)***.

2.5. Virtual Pockets

Kangueane and Sakharkar ***(32)*** and Zhao et al. ***(16)*** developed a method using VPs for each residue pockets gleaned from known structural data. In this method, nine VPs are defined, and the binding affinity between HLA and peptide is given by the sum of residue–residue compatibility between peptide residues and corresponding VPs. The quantification of the interaction between the HLA and peptide residue pair is calculated by the application of the Q matrix described by Mathura and Braun ***(33)***.

2.6. Computational Combinatorial Ligand Design

The computational combinatorial ligand design (CCLD) method uses the 3D information from the crystal structure of the molecule. Zeng et al. ***(34)*** applied CCLD for the prediction of peptides that bind HLA molecule with known structure. Using chemical fragments as models for amino acid residues, a set of peptides predicted to bind the HLA–peptide-binding groove were produced. The results showed that CCLD is sensitive to capture important features of peptide binding in sequence and structure.

2.7. Three-Dimensional Quantitative Structure Activity Relationship

Three-dimensional quantitative structure activity relationship (3D QSAR) studies have been applied to explore the molecular interactions between HLA and peptides. They provide coefficient contour maps identifying areas of the peptides that require a particular physicochemical property to increase binding. Flower and Doytchinova ***(35,36)*** applied comparative molecular similarity indices analysis (CoMSIA) for HLA–peptide-binding prediction. CoMSIA uses the interaction potential around aligned sets of 3D peptide structures to describe the contributions to binding. Five types of similarity index (steric bulk, electrostatic potential, local hydrophobicity, and hydrogen-bond donor and hydrogen-bond acceptor abilities) were calculated, using a common probe atom with 1 Å radius, charge +1, hydrophobicity +1, and hydrogen-bond donor and acceptor properties +1 in this method. CoMSIA can predict the binding affinity of a peptide with a residue not presented in the initial training set. However, it cannot assess the contribution of residues at each position and the interactions between them. CoMSIA also returns 3D representations for visual inspection. This approach is partly data-driven and is dependent on the quality of binding data.

3. Conclusion

The use of MDS, threading, SCEO, VM, VP, FRESNO, CCLD, and CoMSIA is discussed for HLA–peptide-binding prediction. The complete mapping of HLA-specific short antigenic peptides in a given viral/bacterial genome will ultimately result in the rational identification of immunogenic epitopes. Although, peptide–HLA specificity plays an important role in the generation of the cell-mediated immune response, other coupled and undoubtedly important mechanisms such as antigen processing, peptide transport, loading of peptides to HLA molecules, and the phenomenon of T-cell receptor (TCR) repertoires need to be considered. A better understanding of HLA specificity and the multitude of variables associated with the cell-mediated immune response will lead to the development of a methodology for the generation of a library of epitopes in silico that could be tested for HLA–peptide binding and immunogenicity. The successful sampling of short antigenic peptides from a pool of viral/bacterial genome sequence using computational tools will aid in faster and cost-effective means of identifying immunogenic peptides that could be further tested using in vivo models, for consideration as vaccines and therapeutics. The application of mathematical models using computational tools is rapidly advancing to uncover the hidden mystery behind cell-mediated immune response. However, caveats regarding model refinement and cross-validation have to be addressed in future.

References

1. Buus S: Description and prediction of peptide-MHC binding: the "human MHC project." *Curr Opin Immunol* 11:209, 1999.
2. Pinilla C, Martin R, Gran B, Appel JR, Boggiano C, Wilson DB, Houghten RA: Exploring immunological specificity using synthetic peptide combinatorial libraries. *Curr Opin Immunol* 11:193, 1999.
3. Rammensee HG, Friede T, Stevanoviic S: MHC ligands and peptide motifs: first listing. *Immunogenetics* 41:178, 1995.
4. Milik M, Sauer D, Brunmark AP, Yuan L, Vitiello A, Jackson MR, Peterson PA, Skolnick J, Glass CA: Application of an artificial neural network to predict specific class I MHC binding peptide sequences. *Nat Biotechnol* 16:753, 1998.
5. Honeyman MC, Brusic V, Stone NL, Harrison LC: Neural network-based prediction of candidate T-cell epitopes. *Nat Biotechnol* 16:966, 1998.
6. Mamitsuka H: Predicting peptides that bind to MHC molecules using supervised learning of hidden Markov models. *Proteins* 33:460, 1998.
7. Parker KC, Shields M, DiBrino M, Brooks A, Coligan JE: Peptide binding to MHC class I molecules: implications for antigenic peptide prediction. *Immunol Res* 14:34, 1995.
8. Schafer JR, Jesdale BM, George JA, Kouttab NM, De Groot AS: Prediction of well-conserved HIV-1 ligands using a matrix-based algorithm, EpiMatrix. *Vaccine* 16:1880, 1998.
9. Jones DT, Thornton JM: Potential energy functions for threading. *Curr Opin Struct Biol* 6:210, 1996.
10. Altuvia Y, Sette A, Sidney J, Southwood S, Margalit H: A structure-based algorithm to predict potential binding peptides to MHC molecules with hydrophobic binding pockets. *Hum Immunol* 58:1, 1997.
11. Schueler-Furman O, Elber R, Margalit H: Knowledge-based structure prediction of MHC class I bound peptides: a study of 23 complexes. *Fold Des* 3:549, 1998.

12. Kangueane P, Sakharkar MK, Lim KS, Hao H, Lin K, Chee RE, Kolatkar PR: Knowledge-based grouping of modeled HLA peptide complexes. *Hum Immunol* 6:460, 2000.
13. Sette A, Sidney J, del Guercio MF, Southwood S, Ruppert J, Dahlberg C, Grey HM, Kubo RT: Peptide binding to the most frequent HLA-A class I alleles measured by quantitative molecular binding assays. *Mol Immunol* 31:813, 1994.
14. Rognan D, Lauemoller SL, Holm A, Buus S, Tschinke V: Predicting binding affinities of protein ligands from three-dimensional models: application to peptide binding to class I major histocompatibility proteins. *J Med Chem* 42:4650, 1999.
15. Batalia MA, Collins EJ: Peptide binding by class I and class II MHC molecules. *Biopolymers* 43:281, 1997.
16. Zhao B, Mathura VS, Rajaseger G, Moochhala S, Sakharkar MK, Kangueane P: A novel MHCp binding prediction model. *Hum Immunol* 64(12):1123–1143, 2003.
17. Rognan D, Scapozza L, Folkers G, Daser A: Molecular dynamics simulation of MHC-peptide complexes as a tool for predicting potential T cell epitopes. *Biochemistry* 33:11476, 1994.
18. Jernigan RL, Bahar I: Structure-derived potentials and protein simulations. *Curr Opin Struct Biol* 6:195, 1996.
19. Skolnick J, Jaroszewski L, Kolinski A, Godzik A: Derivation and testing of pair potentials for protein folding. When is the quasi-chemical approximation correct? *Protein Sci* 6:676, 1997.
20. Altuvia Y, Schueler O, Margalit H: Ranking potential binding peptides to MHC molecules by a computational threading approach. *J Mol Biol* 249:244, 1995.
21. Schueler-Furman O, Altuvia Y, Sette A, and Margalit H: Structure-based prediction of binding peptides to MHC class I molecules. Application to a broad range of MHC alleles. *Protein Sci* 9:1838, 2000.
22. Miyazawa S, Jernigan RL: Estimation of effective inter-residue contact energies from protein crystal structure, quasi-chemical approximation. *Macromolecules* 18:534, 1985.
23. Miyazawa S, Jernigan RL: Residue-residue potentials with a favorable contact pair term and an unfavorable high packing density term, for simulation and threading. *J Mol Biol* 256:623, 1996.
24. Betancourt MR, Thirumalai D: Pair potentials for protein folding: choice of reference states and sensitivity of predicted native states to variations in the interaction schemes. *Protein Sci* 8:361, 1999.
25. Logean A, Sette A, Rognan D: Customized versus universal scoring functions: application to class I MHC-peptide binding free energy predictions. *Bioorg Med Chem Lett* 11:675, 2001.
26. Logean A, Rognan D: Recovery of known T-cell epitopes by computational scanning of a viral genome. *J Comput Aided Mol Des* 16:229, 2002.
27. Sturniolo T, Bono E, Ding J, Raddrizzani L, Tuereci O, Sahin U, Braxenthaler M, Gallazzi F, Protti MP, Sinigaglia F, Hammer J: Generation of tissue-specific and

promiscuous HLA ligand databases using DNA microarrays and virtual HLA class II matrices. *Nat Biotechnol* 17:555, 1999.
28. Hammer J, Gallazzi F, Bono E, Karr RW, Guenot J, Valsasnini P, Nagy ZA, Sinigaglia F: Peptide binding specificity of HLA-DR4 molecules: correlation with rheumatoid arthritis association. *J Exp Med* 181:1847, 1995.
29. Gross DM, Forsthuber T, Tary-Lehmann M, Etling C, Ito K, Nagy ZA, Field JA, Steere AC, Huber BT: Identification of LFA-1 as a candidate autoantigen in treatment-resistant Lyme arthritis. *Science* 281:703, 1998.
30. Cochlovius B, Stassar M, Christ O, Raddrizzani L, Hammer J, Mytilineos I, Zoller M: In vitro and in vivo induction of a Th cell response toward peptides of the melanoma-associated glycoprotein 100 protein selected by the TEPITOPE program. *J Immunol* 165:4731, 2000.
31. Stassar MJ, Raddrizzani L, Hammer J, Zoller M: T-helper cell-response to MHC class II-binding peptides of the renal cell carcinoma-associated antigen RAGE-1. *Immunobiology* 203:743, 2001.
32. Kangueane, P Sakharkar MK: T-Epitope designer: A HLA-peptide binding prediction server. *Bioinformation* 1(1):21–24, 2005.
33. Venkatarajan MS, Braun W: New quantitative descriptors of amino acids based on multidimensional scaling of a large number of physical–chemical properties. *J Mol Model* 7:445, 2001.
34. Zeng J, Treutlein HR, Rudy GB: Predicting sequences and structures of MHC-binding peptides: a computational combinatorial approach. *J Comput Aided Mol Des* 15:573, 2001.
35. Doytchinova IA, Flower DR: Toward the quantitative prediction of T-cell epitopes: coMFA and coMSIA studies of peptides with affinity for the class I MHC molecule HLA-A*0201. *J Med Chem* 44:3572, 2001.
36. Doytchinova IA, Flower DR: A comparative molecular similarity index analysis (CoMSIA) study identifies an HLA-A2 binding supermotif. *J Comput Aided Mol Des* 16:535, 2002.

22

A Practical Guide to Structure-Based Prediction of MHC-Binding Peptides

Shoba Ranganathan and Joo Chuan Tong

Summary

The binding of bound peptide ligands to major histocompatibility complex (MHC) molecules plays a key role in the activation of normal immune responses and is an intricate theoretical problem that remains unsolved. Geometric and energetic complementarities between an MHC molecule and its corresponding bound peptide ligand are critical in determining the stability of the complex. In this context, the introduction of structural information can greatly facilitate our understanding of how well a peptide ligand can associate with a particular MHC molecule. This chapter introduces the use of structural models as a predictive method to determine whether a peptide sequence can bind to a specific MHC allele.

Key Words: MHC; binding energy; homology modeling; docking; antigens/peptides/epitopes

1. Introduction

In recent years, protein structure prediction has been gaining prominence in the field of structural biology. A useful three-dimensional (3D) model for a receptor–ligand complex of unknown structure can frequently be built using a battery of bioinformatics software. In the context of peptide–MHC (pMHC) complex, the availability of such models allows the prediction of potential immunodominant epitopes at allele-specific level without the need of large experimental data set for training and offers an alternative to the traditional sequence-based predictive techniques. This chapter outlines all the stages in a structure-based pMHC prediction session. The methodology presented here is applicable to the design of both subtype-specific vaccine candidates and promiscuous peptide epitopes.

From: *Methods in Molecular Biology, vol. 409: Immunoinformatics: Predicting Immunogenicity In Silico*
Edited by: D. R. Flower © Humana Press Inc., Totowa, NJ

2. Materials

2.1. Data

1. MHC protein sequences to be modeled (referred to as target sequence) are obtained from Swiss-Prot.
2. Experimental 3D structures of MHC molecules are obtained from the Protein Data Bank (PDB).
3. Experimental binding data of pMHC complexes are obtained from the literature and freely available databases such as MHCPEP (http://wehih.wehi.edu.au/mhcpep/), MHCBN (http://www.imtech.res.in/raghava/mhcbn/), and JENPEP (http://www.jenner.ac.uk/JenPep/).

2.2. Software

1. PSI-BLAST or BLASTP (http://www.ncbi.nlm.nih.gov/blast/) is used for sequence similarity search of the target sequence against available structures in PDB.
2. CLUSTALW *(1)* or JALVIEW *(2)* is used for alignment of target and template structures.
3. MODELLER *(3)*, Internal Coordinates Mechanics (ICM) *(4)*, SWISS-MODEL (http://swissmodel.expasy.org//SWISS-MODEL.html) *(5)*, 3D-JIGSAW *(6)* (http://www.bmm.icnet.uk/servers/3djigsaw), and WHATIF *(7)* are used for comparative modeling of protein 3D structures where experimental 3D structures are not available.
4. ICM *(4)* is used for model refinement and docking simulations.
5. PROCHECK *(8)*, ERRAT *(9)*, PROSA II *(10)*, and WHATIF *(7)* are available in the Biotech structure and model verification server (http://biotech.ebi.ac.uk:8400/) to evaluate the quality of the generated model structures.
6. MATLAB (http://www.mathworks.com/) is used for the calibration of the energy function.

3. Methods

The most important criteria for accurate prediction of MHC-binding peptides using a structure-based approach are the availability of reliable pMHC models and a good scoring technique to effectively discriminate binding peptides from the background of nonbinders. The former requires an accurate prediction of both the receptor-binding site and the bound conformation and orientation of target ligand where experimental 3D structures are not available, whereas the latter is necessary for finding hits with reasonable accuracy.

3.1. Modeling an MHC Receptor of Unknown Structure

The comparative protein structure modeling process involves a series of consecutive steps, with each step depending on the success of the preceding one: (i) search for templates, (ii) template(s) selection, (iii) target-template alignment, (iv) model building, (v) model refinement, and (vi) model evaluation.

3.1.1. Search for Templates

Comparative modeling generally starts by performing a sequence similarity search of the target sequence against available structures in PDB using BLAST (PSI-BLAST or BLASTP) to identify suitable template structures for model building. The template structure is a sequence with known structure that is significantly similar to the target sequence.

3.1.2. Template(s) Selection

Based on the list of potential templates obtained from BLAST, it is necessary to identify one or more suitable templates for modeling the target structure. Criteria for template(s) selection include (i) sequence similarity, (ii) resolution and R-factor of crystallographic structure and the number of restraints per residue for a nuclear magnetic resonance structure, and (iii) presence of missing residues within the binding site. The best template for our problem would have the highest sequence similarity, with the best crystallographic structure (with the least value for resolution) and no missing residues in the critical peptide-binding groove.

3.1.3. Target-Template Alignment

This critical step determines the quality and nature of the model structure. Here, the target sequence is aligned with the template sequence to maximize the structural similarity using either a local-similarity dynamic programming approach ***(11)*** or a global-similarity approach ***(12)***. Although the alignment between the query and template sequences is fairly constant in the case of homologous sequences (> 40% identity), care is required below this value as it is common for standard alignment methods to produce structurally incorrect results ***(13)***. Manual inspection of automatically generated alignments must be performed to detect and correct alignment errors. In most globular proteins, gap regions usually occur in loops. Thus, any gaps occurring in secondary structure elements should be "moved" to the closest loop region by manual editing.

3.1.4. Model Building

Several model building methods are available for constructing a 3D model for the target protein, both as web servers (WHATIF, SWISS-MODEL, and 3D-JIGSAW) and as programs (MODELLER and ICM). These programs are either automated (SWISS-MODEL and 3D-JIGSAW) requiring only the target sequence as input or manual (WHATIF, MODELLER, and ICM) requiring the input of (i) the target sequence, (ii) the template structure(s), and (iii) an alignment of the target and template sequences.

3.1.5. Model Refinement

Once the initial model has been built, geometrical improvements to the structure and the removal of unfavorable nonbonded contacts can be performed by simple energy minimization or molecular dynamics. These are available in all the model building programs listed above.

3.1.6. Model Evaluation

After the initial model has been built, it is important to check for possible errors during the modeling simulation. Evaluation of the quality of the initial model can be performed with the support of quality evaluation programs such as PROCHECK, ERRAT, PROSA II, and WHATIF. These programs can check the energy profile, stereochemical quality, as well as unfavorable side-chain environments (usually a good indicator of incorrectly folded protein structures). A few iterations of model refinement and evaluation can be performed to improve the overall quality of the model.

3.2. Modeling the Bound Conformation of Peptide Ligand

Computer-simulated ligand binding or docking is a useful technique when studying intermolecular interactions or designing new pharmaceutical products. In general, the purpose of docking simulation is twofold: (i) to find the most probable translational, rotational, and conformational juxtaposition of a given ligand–receptor pair and (ii) to evaluate the relative goodness-of-fit for different computed complexes. In this section, we introduce the use of an incremental docking technique to construct the bound conformation of peptide to MHC molecules.

3.2.1. Selection of Probe Residue

The starting point involves the use of a probe or "base fragment" to sample different regions of the receptor-binding site. As such, the selection of a probe is critical to the quality of the ligand-model structure as well as the computational time needed for simulations. Because such technique is combinatorial in nature, the key issue in docking simulation is to enumerate the number of combinations for two molecules within an enclosed sampling space. There are six degrees of global-rotational and -translational freedom of one molecule relative to the other, as well as one internal dihedral rotation per rotational bond. A full search on the conformational space increases exponentially with increasing molecule size and sampling space, as a 10-residue peptide has $>10^{10}$ conformations. As such, a key challenge in pMHC-docking simulation is to keep the sampling space within manageable limits.

A probe must satisfy two criteria: (i) the anchor must have sufficient contact with the receptor and (ii) the structure of the anchor must be highly conserved. Probes that are too short in length will require the exploration of a larger search space and hence longer computational time, whereas probes that are too long may result in insufficient sampling of the receptor-binding site. Various studies have found that that the backbone conformation of bound peptide at both ends of MHC class I and II binding grooves are highly conserved ***(14)***, thus offering a good starting point for docking simulation.

3.2.2. Rigid Docking of Probe Residues

The first step involves rigid docking of the selected probe residues at both ends of MHC–receptor binding groove using ICM global optimization algorithm ***(15)***. Sampling of each probe is localized to small cubic regions of 1.0 Å radius from experimentally determined binding site using a receptor grid map and appropriate restraints. This will effectively restrict the configurational space that needs to be sampled and reduce the number of false hits generated during the simulation. The side-chain torsions of ligand within the grid map were changed in each random step using a biased Monte Carlo procedure, which pseudo-randomly selects a set of torsion angles in the ligand and subsequently finding the local energy minimum about those angles. New conformations are adopted upon satisfaction of the Metropolis criteria with probability $\min(1,\exp[-\Delta G/RT])$, where R is the universal gas constant and T (usually 300K) is the absolute temperature of the simulation. The optimal energy function used during simulations consisted of the internal energy of the ligand and the intermolecular energy based on the same optimized potential maps used in the docking step:

$$E = E_{Hvw} + E_{Cvw} + 2.16E_{el}^{solv} + 2.53E_{hb} + 4.35E_{hp} + 0.20E_{solv}$$

The internal energy included internal van der Waals interactions, hydrogen bonding and torsion energy calculated with ECEPP/3 parameters, and the Coulomb electrostatic energy with a distance-dependent dielectric constant ($e = 4r$). The configurational entropy of the side chains and the surface-based solvation energy were included in the final energy to select the best-refined solutions.

3.2.3. Construction of the Loop Linking the Probes

In this stage, an initial conformation of the central loop is generated by satisfaction of spatial constraints *(3)* based on the allowed subspace for backbone dihedrals in accordance with the conformations of peptides docked into the ends of the binding groove. This is performed in three steps: (i) distance and dihedral angle restraints on the entire peptide sequence are derived from its alignment with the sequences of probes docked into the binding groove; (ii) the restraints on spatial features of the unknown center residues are derived by extrapolation from the known 3D structures of probes in the alignment, expressed as probability density functions, with stereochemical restraints including bond distances, bond angles, planarity of peptide groups and side-chain rings, chiralities of Cα atoms and side chains, van der Waals contact distances and the bond lengths, bond angles, and dihedral angles of cysteine disulfide bridges; and (iii) spatial restraints on the unknown center residues are satisfied by optimization of the molecular probability density function using a variable target function technique that applies the conjugate gradients algorithm to positions of all nonhydrogen atoms.

3.2.4. Refinement of the Ligand Backbone

To improve the accuracy of the initial ligand model, partial refinement was performed for the ligand backbone, using ICM biased Monte Carlo procedure *(4)*. Preliminary stages of refinements attempt to overcome the penalty derived from the initial rigid docking of terminal residues by introducing partial flexibility to the ligand backbone. Restraints were imposed upon the positional variables of the Cα atoms of probes to keep them close to the starting conformation. The energy function adopted for this refinement step is:

$$E = E_{vw} + E_{hbonds} + E_{torsions} + E_{electr} + E_{solv} + E_{entropy}$$

3.2.5. Construction of Flanking Residues

At this stage, MHC class I ligand models have been fully constructed, and the following task is applicable only to MHC class II ligands. Here, the only construction remaining is the flanking residues that extend out of the MHC class II binding groove. Flanking residues have an important contribution to the overall free binding energy of class II peptides and should be modeled accordingly. It is a common mistake to model only the nonameric core recognition regions for class II peptides and develop a scoring function based on these core regions alone. The entire peptide should be modeled, and flanking residues may be constructed by using the biased Monte Carlo procedure described above, using the conformation of existing MHC-bound class II peptides as a guide.

3.2.6. Refinements of Receptor and Ligand Interacting Side Chains

The final stage of the modeling process involves the refinement of the receptor and ligand side-chain torsions in the vicinity of 4.0 Å of the receptor. This step serves to optimize the conformations of all residues involved in the MHC–peptide interaction.

3.3. Scoring

The accuracy of structure-based predictive model relies heavily on training the adopted scoring function using a small set of experimentally determined binders and nonbinders with known IC_{50} values. In the case of class II peptides, nonameric core recognition sequences should be known and modeled into the binding groove appropriately. Generalized scoring functions tend to produce inferior prediction results with poor correlation to experimental data when confronted with the novel receptor–ligand system. The adopted scoring function must be recalibrated to suit the data set by adjusting the relative weight of the different energy terms to improve the discriminative power of the model. New weights for the energy terms can be generated by performing multiple linear regression on a small training data set using the program MATLAB. The calibrated system is then used predictively on the test data set.

Note

1. Automated model building is not recommended unless there is a very high query-template sequence homology. Although laborious, the quality of a structural model based on a manual step-by-step approach is definitely better as well as biologically viable ***(16,17)***.

References

1. Chenna, R., Sugawara, H., Koike, T., Lopez, R., Gibson, T. J., Higgins, D. G. and Thompson, J. D. (2003) Multiple sequence alignment with the Clustal series of programs. Nucleic Acids Res. **31**, 3497–3500.
2. Clamp, M., Cuff, J., Searle, S. M. and Barton, G. J. (2004) The Jalview Java alignment editor. Bioinformatics **12**, 426–427.
3. Sali, A. and Blundell, T. L. (1993) Comparative protein modeling by satisfaction of spatial restraints. J. Mol. Biol. **234**, 774–815.
4. Abagyan R. and Totrov M. (1999) Ab initio folding of peptides by the optimal-bias Monte Carlo Minimization Procedure. J. Comput. Phys. **151**, 402–421.

5. Schwede, T., Kopp, J., Guex, N. and Peitsch, M. C. (2003) SWISS-MODEL: an automated protein homology-modeling server. Nucleic Acids Res. **31**, 3381–3385.
6. Bates, P. A., Kelley, L. A., MacCallum, R. M. and Sternberg, M. J. E. (2001) Enhancement of protein modelling by human intervention in applying the automatic programs 3D-JIGSAW and 3D-PSSM. Proteins **Suppl. 5**, 39–46.
7. Vriend, G. (1990) WHAT IF: a molecular modeling and drug design program. J. Mol. Graph. **8**, 52–56.
8. Laskowski, R. A., MacArthur, M. W., Moss, D. S. and Thornton, J. M. (1993) PROCHECK: a program to check the stereochemical quality of protein structures. J. Appl. Cryst. **26**, 283–291.
9. Colovos, C. and Yeates, T. O. (1993) Verification of protein structures: patterns of nonbonded atomic interactions. Protein Sci. **2**, 1511–1519.
10. Sippl, M. J. (1993) Recognition of errors in three-dimensional structures of proteins. Proteins **17**, 355–362.
11. Smith, T. F. and Waterman, M. S. (1981) Identification of common molecular subsequences. J. Mol. Biol. **147**, 195–197.
12. Needleman, S. B. and Wunsch, C. D. (1970) A general method applicable to the search for similarities in the amino acid sequence of two proteins. J. Mol. Biol. **48**, 443–453.
13. Read, J., Braye, G., Jurek, L. and James M. N. G. (1984) Critical evaluation of comparative model building of Streptomyces griseus trypsin. Biochemistry **23**, 6570–6575.
14. Tong, J. C., Tan, T. W. and Ranganathan, S. (2004) Modeling the structure of bound peptide ligands to major histocompatibility complex. Protein Sci. **13**, 2523–2532.
15. Fernández-Recio, J., Totrov, M. and Abagyan, R. (2002) Soft protein-protein docking in internal coordinates. Protein Sci. **11**, 280–291.
16. Ranganathan, S. (2001) Molecular Modeling on the web. Biotechniques, **30**, 50–52.
17. Ranganathan, S. (2003) Molecular Modeling on the web, Biocomputing: Computer Tools for Biologists, ed. Stuart M. Brown, Biotechniques press, Eaton Publishing, Westborough, USA, Chap. 49, pp. 411–417.

23

Static Energy Analysis of MHC Class I and Class II Peptide-Binding Affinity

Matthew N. Davies and Darren R. Flower

Summary

Antigenic peptide is presented to a T-cell receptor (TCR) through the formation of a stable complex with a major histocompatibility complex (MHC) molecule. Various predictive algorithms have been developed to estimate a peptide's capacity to form a stable complex with a given MHC class II allele, a technique integral to the strategy of vaccine design. These have previously incorporated such computational techniques as quantitative matrices and neural networks. A novel predictive technique is described, which uses molecular modeling of predetermined crystal structures to estimate the stability of an MHC class II–peptide complex. The structures are remodeled, energy minimized, and annealed before the energetic interaction is calculated.

Key Words: MHC; antigenic peptides; energy minimization; simulated annealing

1. Introduction

Major histocompatibility complex (MHC) glycoprotein molecules play a central role in the adaptive immune system, forming a complex with foreign antigenic peptides and displaying them to T-cell receptors (TCRs) on the cell surface of antigen-presenting cells (APCs). The array of MHC–peptide complexes presented by APCs shape the specificity of the T-cell response. Two different types of MHC molecules, class I and class II, are recognized by distinct sets of T cells, CD8 and CD4, respectively. MHC class I molecules, which present antigenic peptides derived from the cytosol, are composed of an α-heavy chain, the small subunit β_2-microglobulin (β_2m), and an antigenic peptide between 8 and 11 amino acids in length. The peptide is bound with a

From: *Methods in Molecular Biology, vol. 409: Immunoinformatics: Predicting Immunogenicity In Silico*
Edited by: D. R. Flower © Humana Press Inc., Totowa, NJ

groove formed by the heavy chain α1 and α2 domains. Peptide-binding motifs have previously been identified based on promiscuous epitopes with binding affinities that transcend the specificities of individual alleles *(1)*. In an attempt to discover novel T-cell epitopes, various techniques have been developed to calculate the affinity of a peptide for a given MHC class I molecule. Empirical methods such as EpiVax *(2)*, Artificial Neural Networks *(3)*, Hidden Markov Models *(4)*, Support Vector Machines *(5)* and Profiles *(6)*, and the Quantitative Structure-Activity Relationship (QSAR)-based additive technique *(7–12)* have been developed as means to calculate the affinity of a given peptide–MHC interaction. Here, we describe how a molecular modeling approach can be used to create a predictive system *(13–14)* by carrying out molecular dynamics (MD) simulations and static energy analysis on the MHC–peptide complexes.

2. Energy Minimization Theory

Molecular modeling simulations estimate the time-dependent behavior of a molecular system on a microscopic scale. They provide detailed information on the fluctuations and conformational changes of proteins and nucleic acids and are routinely used to investigate the structure, dynamics, and thermodynamics of biological molecules. From this it is possible to analyze the interrelationships of molecules and complexes in a theoretical environment. The technique allows detailed information to be generated on the fluctuations and conformational changes of proteins and nucleic acids by simulating the dynamics of experimentally determined structures. These simulations can vary from local motions (atomic fluctuations, side chain, and loop motions) to rigid body motions (helix or subunit motions) to more large-scale motions such as helix coil association/dissociation reactions or the folding and unfolding of proteins. A force field is applied to the system that contains parameters for all covalent and noncovalent interactions between all the atoms within the system. From this it is possible to determine the position and velocity of every represented atom after a given period of time. The method is deterministic in that the movement of the system can be propagated forward or backward from a given time step. Simulations aid our understanding of biochemical processes and give a dynamic dimension to structural data. Time-dependent molecule properties may be determined from the system that approach experimentally measurable ensemble averages. The technique may also be used to explore which conformations of a molecule are thermodynamically plausible. An exponential increase in computer power over the last three decades combined with the ever-increasing number of structures within the Protein Data Bank has allowed modelers to be increasingly imaginative with the applications of the software. Over the past decade, developments in X-ray diffraction techniques and computational methods have allowed for an exponential increase in the number of macromolecular structures being solved. Simulations have moved from the interactions of several atoms to vast biomolecular systems incorporating large multimeric structures and trans-membrane elements. The quality of the simulations has also improved more realistic boundary conditions and better sampling times. Molecule dynamics has also moved from being based in classical physics to begin to incorporate quantum mechanics into force fields. As the quality of the force fields continues to improve and the scope of the simulations graduates from the atomic to the subcellular, there remains a huge potential for the use of the technique in computational science.

3. Methodology

3.1. PDB Structure

In order to create a structure of a given MHC–peptide complex, the residues of the bound peptide must be remodeled using the crystallographic modeling program “O” *(15)*. Mutating a residue consists of two steps: use Mutate_replace to assign the correct residue type. “O” will put it in as the most common side-chain rotamer. The example below would generate the MHC class II CLIP peptide, MRMATPLLM.

```
O > mut_repl
Mut>  Mutate a molecule by replacing one residue type
Mut>  by another.
Mut> Molecule ([M1]) :
Mut> Residue name and new type ( to end) : c1 met
Mut> Residue name and new type ( to end) : c2 arg
Mut> Residue name and new type ( to end) : c3 met
Mut> Residue name and new type ( to end) : c4 ala
Mut> Residue name and new type ( to end) : c5 thr
Mut> Residue name and new type ( to end) : c6 pro
Mut> Residue name and new type ( to end) : c7 leu
Mut> Residue name and new type ( to end) : c8 leu
Mut> Residue name and new type ( to end) : c9 met
Mut> Residue name and new type ( to end) :
Mut>  There are 9 mutations
Mut>  The Rotamer_DB is now being loaded.
```

Then use the normal rebuilding tools to fit the density (lego_side_ch) to orient the side chain into a common rotamer.

```
O > zo ; end
O > le_si_ch cl etc.
```

The structure may then be outputted as a *pdb* file and may be used as the basis for a molecule dynamics simulation.

3.2. AMBER

All MD simulations were carried out using AMBER Version 8 ***(16)***. AMBER is the collective name for a suite of programs that allow users to perform various sorts of molecular modeling ***(16)***. The individual programs can be used together to prepare and run an MD simulation. The term *amber* is also sometimes used to refer to the empirical force fields that are implemented by the programs ***(17)***.

3.2.1. LEaP

The pdb file must first be loaded into the leap program in order to generate the topology and coordinate files necessary to begin the simulation.

```
tleap -s -f leaprc.ff03
```

Activates the leap program (*tleap* is a command line based version of *leap*; *xleap* provides a graphical interface in order to visualize the system)

```
loadamberparams $AMBERHOME/dat/leap/parm/gaff.dat
```

The above command loads the AMBER force field (GAFF) into the leap program. The force field incorporates into its parameters natural processes such as the stretching of bonds, variations in bond angles, and rotations around a single bond. Although various complex factors can be incorporated into the force field, the mechanics can be summarized into four essential components. These are bonds lengths, bond angles, torsion angles, and nonbonded interactions. Deviations from a bond or an angle's reference value are calculated as energetic penalties in the system. The reference or natural bond length does not represent the bond's condition when the system is in equilibrium but the minimum potential energy level when all other terms in the force field are set to zero. The interaction of the various atomic forces cause the system to deviate from the reference values in order to compensate for other forces being placed upon individual bond or angles. GAFF has been specifically designed to cover most pharmaceutical molecules and is compatible with the traditional AMBER force fields in such a way that the two can be mixed during a simulation. Like the traditional AMBER force fields, GAFF uses a simple harmonic function form for bonds and angles, but unlike the traditional protein- and DNA-orientated AMBER force fields, the atom types used in GAFF are much more general such that they cover most of the organic chemical space. The current implementation of the GAFF force field consists of 33 basic atom types and 22 special atom types. It should be recognized, however, that the code and force field are separate: several other computer packages have implemented the *amber* force fields, and other force fields can be implemented with the *amber* programs.

```
x = loadpdb mhc.pdb
```

The above command loads the pdb file you have created into the program. It now exists as the unit *x*. The full structure of the HLA-A*0201–peptide complex was explicitly represented within the simulation.

```
solvatebox x TIP3PBOX 6
```

Hydrogen atoms were added to the structure, and the system was fully solvated using TIP3 waters. The command

```
check x
```

may also be used here to check for inconsistencies that may cause problems with a simulation.

```
saveamberparm x mhc.top mhc.crd
```

The above command saves the topology and coordinate files as mhc.top and mhc.crd, respectively. This command will cause LEaP to search its list of PARMSETs for parameters defining all of the interactions between the ATOMs within the UNIT. The output of this operation can be used for minimizations, dynamics, and thermodynamic integration calculations.

3.2.2. Sander

The energy of the solvated molecular complex was minimized using a steepest descent method that continued for 20,000 1 fs time steps or until the root mean square deviation between successive time steps had fallen below 0.01 Å.

3.2.2.1. Energy Minimization

```
sander -i minimise.inp -o MHC.out -p MHC.top -c MHC.crd -ref
                    MHC.crd -r MHC.restrt
```

The minimize.inp (Appendix 1) provides the parameters for the simulation. MHC.out provides the details of each simulation, and MHC.restrt contains the new coordinate file generated by the program. The system is now at a point of sufficient equilibrium to begin an MD simulation.

3.2.2.2. Molecular Dynamics

```
sander -i md.inp -o MHC_MD.out -p MHC.top -c MHC.restrt -ref
        MHC.restrt -r MHC_MD.restrt -x MHC_MD.traj
```

md.inp (Appendix 2) provides the parameters for the simulation. The simulation was run for a minute and a new coordinate file, MHC_MD.restrt, was generated.

3.2.3. Anal

Following MD, static energy analysis is carried out on the protein–water system using the *anal* program. *Anal* calculates the group–group interaction energies between different parts of the system based on the position of their composite atoms.

```
anal -i anal.inp -o analout -p MHC.top -c MHC_MD.restrt
```

The input file, anal.inp, is presented in Appendix 3. The interaction energies, which are measured in kcal mol^{-1}, may be calculated between all residues comprising the peptide and MHC molecule or they may be calculated for the individual interaction between composite residues. Three values are generated for each interaction: the electrostatic interaction energy, the van der Waals interaction energy, and the total interaction energy (which is the sum of the first two terms). The interaction energies reflect the affinity between the peptide and the MHC molecule and as such may be used as the basis of a predictive system.

3.3. Correlation

The degree of correlation can be calculated by comparison with the experimentally determined IC_{50} or BL_{50} data. The AntiJen database is a comprehensive of all the binding data that are publicly available *(18–20)*. The data set may be broken down into a training and test set; the former can be used to optimize the parameters of the simulation. It is not necessary for the MHC–peptide complex to undergo large conformational shifts to reach the global energy minimum, and therefore, it is not necessary to expose system to extreme temperatures. Affinity can then be calculated for the test set and compared with the experimental values. An overall interaction energy may be calculated between the receptor and ligand, which may be used as the basis for correlation coefficient and receiver operating characteristic (ROC) analysis *(13)*. A more sophisticated approach, however, is to break the interaction energies into the component residues and use QSAR analysis to search for key residues that may act as descriptors in a predictive model *(14)*. This is similar to the COMBINE method, where structural data are combined with QSAR analysis to predict receptor–ligand affinity, which has previously been used successfully on various complexes *(21–23)*. Both MHC Class I and Class II predictive algorithms generated that were either comparable with or surpassed other predictive techniques. However, the molecular dynamics technique are far more computationally expensive than other available techniques.

4. Discussion

All these prediction techniques have been limited in their accuracy by the quality of their scoring function. The determinants of binding include van der Waals interaction energy, electrostatic interaction energy, and the hydrophobic component. It is, however, possible that static energy analysis overlooks a significant energetic interaction that takes place during the binding. The actual thermodynamic property that we are trying to estimate with the scoring function is the Gibbs free energy of binding, ΔG, which is the energy that is released when ligand and receptor bind. It may be represented as $\Delta G = \Delta H - T\Delta S$, where ΔH is the enthalpic (internal) energy, and $T\Delta S$ is an entropy term, which is indicative of the relative gain or loss of disorder when ligand and receptor bind. The ΔH is roughly approximated by the calculation of the internal energy interactions but does not incorporate the entropic contribution, in particular, the behavior of the water molecules in the active site that are displaced when binding occurs. This is unfortunate as there is much evidence to suggest that the binding of a ligand to a receptor is as much driven by entropic energy contributions as enthalpic. A transition state exists within the molecule between the peptide-bound and peptide-unbound states *(24)*. It is likely that the movement from free peptide to this transition state is favored entropically but not enthalpically. However, the change from the transition state to the bound complex is both entropically and enthalpically favorable. The binding groove is hydrophobic in character, particularly in the region of the anchor residues, and the calculated energy interactions reflect this, particularly with the high incidence of charge–charge interactions within the groove. It is necessary for the solvent entropy from the burial of hydrophobic groups to offset the reduction in peptide conformational entropy that occurs upon binding. The measured interactions reflect the formation of hydrogen bonds and salt bridges that occurs when the peptide moves from the loosely packed, partially hydrated interface to stable complex. Water is stripped from partial and fully charged MHC and peptide residues and there is a reduction in the favorable hydrogen bond enthalpy associated with hydrophobically oriented water. It is also the entropic contribution that lowers the activation barrier necessary for the dissociation of the peptide. Hence, both the association and dissociation of the peptide is essentially an entropically driven process. In taking the work forward, it is therefore necessary to incorporate the entropic contribution to the energetic calculation into the descriptors in order to create a more accurate representation of the free-energy change between the MHC molecules bound and unbound states.

References

1. De Groot, A. S., H. Sbai, C. S. Aubin, J. McMurry and W. Martin. 2002. Immunoinformatics: mining genomes for vaccine components. *Immunol. Cell Biol.* 80:255.
2. Brusic, V., G. Rudy and L. C. Harrison. 1994. Prediction of MHC binding peptides using artificial neural networks. In *Complex Systems: Mechanism of Adaptation*, 1st edn. R. J. Stonier and X. S. Yu, eds. IOS Press, Amsterdam; OHMSHA Tokyo, p. 253.
3. Udaka, K., H. Mamitsuka, Y. Nakaseko and N. Abe. 2002. Prediction of MHC class I binding peptides by a query learning algorithm based on hidden Markov models. *J. Biol. Phys.* 28:183.
4. Donnes, P. and A. Elofsson. 2002. Prediction of MHC class I binding peptides, using SVMHC. *BMC Bioinformatics.* 3:25.
5. Reche, P. A., J. P. Glutting and E. L. Reinherz. 2002. Prediction of MHC class I binding peptides using profile motifs. *Hum. Immunol.* 63:701.
6. Sette, A. and J. Sidney. 1999. Nine major HLA class I supertypes account for the vast preponderance of HLA-A and -B polymorphism. *Immunogenetics.* 50(3–4):201–12.

7. Doytchinova, I.A., P. Guan and D.R. Flower. 2004. Identifying human MHC supertypes using bioinformatic methods. *J. Immunol.* 172(7):4314–23.
8. Hattotuwagama, C.K., C.P. Toseland, P. Guan, P.J. Taylor, S.L. Hemsley, I. A. Doytchinova and D. R. Flower. 2006. Toward Prediction of Class II Mouse Major Histocompatibility Complex Peptide Binding Affinity: in Silico Bioinformatic Evaluation Using Partial Least Squares, a Robust Multivariate Statistical Technique. *J. Chem. Inf. Model.* 46:1491–1502.
9. Hattotuwagama, C.K., I.A. Doytchinova and D.R. Flower. 2005. *In silico* prediction of peptide binding affinity to class I mouse major histocompatibility complexes: A Comparative Molecular Similarity Index Analysis (CoMSIA) study. *J. Chem. Inf. Mod.* 45:1415–1423.
10. Hattotuwagama, C. K., P. Guan, I. A. Doytchinova and D. R. Flower. 2007. In silico QSAR-based predictions of class I and class II MHC epitopes. *Immunoinformatics: Opportunities and Challenges of Bridging Immunology with Computer and Information Sciences.* (in press).
11. Hattotuwagama, C. K., P. Guan, I. A. Doytchinova and D. R. Flower. 2004. New horizons in mouse immunoinformatics: reliable *in silico* prediction of mouse class I histocompatibility major complex peptide binding affinity. *Org. Biomol. Chem.* 2:3274–83.
12. Hattotuwagama, C. K., P. Guan, I. A. Doytchinova, C. Zygouri and D. R. Flower. 2004. Quantitative online prediction of peptide binding to the major histocompatibility complex. *J. Mol. Graph. Model.* 22(3):195–207.
13. Davies, M. N., C. Sansom, C. Beazley and D. S. Moss. 2003. A novel predictive technique for the MHC class II peptide-binding interaction. *Mol. Med.* 9 (9–12):220–5.
14. Davies, M. N., C. K. Hattotuwagama, D. S. Moss, M. G. B. Drew and D. R. Flower. 2006. Statistical deconvolution of enthalpic energetic contributions to MHC-peptide binding affinity BMC Struct Biol. 6:5–18.
15. Kleywegt, G. J. and T. A. Jones. 1997. Model-building and refinement practice. *Methods Enzymol.* 277:208–30.
16. Pearlman, D. A., D. A. Case, J. W. Caldwell, W. S. Ross, T. E. Cheatham, III, S. DeBolt, D. Ferguson, G. Seibel and P. Kollman. 1995. AMBER, a package of computer programs for applying molecular mechanics, normal mode analysis, molecular dynamics and free energy calculations to simulate the structural and energetic properties of molecules. *Comp. Phys. Commun.* 91:1–41.
17. J. W. Ponder and D. A. Case. 2003. Force fields for protein simulations. *Adv. Protein Chem.* 66:27–85.
18. Blythe, M. J., I. A. Doytchinova and D. R. Flower. 2002. JenPep: a database of quantitative functional peptide data for immunology. *Bioinformatics.* 18(3): 434–9.
19. Sette, A., J. Sidney, M.-F. del Guercio, S. Southwood, J. Ruppert, C. Dalberg, H. M. Grey and R. T. Kubo. 1994. Peptide binding to the most frequent HLA-A class I alleles measured by quantitative molecular binding assays. *Mol. Immunol.* 31:813–22.

20. McSparron, H., M. J. Blythe, C. Zygouri, I. A. Doytchinova and D. R. Flower. 2003. JenPep: A novel computational information resource for immunology and vaccinology. *J. Chem. Inf. Comput. Sci.* 43:1276–87.
21. Wang, R. and R. Wade. 2002. Comparative binding energy (COMBINE) analysis of OppA-peptide complexes to relate structure to binding thermodynamics. *J. Med. Chem.* 45(22):4828–37.
22. Wang, R. and R. Wade. 2001. Comparative binding energy (COMBINE) analysis of influenza neuraminidase-inhibitor complexes. *J. Med. Chem.* 6:961–71.
23. Tokarski, J. S. and A. J. Hopfinger. 1997. Prediction of ligand-receptor binding thermodynamics by free energy force field (FEFF) 3D-QSAR analysis: application to a set of peptidometic renin inhibitors. *J. Chem. Inf. Comput. Sci.* 37(4):792–811.
24. Binz, A. K., R. C. Rodriguez, W. E. Biddison and B. M. Baker. 2003. Thermodynamic and kinetic analysis of a peptide-class I MHC interaction highlights the noncovalent nature and conformational dynamics of the class I heterotrimer. *Biochemistry.* 42(17):4954–61.

Appendix 1: Minimize.inp

```
# do minimization of all only to 0.01Ang, using EWALD:
&cntrl
imin = 1,
nmropt = 0,
ntx = 1,
irest = 0,
ntrx = 1,
ntxo = 1,
ntpr = 50,
ntave = 10,
iwrap = 0,
ioutfm = 0,
ntf = 1,
ntb = 1,
ntwx = 2000,
dielc = 1,
cut = 8.0,
scnb = 2,
scee = 1.2,
ibelly = 0,
ntr = 0,
maxcyc = 3000000,
```

```
ntmin = 1,
ncyc = 100,
dx0 = 0.01,
drms = 0.01.
```

Appendix 2: md.inp

```
# do minimization of all only to 0.01Ang, using EWALD:
&cntrl
imin = 0,
irest = 0,
ntx = 1,
ntt = 1,
temp0 = 300.0,
tautp = 0.2,
ntp = 1,
taup = 2.0,
ntb = 2,
ntc = 2,
ntf = 2,
nstlim = 500000,
ntwe = 100,
ntwx = 100,
ntpr = 100,
cut = 12.
```

Appendix 3: Anal.inp

```
Pocket 1 Energy Analysis
1 0 0 500 0 1
0 0.0 0.0 0.0 0.0
1 0 1 0 50 0
25.0 2.0 2.0 1.0
1 20.0 20.0 20.0 100.0 50.0 10000. 200. 200. 0.
PDB
ENERGY
MHC residues
RES -1 60
END
```

```
Peptide residues
RES -377 385
END
END
TELL
TORSION
N CA C O
N C CA CB
O C CA CB
N CA CB CG
```

24

Molecular Dynamics Simulations

Bring Biomolecular Structures Alive on a Computer

Shunzhou Wan, Peter V. Coveney, and Darren R. Flower

Summary

The molecular dynamics (MD) simulations play a very important role in science today. They have been used successfully in binding free-energy calculations and rational design of drugs and vaccines. MD simulations can help visualize and understand structures and dynamics at an atomistic level when combined with molecular graphics programs. The molecular and atomistic properties can be displayed on a computer in a time-dependent way, which opens a road toward a better understanding of the relationship of structure, dynamics, and function. In this chapter, the basics of MD are explained, together with a step-by-step description of setup and running an MD simulation.

Key Words: molecular dynamics simulation; visualization; major histocompatibility complex

1. Introduction

Molecular dynamics (MD) is a computer simulation technique where the time evolution of a set of interacting atoms is followed by integrating their equations of motion. MD simulations can, in principle, provide the ultimate details of motional phenomena, which can enhance our understanding of biological function through the structure, dynamics, and function connection.

The MD studies in biology have emerged rapidly in recent years as an important complement to experiment. MD simulations find three major areas of application for the study of macromolecules of biological interest *(1)*. First, MD simulation is used as a means of sampling configuration space, which can determine or refine structures with data obtained from X-ray or nuclear magnetic resonance (NMR) experiments. Second, it is used to determine equilibrium

From: *Methods in Molecular Biology, vol. 409: Immunoinformatics: Predicting Immunogenicity In Silico*
Edited by: D. R. Flower © Humana Press Inc., Totowa, NJ

averages of molecular properties that approach the experimentally measurable ensemble averages. Third, it is used to examine the actual dynamics, giving insights into the properties on different timescales for biomolecules in solvation.

The number of simulation techniques and their applications has greatly expanded. Today in the literature, there exist many specialized techniques for particular problems, including mixed quantum mechanical–molecular mechanical (QM/MM) simulations that are being employed to study enzymatic reactions in the context of the full protein *(2)* and massively parallel MD simulations that mimic real-time MD of water penetration through a membrane protein *(3)*.

Given these many sources for the application of MD simulations, the focus in this chapter is on providing a hand-on session of classical MD simulation of the peptide–major histocompatibility complex (pMHC). The reader is advised to complement these aspects with the more standard topics available elsewhere.

2. MD Approach

2.1. Theory

Structure determination is clearly a critical step toward understanding biological function. However, protein structures solved by X-ray crystallography are heterogeneous and inaccurate *(4)*. In addition, protein function requires motion. MD is the link between structure and function. To bring biomolecular structures alive on a computer, we need to employ a few techniques that manipulate the structure, {**R**}, given the potential energy, V({**R**}). The CHARMM format energy function has the form *(5)*:

$$\begin{aligned} V(\{\mathbf{R}\}) = {} & \sum_{\text{bonds}} K_b(b-b_0)^2 + \sum_{\text{UB}} K_{\text{UB}}(S-S_0)^2 + \sum_{\text{angles}} K_\theta(\theta-\theta_0)^2 \\ & + \sum_{\text{dihedrals}} K_\chi\,[1+\cos(n\chi-\chi_0)] + \sum_{\text{impropers}} K_\varphi(\varphi-\varphi_0)^2 \\ & + \sum_{\text{nonbond}} \left[\left(\frac{A_{ij}}{r_{ij}}\right)^{12} - \left(\frac{B_{ij}}{r_{ij}}\right)^{6} + \frac{q_i q_j}{\varepsilon r_{ij}} \right] \end{aligned} \tag{1}$$

where K_b, K_{UB}, K_θ, K_χ, and K_φ are the bond, Urey–Bradley, angle, dihedral angle, and improper dihedral angle force constants, respectively; b, S, θ, χ, and ϕ are the bond length, Urey–Bradley 1,3-distance, bond angle, dihedral angle, and improper torsion angle, respectively, with the subscript zero representing the equilibrium values; n is the coefficient of symmetry ($n = 1, 2, 3$) of the dihedrals. The van der Waals (vdW) interaction is modeled using the Lennard-Jones 6-12 potential that expresses the interaction energy using the atom-type dependent constants A_{ij} and B_{ij}. q_i is the partial charge of atom i, ε is the effective dielectric constant for the medium, and r_{ij} is the distance between atoms i and j.

The MD simulation method is based on Newton's equation of motion:

$$\mathbf{F} = m\frac{\mathrm{d}^2\mathbf{r}}{\mathrm{d}t^2} \tag{2}$$

where $\mathbf{F}$ is the force exerted on the particle, m is its mass, and $\mathrm{d}^2\mathbf{r}/\mathrm{d}t^2$ is its acceleration. The force can be expressed as the gradient of the potential energy (Eq. 1):

$$\mathbf{F} = -\nabla V(\{\mathbf{R}\}) \tag{3}$$

Combining Eqs. 2 and 3, the derivative of the potential energy can be related to the acceleration as a function of time. Numerical integration of the differential equations of motion then yields a trajectory that describes the evolution of the system's positions and momenta through time. From this trajectory, the average values of properties can be determined.

2.2. Ingredients of a Molecular Simulation

To set up a macromolecular MD simulation, some basic ingredients are needed: a description of the structure, a set of atomic coordinates, and an empirical energy function. A description of the structure, including all atoms in the model, their covalent connectivity, and all the energetic interactions to be calculated, is specified in the Protein Structure File (PSF). An X-ray crystal structure or an NMR structure from the Brookhaven Protein Databank (http://www.rcsb.org/pdb/) is usually used as the initial structure in simulations of biomolecules. An empirical energy function (Eq. 1) is needed to describe the energy of the system for any configuration of the atomic coordinates. In the following simulation, atomic charges, vdW, and stereochemical force-field parameters for the proteins are taken from the CHARMM22 all-atom force field *(5)*.

2.3. Computer Software

The CHARMM *(6)* (http://www.charmm.org/) package is a general purpose MD simulation program. It contains a comprehensive analysis facility that enables the user to examine both static and dynamic properties of a system. It will be used here for preparing the initial molecular model and analyzing the final results.

The NAMD *(7)* (http://www.ks.uiuc.edu/Research/namd/) code is a highly scalable massively parallel MD code designed for high-performance simulation of large biomolecular systems. It shows very substantial acceleration relative to single processor runs for a large system (>10,000 atoms). It will be used for the heating and equilibrium, and the production runs of our system.

The VMD package *(8)* (http://www.ks.uiuc.edu/Research/vmd/) is a molecular visualization program for displaying and animating large biomolecular systems using 3D graphics. It can also be used to build a model and analyze the results using its built-in scripting. It will be used here for visualizing the trajectory.

The CHARMM and NAMD, which we used, are compiled with MPI on Unix/Linux machines, although they support many other platforms. It should be noted that there are other MD simulation programs. Although the detailed commands may change, the general approach taken to setup and execute an MD simulation is the same. The example presented below can be easily adapted to other MD simulation programs. All example CHARMM and NAMD input files can be found in Appendices.

3. MD Simulation of pMHC

In this section, we will describe in some detail the steps of building a model, solvating, minimizing, heating and equilibrating the system, running production simulation, and analyzing the trajectory of Tax/HLA-A2 complex using the CHARMM and NAMD programs. It is assumed that you have the CHARMM, NAMD, and VMD programs installed on your local machine.

3.1. Model Building

To begin an MD simulation, an initial configuration of the system must be chosen first. The X-ray structures of the Tax/HLA-A2 complex (PDB id: 1DUZ) *(9)* are used as the initial structure for the simulation. The first step is to generate the PSF, load the initial coordinates, and place the missing hydrogen atoms. This task is controlled by executing appropriate CHARMM commands sequentially. All the CHARMM commands are put in a file named build.inp (see Appendix 1), and this file is directed into CHARMM by typing the following command at the UNIX prompt:

```
mpirun –np 1 charmm < build.inp > build.out
```

The build.inp file holds the CHARMM commands for reading in the basic chemical units, the parameters for all energy terms, the sequence information, and the Cartesian coordinates, for generating the PSF, and for placing the missing atoms.

3.2. Add Water Molecules and Counterions

The pMHC system is not neutral. It is advised to put excess counterions (Na^+) into the system so that the entire model is neutral. The counterions can be put randomly or on an optimizing position according to the electrostatic energy. Then, explicit water molecules are added to solvate the protein. Some water molecules are already present in the X-ray crystal structure; they are kept in the built model (see build.inp in Appendix 1). The solvating water molecules are usually obtained from a suitable large box of water that has been previously equilibrated. The entire box of water is overlapped onto the protein, and those water molecules that overlap the protein are removed. The distance between the edges of the solvent box and the closest atom in the protein is 10 Å. The CHARMM commands are put in setup_box.inp file, which holds the commands for reading in the PSF file, the coordinate file, for adding counterions and water molecules. The final model is constructed following appropriate removal of overlapping water molecules and trim of the system to desired size. The final model had 45,355 atoms (see Fig. 1). The setup_box.inp file is directed into CHARMM by typing:

```
mpirun -np 1 charmm < setup_box.inp > setup_box.out
```

3.3. Minimization

Before starting an MD simulation, it is advisable to do an energy minimization of the structure. This removes any specious high-energy contacts that may exist. The energy minimization will be done first with the protein heavy atoms fixed in their X-ray positions. This allows the water molecules, the counterions, and the protein's hydrogen atoms to adjust to the protein's heavy atoms. This is followed by a loop of energy minimizations with the backbone and side-chain heavy atoms harmonically restrained at their X-ray positions. The restraint force constants are gradually reduced after each loop. At last, a free energy minimization is performed. All the commands are put in minimization.inp file (Appendix 3). In this file, the topology file, the parameter file, and the PSF are first read into CHARMM; then, a series of energy minimizations are performed after setting respective constraints on a set of atoms. Run the minimization.inp script file as before:

```
mpirun -np 1 charmm < minimization.inp > minimization.out
```

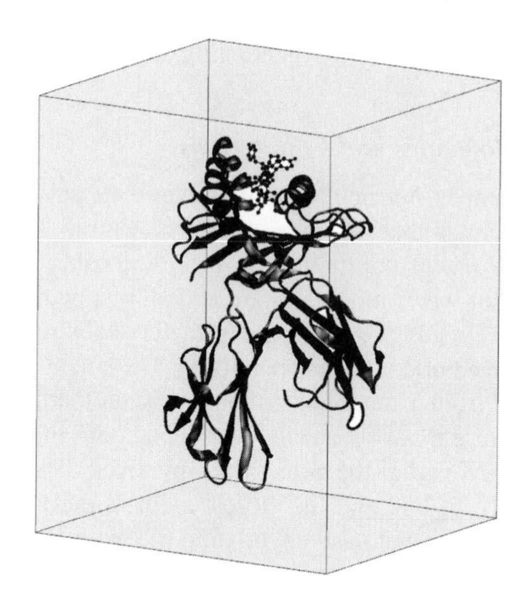

Fig. 1. The peptide–major histocompatibility complex (pMHC) in a box of water. The MHC is represented as ribbon, and Tax peptide as ball-and-stick. Water molecules are not shown for clarity. The picture was created using VMD *(8)*.

3.4. Heating and Equilibrium the System

Heating is the process of increasing the kinetic energy of the system up to a desired temperature. Because MD simulations can often require a significant amount of CPU time, the following simulations will be done using NAMD on multiprocessors. The initial velocities are assigned randomly based on a Maxwellian distribution at 100 K. In our example, new velocities are assigned every 100 steps at a slightly higher temperature (temperature increment is 20 K). This is repeated until the desired temperature (300 K) is reached.

Equilibration is the process where the kinetic energy and the potential energy of the system evenly distribute themselves throughout the system. Once the desired temperature is reached, the simulation continues at this temperature. During this phase, the kinetic and potential energies are monitored. The point of the equilibration phase is to run the simulation until both terms appear to be stable with respect to time.

To perform the above tasks, some commands are needed: reading the coordinates and PSF files, specifying the force field and nonbonded approximations, setting constraint for all bonds involving hydrogen atoms and the internal geometry of the water molecules, setting periodic boundary conditions, using temperature and pressure control, and setting time step and number of steps. All the commands are in md0.inp file (Appendix 4). It is directed to NAMD, which run on 32 processors here as:

```
mpirun -np 32 namd2 md0.inp > md0.out
```

The potential and kinetic energies are extracted from the md0.out. They are shown in Fig. 2 .

3.5. Production Run

The NAMD command file md1.inp (Appendix 5) provides the commands necessary to continue the simulation. In general, these commands are the same as those carried out for heating and equilibrium, except that there is no velocity reassigning during the production phase. The system is allowed to propagate in time without any further intervention. The trajectory (md1.dcd) is collected at this phase for further analysis. This is done by NAMD on 32 processors:

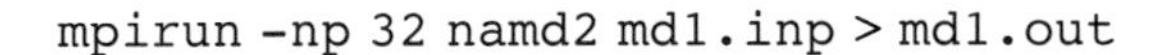

```
mpirun -np 32 namd2 md1.inp > md1.out
```

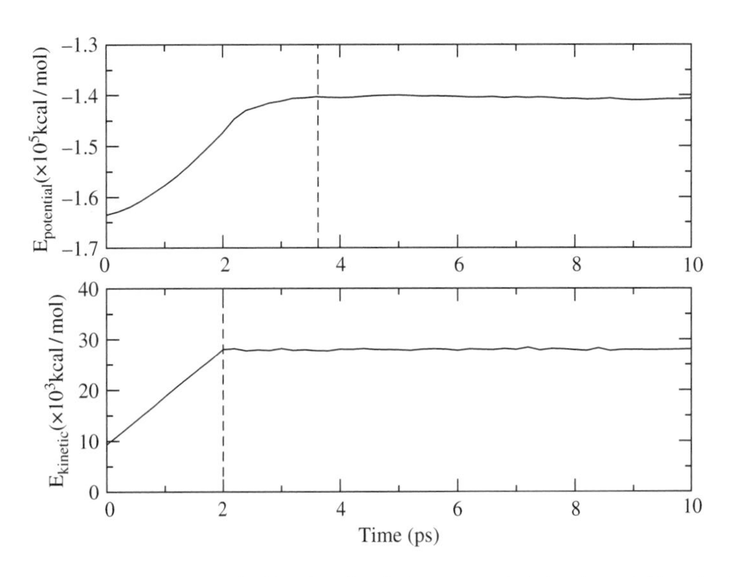

Fig. 2. The time evolution (solid lines) of the potential (top) and kinetic (bottom) energies. The heating process is during 0–2 ps, whereas the equilibration process lasts until 3.6 ps (broken lines).

3.6. Analysis and Visualization

An essential part of any simulation is the analysis of the trajectory, which extracts the structural and energetic information from the production run. The analysis aims at gaining structural and dynamical insights, predicting meaningful statistical properties, and relating structure to function.

It is always helpful to actually see what happens during the simulation. So, we will first use VMD to visualize the trajectory. To start VMD, type *vmd* on the command line of your shell. Load the PSF file first and then md1.dcd file into the same molecule. This will read the DCD trajectory frames and animate it.

The energy information can be extracted from the md1.out file, as we did above for the heating and equilibrium phase.

We will then quantify the changes in protein structures by calculating the root-mean square (RMS) values. The RMS deviations are analyzed after least square fitting of main chain atoms to their X-ray-defined coordinates at the antigen-binding site. The CHARMM commands needed for this task are reading in the topology and parameter files, the PSF, the X-ray coordinates, and the trajectory files; aligning each frame to the X-ray structure; and calculating the

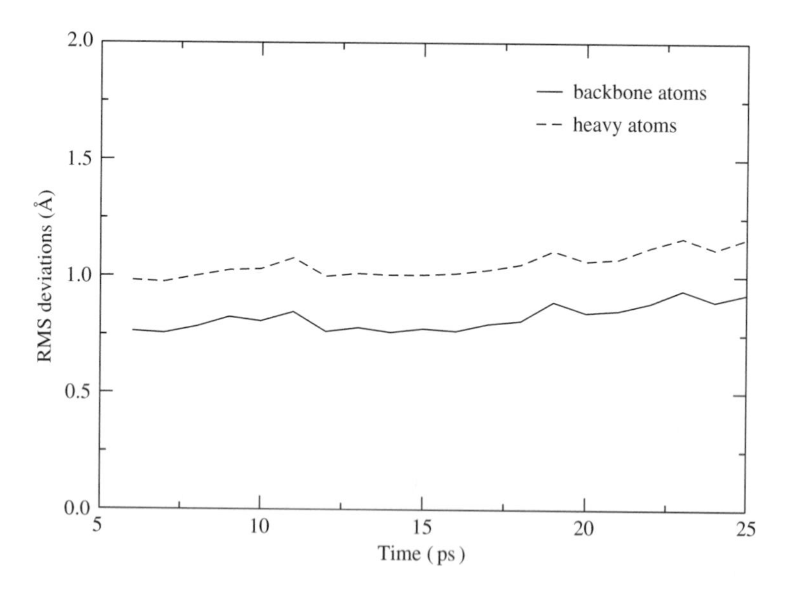

Fig. 3. The time evolution of the root-mean square (RMS) deviations between snapshots of the simulations and the X-ray structure. Broken line represents all heavy atoms and solid line represents backbone atoms. Note that longer simulation is needed to get a stable structure.

RMS deviations for the backbone and all heavy atoms. All commands are in rmsd-t.inp (Appendix 6) and directed to CHARMM by typing:

```
mpirun -np 1 charmm < rmsd-t.inp > rmsd-t.out
```

Figure 3 shows the time evolution of the RMS deviations of the backbone and all heavy atoms from the X-ray structures.

There are more properties that can be obtained by analyzing a longer trajectory. These include the conformational changes, the dynamical properties such as time correlation functions and transport coefficients, and the energetic properties such as binding free energy. Some of these can be related to experimental measurements. We will not cover these aspects here.

References

1. Karplus, M., 2002. Molecular dynamics simulations of biomolecules. *Acc. Chem. Res.* 35: 321–323.
2. Cheng, Y., Zhang, Y. and McCammon, J. A. 2005. How does the cAMP-dependent protein kinase catalyze the phosphorylation reaction: An ab initio QM/MM study. *J. Am. Chem. Soc.* 127: 1553–1562.
3. Tajkhorshid, E., Nollert, P., Jensen, M. Ø., Miercke, L. J. W., O'Connell, J., Stroud, R. M. and Schulten, K. 2002. Control of the selectivity of the aquaporin water channel family by global orientational tuning. *Science* 296: 525–530.
4. DePristo, M. A., de Bakker, P. I. W. and Blundell, T. L. 2004. Heterogeneity and inaccuracy in protein structures solved by x-ray crystallography. *Structure* 12: 831–838.
5. MacKerell, Jr. A. D., Bashford, D., Bellott, M., Dunbrack, Jr. R. L., Evanseck, J. D., Field, M. J., Fischer, S., Gao, J., Guo, H., Ha, S., Joseph-McCarthy, D., Kuchnir, L., Kuczera, K., Lau, F. T. K., Mattos, C., Michnick, S., Ngo, T., Nguyen, D. T., Prodhom, B., Reiher, III, W. E., Roux, B., Schlenkrich, M., Smith, J. C., Stote, R., Straub, J., Watanabe, M., Wiorkiewicz-Kuczera, J., Yin, D. and Karplus, M. 1998. All-atom empirical potential for molecular modeling and dynamics studies of proteins. *J. Phys. Chem. B* 102: 3586–3616.
6. Brooks, B. R., Bruccoleri, R. E., Olafson, B. D., States, D. J., Swaminathan, S. and Karplus, M. 1983. CHARMM: a program for macromolecular energy, minimization, and dynamics calculations. *J. Comput. Chem.* 4: 187–217.
7. Kalé, L., Skeel, R., Bhandarkar, M., Brunner, R., Gursoy, A., Krawetz, N., Phillips, J., Shinozaki, A., Varadarajan, K. and Schulten, K. 1999. NAMD2: greater scalability for parallel molecular dynamics. *J. Comput. Phys.* 151: 283–312.
8. Humphrey, W., Dalke, A. and Schulten, K. 1996. VMD - visual molecular dynamics. *J. Mol. Graph.* 14: 33–38.
9. Khan, A. R., Baker, B. M., Ghosh, P., Biddison, W. E. and Wiley, D. C. 2000. The structure and stability of an HLA-A*0201/octameric tax peptide complex with an empty conserved peptide-N-terminal binding site. *J. Immunol.* 164: 6398–6405.

Appendix 1: build.inp for CHARMM (Section 3.1)

```
* build.inp: generate CHARMM PSF for pMHC (PDB id: 1DUZ);
* Using chain A (HLA), B (B2M), C (TAX) and F (water);
* Some adaption needed for PDB format to fit the CHARMM notation.
*

! read in topology (basic chemical units) and parameter files
open unit 1 card read name top_all27_prot_na.rtf
read rtf card unit 1
close unit 1

open unit 1 card read name par_all27_prot_na.prm
read para card unit 1
close unit 1

! read sequences from the PDB coordinate files;
! separate the original PDB to files for each chain;
! generate one segment for each chain;
open unit 1 card read name hla.pdb
read sequ pdb unit 1
generate hla setu

open unit 1 card read name b2m.pdb
read sequ pdb unit 1
generate b2m setu
open unit 1 card read name tax.pdb
read sequ pdb unit 1
generate tax setu

open unit 1 card read name water.pdb
read sequ pdb unit 1
generate wat noangle nodihedral

! add disulfide bonds using the patch command
patch disu hla 101 hla 164
patch disu hla 203 hla 259
patch disu b2m 26 b2m 81

! the CD1 of ILE (PDB notation) is denoted as CD in the CHARMM,
! carboxyl terminal oxygens are referred to as O and OXT in the PDB;
! they are OT1 and OT2 in CHARMM.
! rename the carboxyl terminal oxygens OT1 to O, and OT2 to OXT.
rename atom CD1 select resname ILE .and. type CD end
rename atom O select type OT1 end
rename atom OXT select type OT2 end
```

```
! read coordinates from the PDB files
open unit 1 card read name hla.pdb
read coor pdb unit 1

open unit 1 card read name b2m.pdb
read coor pdb append unit 1

open unit 1 card read name tax.pdb
read coor pdb append unit 1

open unit 1 card read name water.pdb
read coor pdb append unit 1

! build in hydrogens if using a crystal structure
hbuild sele all end

! delete water molecules far (>=5Å) from protein
delete atom sort -
           select .byres. (segid wat .AND. type oh2 .and. .not. -
           ((.not. (segid wat .OR. hydrogen)) .around. 5)) end

join wat renumber

! write the protein structure file (psf) and coordinate file
open write formatted unit 27 name pmhc.psf
write psf card unit 27
* PSF for pMHC (1DUZ)
*

open unit 1 card write name pmhc.crd
write coor card unit 1
* pMHC (1DUZ): Coordinate with hydrogens
*

close unit 1

stop
```

Appendix 2: setup_box.inp for CHARMM (Section 3.2)

```
* setup_box.inp: add counterions and water molecules
*

! read in topology (basic chemical units) and parameter files
open unit 1 card read name top_all27_prot_na.rtf
read rtf card unit 1
close unit 1
```

```
open unit 1 card read name par_all27_prot_na.prm
read para card unit 1
close unit 1

! read pMHC's psf and coordinate files
open read formatted unit 27 name pmhc.psf
read psf card unit 27
close unit 27

open unit 1 card read name pmhc.crd
read coor card unit 1
close unit 1

! add counterions to make system neutral
open unit 1 card read name na.pdb
read sequ pdb unit 1
generate na setu

open unit 1 card read name na.pdb
read coor pdb append unit 1

! some useful selection definitions
define prot sele .not. ( resname tip3 .or. resname sod ) end
define noh sele .not. hydrogen end

! orientation and translation:
! the geometric centre of selected atoms (protein) is at the origin;
! the principle geometric axis coincides with the {x,y,z} axis.
coor orien sele prot .and. noh end
coor stat sele prot .and. noh end
calc xx = ?xmax + ?xmin
calc yy = ?ymax + ?ymin
calc zz = ?zmax + ?zmin
coor tran sele all end xdir @xx ydir @yy zdir @zz fact -0.5
coor stat sele prot .and. noh end

! add water molecules
! a large water box is used with four segments
read sequence tip3 7560
gene wat1 noangle nodihe
read sequence tip3 7560
gene wat2 noangle nodihe
read sequence tip3 7560
gene wat3 noangle nodihe
read sequence tip3 7560
gene wat4 noangle nodihe
```

```
open unit 1 card read name watbox654.crd
read coor card append unit 1

! keep all crystal water (segment name wat),
! delete other water molecules which are within 2.8Å of
! { protein + crystal water + counterion }
delete atom sort -
            select .byres. ( ( resname tip3 .and. .not segid wat ) -
            .and. type oh2 .and. -
            ( ( ( prot .or. segid wat .or. segid sod ) .and. noh ) -
            .around. 2.8 ) ) end

! trim down system to desired size,
! calculate the desired size first,
! distance between the wall of the box and the closest protein is 10Å
coor stat sele prot .and. noh end
calc xxmin = ?xmin - 10
calc xxmax = ?xmax + 10
calc yymin = ?ymin - 10
calc yymax = ?ymax + 10
calc zzmin = ?zmin - 10
calc zzmax = ?zmax + 10
delete atom sort select .byres. ( ( -
            prop x .lt. @xxmin .or. prop x .gt. @xxmax .or. -
            prop y .lt. @yymin .or. prop y .gt. @yymax .or. -
            prop z .lt. @zzmin .or. prop z .gt. @zzmax ) .and. -
            resname tip3 .and. type oh2 ) end

! write the psf (CHARMM and XPLOR formats) and coordinate (crd & pdb)
open write formatted unit 27 name setup_box.psf
write psf card unit 27
* PSF for pMHC in solvation, CHARMM format
*

open write formatted unit 27 name setup_box_xplor.psf
write psf card xplor unit 27
* PSF for pMHC in solvation, XPLOR format
*

open unit 1 card write name setup_box.pdb
write coor pdb unit 1
* coordinates of pMHC in solvation, pdb format
*

open unit 1 card write name setup_box.crd
write coor card unit 1
```

```
* coordinates of pMHC in solvation, CHARMM format
*

stop
```

Appendix 3: minimization.inp for CHARMM (Section 3.3)

```
* minimization.inp: minimization
* protein is first fixed, then harmonically restrained, and
* at last there are no any constraints.
*

! Read in Topology (basic chemical units) and Parameter files
open unit 1 card read name top_all27_prot_na.rtf
read rtf card unit 1
close unit 1

open unit 1 card read name par_all27_prot_na.prm
read para card unit 1
close unit 1

! Read pMHC's psf and coordinate files
open read formatted unit 27 name setup_box.psf
read psf card unit 27
close unit 27

open unit 1 card read name setup_box.crd
read coor card unit 1
close unit 1

! keep a copy
coor copy comp

! some useful selection definitions
define prot sele ( .not. (resname tip3 .or. resname sod)) end
define noh  sele ( .not. hydrogen ) end
define back sele ( type n .or. type ca .or. type c ) end
define side sele ( (.not. back) .and. prot .and. noh ) end

! fix the heavy atoms of protein
! relax the water molecules, counterions, and protein's hydrogens
cons fix purge sele prot .and. noh end
minimize sd nstep 500 nprint 10 ihbfrq 0 inbfrq 10

! specify energy minimization inside command loop
! constraints on backbone and sidechain with different constants
! Reduce the constraints after each loop
```

```
set 1    0   ! step count
set 2  100   ! step increment (no. of minimization steps each pass)
set 3 1000   ! step limit
set 4   10   ! print frequency

! initialize harmonic constraint potential.
cons harm exponent 2 force 1

! tight on backbone atoms
scalar constraint set 50. select ( prot .and. back ) end
! looser on sidechain atoms
scalar constraint set 25. select ( prot .and. side ) end

label mini

   minimize sd nstep @2 nprint @4 ihbfrq 0 inbfrq 10
   incr 1 by @2

! reduce the harmonic force constants
   scalar constraint multiply 0.5

if 1 lt @3 goto mini ! check for step count

! remove all constraints and minimize
cons harm clear
minimize sd nstep 500 nprint 10 ihbfrq 0 inbfrq 10

! write out the coordinate file
open unit 1 card write name pmhc_min.pdb
write coor pdb unit 1
* Minimization of pMHC (1DUZ) in solvation
*

stop
```

Appendix 4: md0 for NAMD (Section 3.4)

```
# NAMD CONFIGURATION FILE FOR pMHC

# molecular system: coordinate and PSF
coordinates      pmhc_min.pdb
structure        setup_box_xplor.psf

# force field
paraTypeCharmm  on
parameters       par_all27_prot_na.prm
```

```
exclude          scaled1-4
1-4scaling       1.0

# nonbonded interaction approximations
switching        on
switchdist       10
cutoff           12
pairlistdist     13.5
margin           0
stepspercycle    20
fullElectFrequency 4

# PME: long-range electrostatic interaction
pme              on
pmegridsizex     96
pmegridsizey     75
pmegridsizez     72

# SHAKE: all bonds involving hydrogens are to be fixed
rigidbonds       all
rigidtolerance   0.00001
rigiditerations  100

# boundary
cellBasisVector1    94.287   0.0     0.0
cellBasisVector2     0.0    75.951   0.0
cellBasisVector3     0.0     0.0    71.099

# output frequency, output name and format
outputenergies   50
outputtiming     100
outputname       md0
binaryoutput     yes

# temperature control
temperature      100
reassignFreq     100
reassignIncr     20
reassignHold     300
seed             31415926

# pressure control (1 atm)
useGroupPressure     on
BerendsenPressure                    on
BerendsenPressureTarget               1.01325
BerendsenPressureCompressibility 0.0000446
```

```
BerendsenPressureRelaxationTime 200.0
BerendsenPressureFreq           4

# protocol: timestep (2fs) and number of steps (0-10ps)
timestep      2.0
numsteps      5000
```

Appendix 5: md1.inp for NAMD (Section 3.5)

```
# NAMD CONFIGURATION FILE FOR pMHC

# molecular system: coordinate, velocities and PSF
coordinates     pmhc_min.pdb
bincoordinates  md0.coor
binvelocities   md0.vel
structure       setup_box_xplor.psf

# force field
paraTypeCharmm  on
parameters      par_all27_prot_na.prm
exclude         scaled1-4
1-4scaling      1.0

# nonbonded interaction approximations
switching       on
switchdist      10
cutoff          12
pairlistdist    13.5
margin          0
stepspercycle   20
fullElectFrequency 4

# PME: long-range electrostatic interaction
pme             on
pmegridsizex    96
pmegridsizey    75
pmegridsizez    72

# SHAKE: all bonds involving hydrogens are to be fixed
rigidbonds      all
rigidtolerance  0.00001
rigiditerations 100

# boundary
extendedSystem  md0.xsc
wrapall         on
```

```
# output frequency, output name and formatoutputenergies 100
outputtiming    1000
outputname      md1
binaryoutput    yes
restartname     md1_rst
restartfreq     5000
DCDfile         md1.dcd
DCDfreq         500
DCDUnitCell     yes

# temperature control
tcouple         on
tcoupletemp     300
tcouplefile     pmhc_min.pdb
tcouplecol      O

# pressure control
useGroupPressure                    on
BerendsenPressure                   on
BerendsenPressureTarget              1.01325
BerendsenPressureCompressibility     0.0000446
BerendsenPressureRelaxationTime    100.0
BerendsenPressureFreq               4
useFlexibleCell                     yes

# protocol: timestep (2fs) and number of steps (10-30ps)
timestep        2.0
numsteps        15000
firsttimestep   5000
```

Appendix 6: rmsd-t.inp for CHARMM (Section 3.6)

```
* rmsd-t.inp: compute the time-dependence of rms deviation from
* the x-ray structure.
*

! Read in Topology (basic chemical units) and Parameter files
open unit 1 card read name top_all27_prot_na.rtf
read rtf card unit 1
close unit 1

open unit 1 card read name par_all27_prot_na.prm
read para card unit 1
close unit 1

! Read pMHC's psf and coordinate files
```

```
open read formatted unit 27 name setup_box.psf
read psf card unit 27
close unit 27

open unit 1 card read name setup_box.crd
read coor card unit 1
close unit 1

! keep a copy
coor copy comp

! some useful selection definitions
define noh sele .not. hydrogen end
define pro sele segid hla .or. segid b2m .or. segid tax end
define bb sele pro .and. ( type n .or. type ca .or. type c ) end

! read in trajectory
open unit 11 read unformatted name md1.dcd
traj iread 11 nread 1

! read in every frame
! align each frame to the x-ray structure using backbone atoms
! calculate the RMS deviations of backbone and all heavy atoms
set 8 0
label loop
     incr 8 by 1
     traj read
     coor orie rms sele bb .and. pro end
     coor rms sele noh .and. pro end
if 8 lt 20 goto loop

stop
```

25

An Iterative Approach to Class II Predictions

Ronna Reuben Mallios

Summary

An iterative approach to resolving protein–peptide binding motifs is appropriate when the length of the binding protein is variable and a variety of amino acid residues may successfully occupy multiple positions. This chapter describes an iterative algorithm that first aligns binding peptides of variable lengths and then extracts a quantitative motif from the resulting alignment. Numerous examples are presented to illustrate the utility of the iterative process.

Key Words: binding-motif prediction; iterative algorithms; discriminant analysis

1. Introduction

The prediction of class II major histocompatibility complex (MHC)–peptide binding affinity presents a double challenge. It requires both peptide alignment and motif extraction. Although the binding groove in class I MHC molecules is closed at one end and provides a backstop for binding peptides that are 9–10 amino acids long, the binding groove in class II MHC molecules is open at both ends and accommodates peptides varying from 9 to 25 amino acids long *(1,2)*. Consequently, it is necessary to identify the segment of the peptide that participates in the binding before binding attributes can be ascertained.

The iterative algorithm described in this chapter is primarily designed to identify the subsequence of amino acids that participate in binding to a given class II MHC molecule. However, because the resulting model is instrumental in identifying the binding region, it often effectively predicts MHC–peptide binding as well.

From: *Methods in Molecular Biology, vol. 409: Immunoinformatics: Predicting Immunogenicity In Silico*
Edited by: D. R. Flower © Humana Press Inc., Totowa, NJ

2. Methods

2.1. Iterative Algorithm Components

2.1.1. Databases and Data Sets

As their names suggest, data-driven algorithms and models are highly dependent upon the data that serve as input. Standardized measuring techniques and laboratory conditions are necessary to assure the integrity of a database. Additionally, the degree to which the data span the entire sequence space impacts the validity of the resulting model. The term data set denotes a subset of a database that pertains to a given MHC allele. A member of a data set consists of a peptide sequence and the affinity level associated with the peptide and allele under consideration.

For the past decade, the two best sources for publicly available online MHC–peptide affinity data have been MHCBN *(3)* at http://www.imtech.res.in/raghava/mhcbn/ and MHCPEP *(4)* at http://wehih.wehi.edu.au/mhcpep/. For a given MHC allele, both web sites provide a set of peptides with categorical affinities of high, moderate, low, or none. Peptide sequences are reported in standard format with uppercase letters denoting amino acids residues. It should be noted that it is important to include nonbinding peptides as well as peptides with positive binding affinities. The current algorithm was developed using these databases and consequently utilizes categorical binding affinities.

More recently, AntiJen *(5,6)* at http://www.jenner.ac.uk/AntiJen/antijenhomepage.htm has offered several quantitative measures of MHC–peptide binding affinity. Additionally, the National Institute of Allergies and Infectious Diseases (NIAID) is currently contracting the development and maintenance of an integrated, web-based searchable database of antibody-binding sites and antigenic MHC-binding peptides for a wide variety of infectious agents and immune-mediated diseases *(7)*. These databases will be valuable tools in developing a similar iterative algorithm for continuous binding affinities.

2.1.2. Outcome and Potential Predictor Variables

MHC–peptide binding is coded as a categorical variable according to the number of affinity levels being considered. The standard set of potential predictors consists of a binary-valued variable for each amino acid residue and position of interest. Thus, if subsequences of length 9 are under consideration, Y1, Y2, …, Y9 assume a value of 1 when there is a tyrosine present in the indicated position, and a 0 otherwise. In the same manner, additional positional variables can denote the presence or absence of acidic, aliphatic, amidic, aromatic, basic, hydroxylic, or sulfur-containing residues. Interactions among variables are formed by multiplying them together, for example, Y1*V2, Y1*ACIDIC2, or Y1*V2*ACIDIC3.

2.1.3. Procedures for Building Classification Models

For this application, a classification model is a mathematical tool that assigns a binding level to a peptide given its primary sequence. There are many computerized procedures available that extract classification models from data sets. These procedures include artificial neural networks ***(8)***, hidden Markov models ***(9)***, support vector machines ***(10)***, logistic regression, and discriminant analysis ***(11)***. Discriminant analysis is the procedure utilized in this chapter to demonstrate the iterative algorithm.

Given observations that are known to be members of m mutually exclusive and exhaustive groups, and a set of potential predictor variables, discriminant analysis determines classification function coefficients that classify observations into one of the m groups. The stepwise feature provides for the entry of only significant predictors (in the sense of a stepwise multiple analysis of covariance procedure), with the most significant entering first. As there is one set of coefficients for each of m binding levels, if i is the number of steps completed in the discriminant analysis, the classification function for level j is

$$u_j = b_{0j} + b_{1j}v_{1j} + b_{2j}v_{2j} + \ldots + b_{ij}v_{ij}, \tag{1}$$

where v_{1j} through v_{ij} are the predictor variables selected by discriminant analysis and b_{1j} through b_{ij} are the corresponding coefficients. Because the value of all predictor variables is either 0 or 1, the value of u_j reduces to the sum of the coefficients of the variables present in a subsequence plus the constant b_{0j}. In Eq. 1, $u = -D^2/2$, where D quantifies the *Mahalanobis distance* ***(11)*** from an observation to the center of each of the m groups. The shorter the *Mahalanobis distance* is the greater the probability that the observation is a member of a particular group.

Classification functions are converted to the probability of group membership by the following relationship:

$$P_j = \frac{e^{u_j}}{\sum_{i=0,m-1} e^{u_i}}, \tag{2}$$

where P_j is probability that a subsequence belongs to binding level j. The predicted classification of a subsequence is determined by selecting the binding level that is associated with the greatest P_j. For a given observation, the sum of the P_js will always equal 1.

To evaluate the accuracy of a model, the actual binding level and the predicted classification level for each observation are compared. Cross-validation is done by a Leave-one-out analysis where each case is classified by the functions derived from all cases other than that case.

It should be noted that there are two assumptions in discriminant function analysis that are often violated in this application: multivariate normality and equal covariance matrices among groups (homoscedasticity). However, the robustness of the analysis mitigates any serious problems ***(11,12)***. This robustness is confirmed with cross-validation.

2.1.4. Computer Software

The models described in Section 3 were produced using SPSS 12.0 for Windows statistical software and Microsoft Excel 2000.

2.2. Iterative Algorithm

The over-arching iterative algorithm is diagrammed in Fig. 1. The first phase builds the initial model. The iterative phase modifies the model until the sequence alignment stabilizes. The last phase allows for further model refinement after alignment.

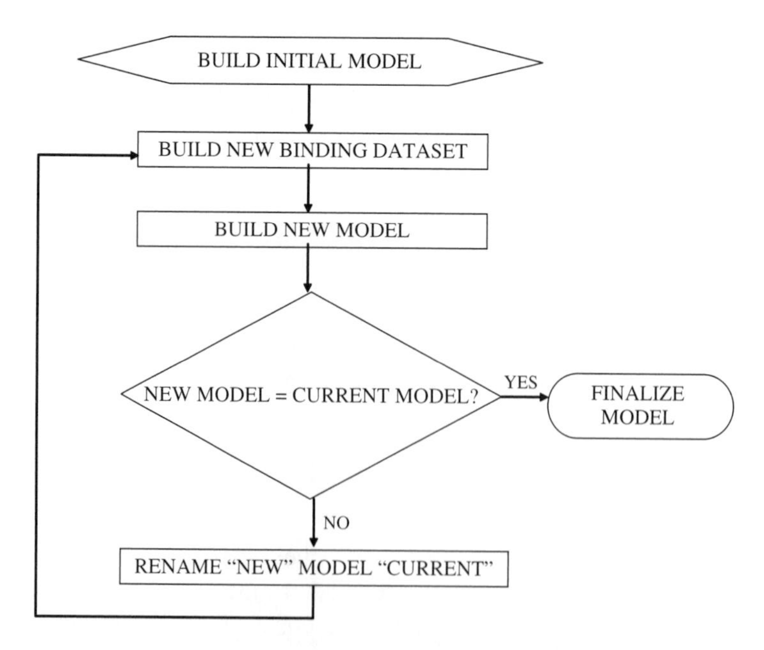

Fig. 1. Diagram of iterative algorithm.

2.2.1. Build Initial Model

Although initial models can be obtained from previous publications or studies, the following method is recommended to avoid bias:

1. Select *n*, the length of subsequences to be considered, and identify the set of potential predictor variables.
2. Populate the nonbinding data set with every subsequence of length *n* from each nonbinding sequence and a binding level of 0. The nonbinding data set will remain unchanged.
3. Populate the initial binding data set with every subsequence of length *n* from each binding sequence and the binding level of the parent sequence.
4. For each subsequence in both data sets, determine the values of all potential predictor variables.
5. Merge the nonbinding data set and the initial binding data set and apply discriminant analysis with the binding level as the dependent variable.
6. Name the classification function coefficients the "current" model.

2.2.2. Iterate Until Alignment Stabilizes

1. Build new binding data set

   a. For each binding sequence, using the "current" model and Eqs. 1 and 2, calculate the probability that each subsequence belongs to the binding level assigned to the parent sequence.
   b. Select for the new binding data set, the subsequence with the maximum probability in Step "a," along with the binding level of the parent sequence.
   c. For each subsequence selected in Step "b," determine the value of all potential predictor variables.

Example: Consider the peptide YVKQNTLKLAT which is known to bind to HLA-DRB1*0101. Suppose the current model is defined by the following classification functions for a two-level prediction of nonbinding (0) and binding (1):

$$u_0 = -2 + 2^*Y1 + 2^*V1 + 2^*K2 + 2^*Q2 + 2^*V2 + 2^*L6 + 2^*A8 + 2^*L9 + 2^*T9 \text{ and}$$

$$u_1 = -12 + 11^*Y1 + 4^*V1 + 5^*K2 + 3^*Q2 + 6^*V2 + 3^*L6 + 7^*A8 + 4^*L9 + 4^*T9.$$

Then, the probabilities of binding (P_1) for the three subsequences of length 9, (a) **YV**KQNTLK**L**, (b) **VK**QNT**L**KLA, and (c) K**Q**NTLKL**AT**, are

a. $u_0 = -2 + 2 + 2 + 2 = 4$; $u_1 = -12 + 11 + 6 + 4 = 9$; $e^{u0} = 54.6$; $e^{u1} = 8103$; $P_1 = 8103/(54.6 + 8103) = 0.99$;
b. $u_0 = 4$; $u_1 = 0$; $e^{u0} = 54.6$; $e^{u1} = 1.0$; $P_1 = 0.02$; and
c. $u_0 = 4$; $u_1 = 2$; $e^{u0} = 54.6$; $e^{u1} = 7.4$; $P_1 = 0.12$.

Because P_1(YVKQNTLKL) = 0.99 is the largest of the three probabilities, YVKQNTLKL is selected as the subsequence most likely to be responsible for binding, and it is entered into the new binding data set. For data analysis, Y1, V2, K3, Q4, N5, T6, L7, K8, and L9 are set to 1, whereas all other potential predictor variables are assigned to 0.

2. Build new model: Merge the permanent nonbinding data set and the new binding data set and apply discriminant analysis with the binding level as the dependent variable.
3. Name the classification function coefficients the "new" model.
4. When the coefficients of the "new" model equal those of the "current" model, the alignment process has stabilized.

2.2.3. Finalize Model

At this point, the model can be utilized as is or can be refined. One method of refinement is to employ receiver operating characteristic (ROC) analysis ***(13)*** to establish a threshold value that separates one level of binding from another. Another refinement collapses binding levels. Additionally, any of the alternative classification methods identified in Section 2.1.3 can be applied.

3. Results: Modeling HLA-DRB1*0101–Peptide Binding Affinity

Four applications of the iterative algorithm for modeling HLA-DRB1*0101 binding are presented. These examples illustrate binary and multilevel classification, predictor variable restriction, and postalignment refinement. In concordance with the goals of vaccine design, the strategy in each case is to identify high-affinity binders. Data sets were retrieved from MHCBN in October 2004 with four binding levels. Affinity levels were originally coded 0 through 3, respectively, denoting none, low, moderate, and high.

3.1. Binary Classification

To reclassify peptides into nonbinding and binding, outcome values 1 through 3 were recoded as 1 and labeled as binding, whereas 0 remained nonbinding. The iterative alignment procedure converged in 11 iterations with 41 predictors. Table 1A displays the dominant standard classification function elements, those with coefficients greater than 3. Within a column, entries are ordered by their contribution to binding.

With two binding levels, classification is determined by which P_j (P_0 or P_1) is greater than 0.5. In addition to accuracy (ACC), binary classification allows for the following evaluation measures: sensitivity (SN), specificity (SP), positive predictive value (PPV), and negative predictive value (NPV). Evaluation of the standard binary converged alignment model is recorded in Table 2A. Note that

Table 1
Binary classification functions (coefficients: binding and nonbinding)

1	2	3	4	5	6	7	8	9
A. Standard converged alignment classification function [Constants (−11.9, −2.5)]								
F(11.9,2.1)	W(6.1,2.2)	M(7.3,1.5)	R(4.1,2.2)	M(5.5,0.9)	I(6.8,2.3)	Q(5.2,1.2)	S(7.0,2.1)	I(8.3,1.8)
Y(11.5,2.1)	V(6.0,2.1)	W(6.2,1.9)			S(4.1,2.0)	P(3.8,1.2)	A(6.8,1.9)	M(7.9,1.9)
I(5.8,2.1)	K(4.5,1.7)	Y(4.2,1.6)			A(3.4,1.7)			L(4.3,1.6)
L(5.3,1.6)	A(4.0,1.8)				L(3.2,1.8)			T(4.0,2.0)
V(3.6,1.7)	Q(3.2,1.7)							F(3.7,1.6)
								V(3.4,1.9)
B. Converged alignment classification function determined by anchor positions [Constants (−7.7, −1.3)]								
Y(10.7,1.2)			L(4.1,1.2)		M(11.1,1.2)			V(3.0,1.4)
F(10.5,1.3)			A(3.3,1.0)		I(10.0,1.5)			I(2.7,0.9)
W(5.5,1.7)			I(3.3,1.4)		A(4.3,1.4)			S(2.7,1.3)
V(4.1,1.6)			D(0.4,1.7)		D(−0.1,1.2)			P(0.4,1.7)
L(3.3,1.3)					M(11.1,1.2)			Q(0.1,1.4)
								H(−0.3,1.8)
								N(−0.4,1.4)
C. Postalignment model after alignment by anchor positions [Constants (−8.8, −1.6)]								
F(11.6,1.6)	V(1.7,0.6)		L(4.6,1.4)	D(−0.2,1.5)	M(11.4,1.3)	S(3.3,1.5)	C(−2.0,0.7)	V(3.4,1.5)
Y(11.5,1.4)	F(−2.8,0.2)		I(3.8,1.5)		I(10.5,1.6)	Q(3.1,1.6)		S(3.2,1.5)
W(5.8,1.6)			A(3.1,0.9)		A(3.8,1.2)	L(3.0,1.6)		I(2.7,0.9)
V(4.6,1.7)			D(0.2,1.7)		D(−0.5,1.1)	A(2.7,1.3)		P(0.0,1.3)
L(3.6,1.5)								H(−0.9,1.6)

Table 2
Evaluation of binary models: binding versus nonbinding

	SN	SP	PPV	NPV	ACC
A. Standard binary alignment model	0.95	0.95	0.91	0.98	0.95
Cross-validation	0.94	0.94	0.90	0.97	0.94
B. Binary anchor alignment model	0.91	0.92	0.86	0.95	0.92
Cross-validation	0.90	0.92	0.86	0.94	0.91
C. Binary anchor postalignment model	0.91	0.93	0.87	0.95	0.92
Cross-validation	0.90	0.93	0.86	0.95	0.92

ACC, accuracy; NPV, negative predictive value; PPV, positive predictive value; SN, sensitivity; SP, specificity.

although it is not known which subsequence is truly responsible for binding, measures of evaluation assume that the subsequence selected for the binding data set is the responsible subsequence.

The mean probability of binding (P_1) was calculated for the selected subsequences of each binding level. The results were none = 0.07, low = 0.88, moderate = 0.92, and high = 0.96. These values suggested that probability of binding and strength of binding were correlated. Consequently, an ROC analysis was performed with P_1 as the test variable, the parent binding level as the state variable, and high as the value of the state variable. The ROC analysis produced an area under the curve (AUC) of 0.922 and indicated an optimal cutoff of 0.94. Thus, if a threshold of 0.94 is applied to P_1, a test is generated for predicting high-affinity binders versus not high-affinity binders. Table 3 displays results of all high-binding versus not high-binding tests devised in this chapter. Row A depicts this binding versus nonbinding model with a 0.94 threshold on P_1.

3.2. Binary Classification Determined by Anchor Positions

This binary model aligned peptides on anchor positions by restricting the potential predictor variables to positions 1, 4, 6, and 9. The iterative procedure converged in 15 iterations with 20 predictors. To refine the model after alignment, discriminant analysis was applied again with potential predictors from all positions. The classification functions for the alignment model are summarized in Table 1B and those for the postalignment model in Table 1C. Model evaluations in Table 2B and 2C reveal that the postalignment model slightly improved the alignment model, but that both did not classify as well as the original binary classification model. Thus, in this instance, focusing on anchor positions appears detrimental rather than advantageous.

Table 3
Evaluation of all high-binding versus not high-binding tests

Model	SN	SP	PPV	NPV	ACC
A. Binding versus nonbinding, 0.94 threshold	0.85	0.86	0.60	0.96	0.85
B. 4-Level collapsed	0.90	0.95	0.82	0.97	0.94
C. 3-Level collapsed	0.96	0.96	0.86	0.99	0.96
D. 3-Level, 0.80 threshold	0.90	0.98	0.91	0.98	0.96

ACC, accuracy; NPV, negative predictive value; PPV, positive predictive value; SN, sensitivity; SP, specificity.

3.3. 4-Level Classification

The MHCBN peptide categorization of none, low-, moderate-, or high-affinity binding naturally lends itself to a 4-level classification model. The iterative procedure converged in 18 iterations with 33 predictors. The dominant classification elements, with coefficients greater than 4, are found in Table 4. The 4-level model correctly classifies 81% of nonbinders, 83% of low binders, 75% of moderate binders, and 90% of high binders, for a total accuracy of 0.83. The accuracy of cross-validation is 0.81.

The 4-level classification functions can be used to separate high binders from non-high binders by collapsing the first three groups into one. Thus, if P_0, P_1, P_2, P_3 are the probabilities of group membership for a given peptide, then the peptide is classified as high binding only if P_3 is the largest of the four probabilities. Otherwise it is classified as non-high binding. Table 3B evaluates the collapsed 4-level model.

3.4. 3-Level Classification

There is justification for condensing low and moderate binders into one group. First, the primary objective is to identify high binders. Second, nonbinders are likely to be very different from the others. Consequently, a 3-level model of nonbinders, low/moderate binders, and high binders was generated. For this 3-level model, the iterative alignment procedure converged in 13 iterations with 38 predictors. The dominant classification elements, with coefficients greater than 3, appear in Table 5. This model correctly classifies 88% of nonbinders, 89% of the

Table 4
4-Level converged alignment classification function coefficients: high, moderate, low, and none. Constants: −12.1, −11.0, −10.7, −2.6.

1	3	4	5	6	7	8
Y(12.7,3.5,−0.5,1.2)	F(7.3,1.7,2.7,1.7)	L(5.5,2.8,2.0,1.5)	M(16.5,4.8,1.8,2.3)	I(12.5,4.4,1.3,2.2)	M(3.8,15.6, −0.0,2.3)	L(5.1,3.3,0.5,1.8)
F(6.5,2.2,1.9,1.1)	M(6.6,2.6,4.4,1.6)	I(2.8,14.6,0.1,1.4)	F(0.6,4.3,3.1,1.3)	T(2.0,4.5,0.3,1.7)	C(−0.4,2.1,7.2,1.3)	S(2.6,5.6,1.2,1.8)
W(6.2,2.5,1.1,1.1)	K(4.6,1.4,0.7,1.2)	E(1.1,1.7,9.8,1.6)	I(0.5,0.8,15.1,1.4)			A(2.4,5.9,−0.4,1.3)
	Y(3.4,11.7,2.2,2.1)					
	V(1.5,1.2,4.6,1.3)					
	L(1.4,0.7,8.7,1.6)					

Table 5
3-Level converged alignment classification function coefficients: high, moderate, low, and none. Constants: −12.1, −9.4, −2.4.

1	2	3	4	5	7	8	9
F(16.3,2.7,2.4)	Y(11.2,0.9,0.9)	M(8.5,2.4,0.9)	I(7.5,2.3,1.8)	M(12.9,0.6,1.5)	A(4.2,2.0,1.5)	L(5.2,0.6,1.0)	I(2.0,13.0,2.2)
Y(10.3,3.5,2.3)			L(5.6,2.8,1.7)	P(−0.1,4.3,2.2)	I(2.2,7.3,2.2)		V(1.7,8.9,1.9)
W(8.7,1.8,2.1)			A(4.0,2.8,1.4)		E(0.9,4.0,1.3)		
K(5.3,1.3,1.7)					C(−0.1,6.2,1.5)		
L(2.3,5.9,1.9)							
E(2.0,4.5,2.4)							

middle group, and 96% of high binders, for a total accuracy of 0.89. The accuracy of cross-validation is 0.87.

Similar to the 4-level model, the collapse of the lower levels produces a high-binding versus not high-binding prediction. Table 3C evaluates the collapsed 3-level model.

The collapsed 3-level model classifies a peptide as high binding only if P_2 is larger than P_0 and P_1. Another strategy, similar to that used with the binary classification model, is to classify a peptide as high binding if P_2 is greater than a given threshold, necessarily greater than 0.5. This strategy has the advantage of increasing the PPV at the expense of SN, which may be desirable in vaccine design. Table 3D illustrates the results when a threshold of 0.80 is applied to P_2 from the 3-level model. A final strategy, often used in practice, is to rank subsequences by the probability of high binding and select top-ranked peptides *(14)*.

4. Conclusion

All of the tests illustrated in Table 3 yield very respectable results for predicting high binding versus non-high binding. In addition, a consensus of the classification functions identifies F1, Y1, W1, I4, L4, A4, I9, and V9 as the most dominant predictors, in agreement with other major studies ***(15–17)***. Thus, the methods presented here have faithfully extracted information from the data ***(18)***, given subsequences of length 9. The accuracy of the modeling process depends upon the true length of the binding region and the reliability and completeness of the data.

References

1. Brown JH, Jardetzky TS, Gorga JC, et al. Three-dimensional structure of the human class II histocompatibility antigen HLA-DR1. *Nature*. 1993;364(6432):33–9.
2. Stern LJ, Brown JH, Jardetzky TS, et al. Crystal structure of the human class II MHC protein HLA-DR1 complexed with an influenza virus peptide. *Nature*. 1994;368(6468):215–21.
3. Bhasin M, Singh H, Raghava GP. MHCBN: a comprehensive database of MHC binding and non-binding peptides. *Bioinformatics*. 2003;19(5):665–6.
4. Brusic V, Rudy G, Harrison LC. MHCPEP, a database of MHC-binding peptides: update 1997. *Nucleic Acids Res*. 1998;26(1):368–71.
5. McSparron H, Blythe MJ, Zygouri C, Doytchinova IA, Flower DR. JenPep: a novel computational information resource for immunobiology and vaccinology. *J Chem Inf Comput Sci*. 2003;43(4):1276–87.
6. Blythe MJ, Doytchinova IA, Flower DR. JenPep: a database of quantitative functional peptide data for immunology. *Bioinformatics*. 2002;18(3):434–9.
7. NIAID. Biodefense Grants and Contracts: FY 2004 Awards http://www2.niaid.nih.gov/biodefense/research/2004awards.

8. Brusic V, Rudy G, Honeyman G, Hammer J, Harrison L. Prediction of MHC class II-binding peptides using an evolutionary algorithm and artificial neural network. *Bioinformatics*. 1998;14(1):121–30.
9. Kato R, Noguchi H, Honda H, Kobayashi T. Hidden Markov model-based approach as the first screening of binding peptides that interact with MHC class II molecules. *Enzyme Microb Technol*. 2003;33(4):472–81.
10. Bhasin M, Raghava GP. SVM based method for predicting HLA-DRB1*0401 binding peptides in an antigen sequence. *Bioinformatics*. 2004;20(3):421–3.
11. Afifi AA, Clark V. *Computer-aided multivariate analysis*. 2nd edn. New York: Van Nostrand Reinhold; 1990.
12. Mallios WS. *Statistical modeling: applications in contemporary issues*. 1st edn. Ames: Iowa State University Press; 1989.
13. Swets JA. Measuring the accuracy of diagnostic systems. *Science*. 1988;240(4857): 1285–93.
14. Mallios RR. A consensus strategy for combining HLA-DR binding algorithms. *Hum Immunol*. 2003;64(9):852–6.
15. Chicz RM, Urban RG, Lane WS, et al. Predominant naturally processed peptides bound to HLA-DR1 are derived from MHC-related molecules and are heterogeneous in size. *Nature*. 1992;358(6389):764–8.
16. Falk K, Rotzschke O, Stevanovic S, Jung G, Rammensee HG. Pool sequencing of natural HLA-DR, DQ, and DP ligands reveals detailed peptide motifs, constraints of processing, and general rules. *Immunogenetics*. 1994;39(4):230–42.
17. Hammer J, Takacs B, Sinigaglia F. Identification of a motif for HLA-DR1 binding peptides using M13 display libraries. *J Exp Med*. 1992;176(4):1007–13.
18. Mallios WS. *The analysis of sports forecasting: modeling parallels between sports gambling and financial markets*. Boston: Kluwer Academic; 2000.

26

Building a Meta-Predictor for MHC Class II-Binding Peptides

Lei Huang*, **Oleksiy Karpenko, Naveen Murugan, and Yang Dai**[1]

Summary

Prediction of class II major histocompatibility complex (MHC)–peptide binding is a challenging task due to variable length of binding peptides. Different computational methods have been developed; however, each has its own strength and weakness. In order to provide reliable prediction, it is important to design a system that enables the integration of outcomes from various predictors. In this chapter, the procedure of building such a meta-predictor based on Naïve Bayesian approach is introduced. The system is designed in such a way that results obtained from any number of individual predictors can be easily incorporated. This meta-predictor is expected to give users more confidence in the prediction.

Key Words: MHC class II binding; epitope prediction; meta-predictor; Naïve Bayesian classifier

1. Introduction

T-cell-mediated immune responses are initiated by the activation of effector T cells. The activation process requires the recognition of the complex formed between an antigen peptide and a major histocompatibility complex (MHC) protein by the T-cell receptor. The identification of peptides that bind to MHC molecules plays a crucial role in understanding the mechanisms of both humoral

* This authors contributed equally.

[1] Address for correspondence: Department of Bioengineering (M/C063), University of Illinois at Chicago, 851 South Morgan Street, Chicago, IL 60607, USA. Tel.: (312) 413-1487; Fax: (312) 413-2018; Email: yandai@uic.edu

From: *Methods in Molecular Biology, vol. 409: Immunoinformatics: Predicting Immunogenicity In Silico*
Edited by: D. R. Flower © Humana Press Inc., Totowa, NJ

and adaptive immunity as well as developing epitope-based vaccines. Experiments for measuring the binding affinities of peptides to MHC molecules are time-consuming and expensive. It is a prohibitive task to identify potential binding peptides from the host and pathogen proteins on a genome-wise scale. Therefore, considerable efforts have been made on the development of computational tools for the identification of MHC-binding peptides *(1,2)*.

Two major types of MHC molecules are involved in the peptide-binding process. MHC class I molecules present endogenous antigens (e.g., viral peptides or tumor antigens synthesized within the cytoplasm of a cell) to CD8+ cytotoxic T cells. MHC class II molecules, on the other hand, present exogenously derived proteins (e.g., bacterial proteins or viral capsid proteins) through antigen-presenting cells (APCs) to CD4+ helper T cells *(3)*. Generally, antigen peptides that bind to both MHC class I and class II molecules are approximately nine amino acid residues long. However, the peptide-binding groove of an MHC class II molecule is open at both ends, which makes it capable of accommodating longer peptides of 10–30 residues *(4–6)*.

The length variability complicates the prediction of peptide–MHC class II binding. However, analyses of the binding motif and the structure of peptide–MHC class II complexes have suggested that a core of nine residues within a peptide is essential for peptide–MHC binding. Computational methods for the prediction include simple binding motifs *(7,8)*, quantitative matrices *(9)*, hidden Markov models *(10)*, artificial neural networks *(11,12)*, and support vector machines *(13)*. Some of these methods require a preprocessing step to align binding sequences with various lengths for the identification of subsequences of the binding cores. Because each method has its own strength and weakness, it is hard for an immunologist to select a single method from the pool of existing predictors. Therefore, a system that produces reliable prediction through the integration of outcomes from major prediction methods is in clear need.

In this chapter, the steps for building such a system based on the Naïve Bayesian *(14)* approach are presented. The Bayesian framework has the flexibility to incorporate any predictor that makes prediction from a computed score correlated with the binding affinity of MHC class II peptides. Here, in order to illustrate the steps of the Bayesian framework, three individual predictors, that is, ProPred, the Gibbs sampler, and the linear programming (LP) model are selected.

ProPred, designed by Singh and Raghava *(15)*, applied the quantitative matrices from 51 HLA-DR alleles for the prediction of MHC class II-binding peptides. These matrices were generated from a pocket profile database described by Sturniolo et al. *(9)* and covered the majority of human HLA-DR specificity.

Nielsen et al. *(16)* proposed an advanced motif sampler method based on the Gibbs sampling technique, which efficiently samples the possible alignment space of binder sequences. For each alignment, a log-odds weight matrix was calculated for the identified binding core subsequences. This matrix serves as the position-specific scoring matrix for the computation of a score for a nonamer.

Motivated by a text mining model designed for building a classifier from labeled and unlabeled examples, Murugan and Dai *(17)* developed an iterative supervised learning model for the prediction of MHC class II-binding peptides. The iterative learning model, based on LP, enables the use of nonbinder information for the detection of the binding cores from a set of putative binding cores and for the construction of the predictor simultaneously. The outcome of this predictor is a position-specific weight matrix that can score amino acids at each position of a nonamer.

2. Materials

1. A data set that includes binding and nonbinding peptides for a specific MHC class II allele. The recommended size of binders is above 100. Any in-house peptide set can be used. If the number of peptides is not sufficient, peptides from databases such as AntiJen *(18)* and MHCBN *(19)* can be added for training. For some alleles, the number of nonbinders may be extremely small. In this case, the random sequences can be added (*see* **Note 1**).
2. Predictors that can score the binding ability for each individual peptide (*see* **Note 2**).

3. Methods

The Bayesian predictor is trained based on the prediction outcome obtained from each individual predictor for a set of training peptides. The system is flexible to incorporate results from any number of predictors. Suppose that the number of predictors is m. In general, the requirement for each predictor is the generation of a score for a given peptide sequence. This score of a peptide is designated as the highest value among all scores that are assigned to the overlapping 9 mer of the peptide by a predictor. A peptide is predicted as a binder/nonbinder if this score is above/below a prescribed threshold value. The steps for building a Bayesian predictor are as follows:

1. Prepare a training data set. Any peptide sequence with length less than nine residues or with undetermined residues in certain positions should be discarded.
2. Reduce the redundancy in the data set. This step is to prevent overestimation of the performance of a predictor. After the reduction, there should be no two peptide

sequences in the set with sequence identity >90% over an alignment of length at least nine residues.

3. Obtain a predictive score for each peptide in the training set (including binding and nonbinding sequences) from each individual predictor. These scores form the input set from which a Bayesian predictor can be built.
4. Determine a set of threshold values that produce distinct pairs of sensitivity and specificity (*see* **Note 3**). This procedure should be performed for each predictor on the training set. Upon the completion of this step, a set of threshold values for predictor j is obtained, say $\delta^j = \left(\delta^j_1, \cdots, \delta^j_{t_j}\right), j = 1, \ldots, m$, where t_j is the number of possible threshold values with the above property for predictor j.
5. Determine the best combination $\delta^* = \left(\delta^{*1}, \cdots, \delta^{*m}\right)$ of threshold values, where each $\delta^{*j} (j = 1, \ldots, m)$ is the selected threshold value for predictor j. This combination can be determined by finding the highest average area under receiver operating characteristic curve A_{ROC} (*see* **Note 3**) value for the Bayesian predictor with a k-fold cross-validation procedure described as follows:

   a. For each combination of threshold values $\delta^1_{i_1}, \cdots, \delta^m_{i_m}$ set up a prediction outcome table for the $(k-1)$-folds of the training peptides (*see* **Note 4**), where $\delta^j_{i_j}$ is the i_jth threshold value for predictor $j, j = 1, \ldots, m$ and $i_j = 1, \ldots, t_j$. This table is of size $n \times m$, where n is the number of peptides in the training folds. The outcome obtained from predictor j for a peptide is denoted by a binary number $f_j : f_j = 1$ if the peptide is predicted as binder, $f_j = 0$ otherwise. Accordingly, the prediction outcome obtained from the m predictors for each peptide will be coded by a binary string $f_1 f_2 \cdots f_m$.
   b. Build the probability table for the Bayesian predictor from the $n \times m$ table described above. Let y_i denote the label of each peptide: $y_i = 1$ if it is a binder and $y_i = -1$ if it is not a binder. The probabilities for each value f_j of the m features for the binder class and the nonbinder class are computed as follows:

$$p(f_j = 1|\text{binder class}) = \frac{\sum_{i:y_i=1} I(f_{ij} = 1)}{\text{total number of binders}}, \quad j = 1, \ldots, m,$$

$$p(f_j = 0|\text{binder class}) = \frac{\sum_{i:y_i=1} I(f_{ij} = 0)}{\text{total number of binders}}, \quad j = 1, \ldots, m,$$

$$p(f_j = 1|\text{nonbinder class}) = \frac{\sum_{i:y_i=-1} I(f_{ij} = 1)}{\text{total number of nonbinders}}, \quad j = 1, \ldots, m, \text{ and}$$

$$p(f_j = 0|\text{nonbinder class}) = \frac{\sum_{i:y_i=-1} I(f_{ij} = 0)}{\text{total number of nonbinders}}, \quad j = 1, \ldots, m,$$

where $I(\cdot) = 1$ if the condition in the parenthesis is true; $I(\cdot) = 0$ otherwise. Note that (i) the total numbers of binders and nonbinders are respectively those in the $(k-1)$ training folds; (ii) the index i in the numerator of each formula runs through all peptides in the $(k-1)$ training folds; and (iii) f_{ij} is the prediction by predictor j for the 9 mer with the highest score from peptide i.

c. For each overlapping 9-mer s_i of a peptide x from the testing fold, compute the ratio of probabilities

$$R_i = \frac{p(f = 1|s_i)}{p(f = 0|s_i)} = \frac{\prod_{j=1}^{m} p(f_{ij}|\text{binder class})}{\prod_{j=1}^{m} p(f_{ij}|\text{nonbinder class})}$$

and select the highest one as the ratio R_x of the peptide x. Here, f_{ij} is the prediction outcome obtained from predictor j for 9-mer s_i. This formula is a straightforward application of the Bayesian rule, without the inclusion of the ratio of prior probabilities p(binder) and p(nonbinder). The influence of prior probabilities on prediction will be implicitly considered through threshold of ratio R_i. With a prescribed threshold δ_B for the Bayesian predictor, the peptide is predicted as a binder if R_x is greater than δ_B, otherwise a nonbinder. Varying the threshold values for δ_B, the A_{ROC} value for the current testing fold can be calculated.

d. Repeat the above steps for the other $k-1$ sets of different training and testing folds and obtain the average A_{ROC} value from the k-testing folds.

e. After obtaining the average A_{ROC} values for all possible combinations of $(\delta^l_{i_l}, \ldots, \delta^m_{i_m})$, identify the best combination $\delta^* = (\delta^{*l}, \ldots, \delta^{*m})$ that corresponds to the highest average A_{ROC} value.

6. Construct the final Bayesian predictor by using the outcome table determined from the best combination of threshold $\delta^* = (\delta^{*l}, \ldots, \delta^{*m})$ for the entire training peptides. That is, build the outcome table following the step 5a with threshold $\delta^* = (\delta^{*l}, \ldots, \delta^{*m})$ and the entire training set. Then compile the probability table as described in the above step 5b. By varying threshold values for δ_B, obtain the corresponding sensitivity and specificity for the entire training set and compute an A_{ROC} value.

The general framework of building a Bayesian predictor is summarized in Fig. 1.

The threshold δ_B for the Bayesian classifier for testing has to be determined based on the requirement for sensitivity and specificity specified by users (*see* **Note 5**). The Bayesian predictor predicts a peptide as a binder if the highest value among the ratios $p(f = 1|s_i)/p(f = 0|s_i)$ for all overlapping 9-mer s_i from the peptide is great than δ_B, otherwise predicts it as nonbinder.

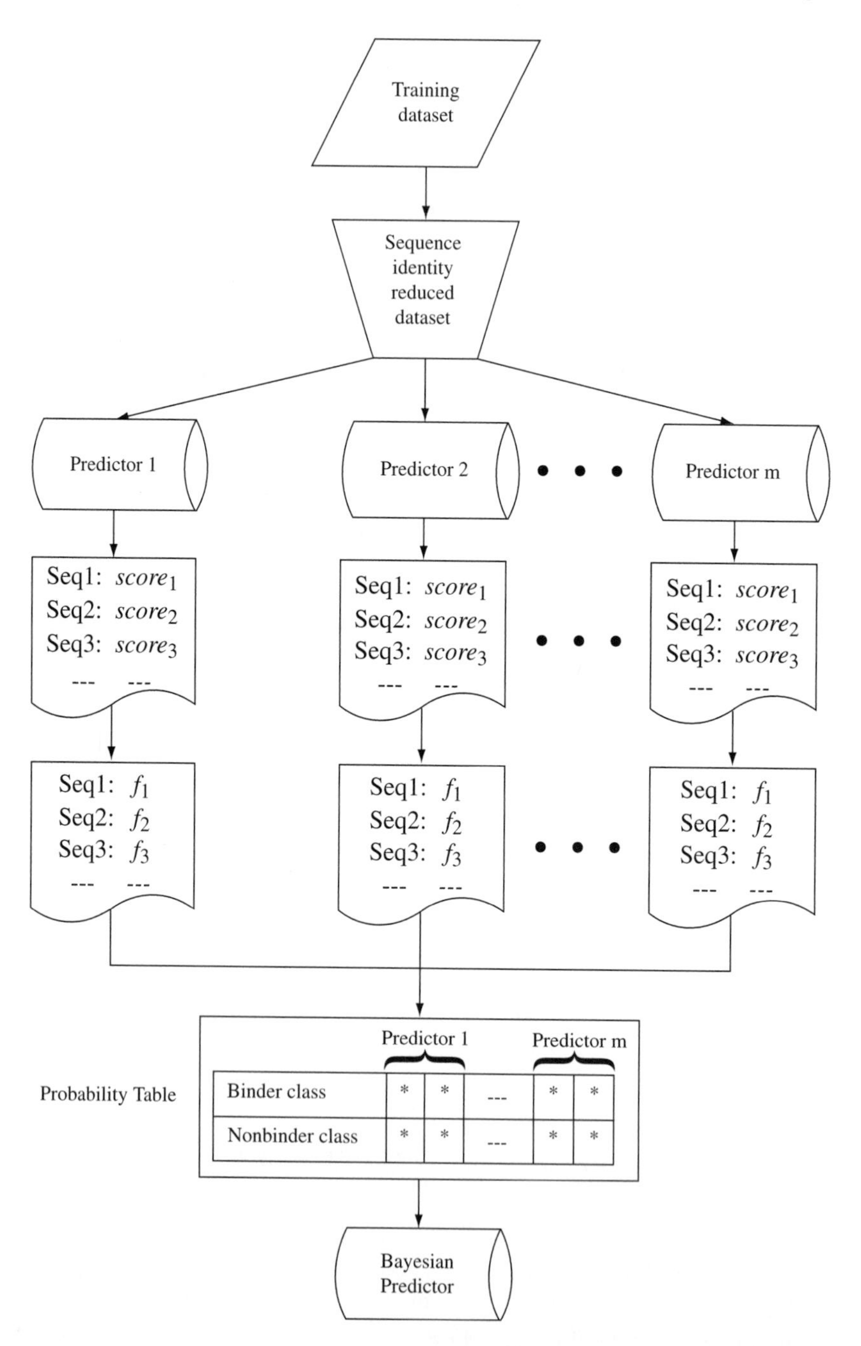

Fig. 1. Illustration of the framework for building a Bayesian predictor.

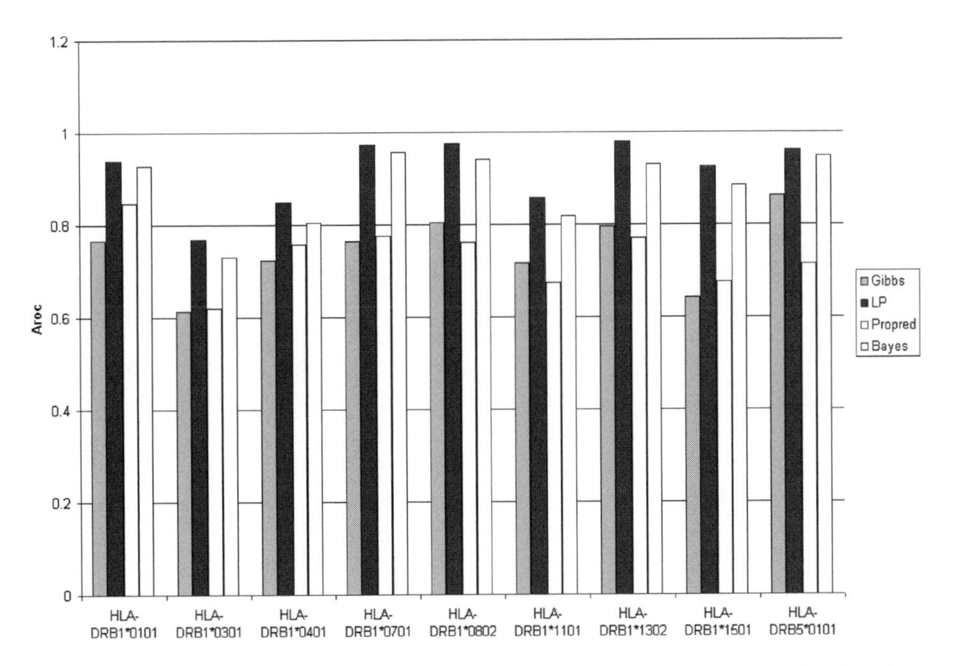

Fig. 2. Comparison of the performance of the Bayesian predictor with the three individual predictors.

For reference the performance of the Bayesian predictor built from the three individual predictors (i) ProPred ***(15)***, (ii) Gibbs sampler ***(16)***, and (iii) the LP predictor ***(17)*** in our illustrative example is shown in Fig. 2. The corresponding web server can be accessed at http://array.bioengr.uic.edu/cgi-in/mhc2srv/testing.web.pl.

Acknowledgments

This work is partially supported by the NIH grant (1 R03 AI069391-01).

Notes

1. In our study, peptide sequences were obtained from two databases: AntiJen ***(18)*** and MHCBN ***(19)***. Considering the size of training set, nine alleles were selected: HLA-DRB1*0101, HLA-DRB1*0301, HLA-DRB1*0401, HLA-DRB1*0701, HLA-DRB1*0802, HLA-DRB1*1101, HLA-DRB1*1302, HLA-DRB1*1501, and HLA-DRB5*0101. Random peptide sequences can be randomly chosen from Swiss-Prot or Entrez database.
2. Each individual predictor may be a position-specific scoring matrix, which is of size 20 by 9. The score of a 9-mer is defined as $\sum_{l=1}^{9} s(l)$, where $s(l)$ is the value in

the lth column of the matrix corresponding to the residue appeared at position l of the 9 mer. The score may not be the actual binding affinity of the 9 mer; however, the magnitude correlates the strength of the binding.

3. The A_{ROC} value is the area under receiver operating characteristic curve ***(20)***, which is determined from a set of values of (1-specificity, sensitivity) derived from different values of threshold of a predictor for a set of binder and nonbinder peptides. The sensitivity and specificity are defined as TP/(TP + FN) and TN/(TN + FP), respectively, where TP and FN are the respective numbers of predicted binders and nonbinders which are true binders; TN and FP are respective numbers of predicted nonbinders and binders which are true nonbinders. An A_{ROC} value close to 1 is desirable for a predictor. A random predictor has an A_{ROC} value of 0.
4. In the k-fold cross-validation, the ratio between the number of binders and the number of nonbinders in all k-folds should be approximately equal. This is important for training.
5. The testing threshold δ_B for the Bayesian predictor is specified by the requirement of the users. In general, the recommended value for δ_B is that the sensitivity and specificity of the predictor are approximately equal. However, it is also possible to select a value for δ_B at which the sensitivity is higher than the specificity; or conversely, δ_B at which the specificity is higher than the sensitivity. The values for δ_B and the corresponding values of the sensitivity and specificity can be obtained for the Bayesian predictor. These values indicate the quality that the predictor may have when the prediction is made for new peptide sequences. In our illustrative example, if one wishes the final predictor to target sensitivity at a level of 0.7, then the proper choice for δ_B should be 1.188 (see Table 1).

Table 1
Threshold values for the Bayesian predictor and the corresponding sensitivity and specificity for HLA-DRB1*0401 allele

Threshold	Sensitivity	Specificity
1.602	0.000	1.000
1.601	0.554	0.876
1.188	0.710	0.817
0.849	0.790	0.774
0.630	0.842	0.651
0.563	0.878	0.505
0.402	0.918	0.339
0.299	1.000	0.000

References

1. Flower, D. R. (2004) Vaccines in silico – the growth and power of immunoinformatics. *The Biochemist* **26**, 17–20.
2. De Groot, A. S. and Berzofsky, J. A. (2004) From genome to vaccine – new immunoinformatics tools for vaccine design. *Methods* **34**, 425–428.
3. Parham, P. (2005) *The Immune System*. Garland Science, New York, NY.
4. Castellino, F., Zhong, G., and Germain, R. N. (1997) Antigen presentation by MHC class II molecules: invariant chain function, protein trafficking, and the molecular basis of diverse determinant capture. *Hum. Immunol.* **54**, 159–169.
5. Sette, A., Buus, S., Appella, E., Smith, J. A., Chesnut, R., Miles, C., Colon, S. M., and Grey, H. M. (1989) Prediction of major histocompatibility complex binding regions of protein antigens by sequence pattern analysis. *Proc. Natl. Acad. Sci. U.S.A.* **86**, 3296–3300.
6. Max, H., Halder, T., Kropshofer, H., Kalbus, M., Muller, C. A., and Kalbacher, H. (1993) Characterization of peptides bound to extracellular and intracellular HLA-DR1 molecules. *Hum. Immunol.* **38**, 193–200.
7. Rammensee, H., Bachmann, J., Emmerich, N. P., Bachor, O. A., and Stevanovic, S. (1999) SYFPEITHI: database for MHC ligands and peptide motifs. *Immunogenetics* **50**, 213–219.
8. Borras-Cuesta, F., Golvano, J., Garcia-Granero, M., Sarobe, P., Riezu-Boj, J., Huarte, E., and Lasarte, J. (2000) Specific and general HLA-DR binding motifs: comparison of algorithms. *Hum. Immunol.* **61**, 266–278.
9. Sturniolo, T., Bono, E., Ding, J., Raddrizzani, L., Tuereci, O., Sahin, U., Braxenthaler, M., Gallazzi, F., Protti, M. P., Sinigaglia, F., and Hammer, J. (1999) Generation of tissue-specific and promiscuous HLA ligand databases using DNA microarrays and virtual HLA class II matrices. *Nat. Biotechnol.* **17**, 555–561.
10. Kato, R., Noguchi, H., Honda, H., and Kobayashi, T. (2003) Hidden Markov model-based approach as the first screening of binding peptides that interact with MHC class II molecules. *Enzyme Microb. Technol.* **33**, 472–481.
11. Brusic, V., Rudy, G., Honeyman, G., Hammer, J., and Harrison, L. (1998) Prediction of MHC class II-binding peptides using an evolutionary algorithm and artificial neural network. *Bioinformatics* **14**, 121–130.
12. Nielsen, M., Lundegaard, C., Worning, P., Lauemoller, S. L., Lamberth, K., Buus, S., Brunak, S., and Lund, O. (2003) Reliable prediction of T-cell epitopes using neural networks with novel sequence representations. *Protein Sci.* **12**, 1007–1017.
13. Bhasin, M. and Raghava, G. P. (2004) SVM based method for predicting HLA-DRB1*0401 binding peptides in an antigen sequence. *Bioinformatics* **20**, 421–423.
14. Theodoridis, S. and Koutroumbas, K. (1999) *Pattern Recognition*. Academic Press, San Diego, CA.

15. Singh, H. and Raghava, G. P. (2001) ProPred: prediction of HLA-DR binding sites. *Bioinformatics* **17**, 1236–1237.
16. Nielsen, M., Lundegaard, C., Worning, P., Hvid, C. S., Lamberth, K., Buus, S., Brunak, S., and Lund, O. (2004) Improved prediction of MHC class I and class II epitopes using a novel Gibbs sampling approach. *Bioinformatics* **20**, 1388–1397.
17. Murugan, N. and Dai, Y. (2005) Prediction of MHC class II binding peptides based on an iterative learning model. *Immunome Res.* **1**, 6.
18. Toseland, C. P., Clayton, D. J., McSparron, H., Hemsley, S. L., Blythe, M. J., Paine, K., Doytchinova, I. A., Guan, P., Hattotuwagama, C. K., and Flower, D. R. (2005) AntiJen: a quantitative immunology database integrating functional, thermodynamic, kinetic, biophysical, and cellular data. *Immunome Res.* **1**, 4.
19. Bhasin, M., Singh, H., and Raghava, G. P. (2003) MHCBN: a comprehensive database of MHC binding and non-binding peptides. *Bioinformatics* **19**, 665–666.
20. Swets, J. A. (1988) Measuring the accuracy of diagnostic systems. *Science* **240**, 1285–1293.

27

Nonlinear Predictive Modeling of MHC Class II–Peptide Binding Using Bayesian Neural Networks

David A. Winkler* **and Frank R. Burden**

Summary

Methods for predicting the binding affinity of peptides to the MHC have become more sophisticated in the past 5–10 years. It is possible to use computational quantitative structure-activity methods to build models of peptide affinity that are truly predictive. Two of the most useful methods for building models are Bayesian regularized neural networks for continuous or discrete (categorical) data and support vector machines (SVMs) for discrete data. We illustrate the application of Bayesian regularized neural networks to modeling MHC class II-binding affinity of peptides. Training data comprised sequences and binding data for nonamer (nine amino acid) peptides. Peptides were characterized by mathematical representations of several types. Independent test data comprised sequences and binding data for peptides of length ≤ 25. We also internally validated the models by using 30% of the data in an internal test set. We obtained robust models, with near-identical statistics for multiple training runs. We determined how predictive our models were using statistical tests and area under the receiver operating characteristic (ROC) graphs (A_{ROC}). Some mathematical representations of the peptides were more efficient than others and were able to generalize to unknown peptides outside of the training space. Bayesian neural networks are robust, efficient "universal approximators" that are well able to tackle the difficult problem of correctly predicting the MHC class II-binding activities of a majority of the test set peptides.

Key Words: Bayesian neural networks; quantitative structure-activity relationships; T-cell epitope; major histocompatibility complex; peptide binding

* Address for correspondence: Centre for Complexity in Drug Discovery, CSIRO Molecular and Health Technologies, Clayton, Australia. Tel.: +61-3-9545-2477; Fax: +61-3-9545-2446; Email: dave.winkler@csiro.au

From: *Methods in Molecular Biology, vol. 409: Immunoinformatics: Predicting Immunogenicity In Silico*
Edited by: D. R. Flower © Humana Press Inc., Totowa, NJ

1. Introduction

Major histocompatibility complex (MHC) proteins are cell-surface glycoproteins present on antigen-presenting cells. They recognize and bind peptides identified by CD4+ T cells and result in activation of the T cell, thus playing a crucial role in initiation, enhancement, and suppression of immune responses. Two classes of MHC molecules exist. Those that bind peptides derived by degradation of intracellular proteins are labeled class I, and those binding peptides derived from extracellular protein degradation are class II. MHC class II-binding peptides, which induce and recall T-cell responses, are called T-cell epitopes.

1.1. Summary of MHC Modeling Research

Identification of T-cell epitopes can be useful for developing disease therapies (e.g., malaria). A relatively small number of research groups have tackled the difficult problem of generating predictive Quantitative Structure-Activity Relationship (QSAR) models for peptide binding to the MHC class II-binding peptides. Buus *(1)* identified privileged binding motifs binding peptides and used these to help build QSAR models of human immune reactivities. Doytchinova and Flower *(2)* employed the 3D QSAR (those that consider the topographical rather than topological properties of the binding peptides) to model the affinity of peptides for the class I MHC HLA-AA*0201 allele. More recently, these workers used an "additive" linear regression method to predict MHC protein–peptide binding *(3)*. They assumed that binding affinity was an additive function of the contributions of amino acids in each position of the peptide, plus some linear interactions between amino acids and their neighbors (*see* **Note 1**). Logean, Sette, and Rognen *(4)* derived a customized free energy scoring function to predict the binding affinity of 26 peptides to the class I MHC HLA-B*2705 protein. Their method ranked the binding affinities and predicted their binding energies within 3–4 kJ/mol. Brusic et al. *(5)* employed backpropagation neural networks to generate a nonlinear model of HLA-A11 binding by nine amino acid peptides. Comparisons between neural networks and alternative QSAR modeling methods for MHC–peptide binding have been reported by Gulukota and coworkers *(6)*. De Hann et al. *(7)* elucidated the relative individual contributions of side-chain hydrogen bonding and flexibility to MHC-binding affinity of peptides using peptoid surrogates.The very efficient support vector machine (SVM) method by Bhasin and Raghava *(8)* was used to classify a relatively large set of peptides binding to HLA-DRB1* 0401 allele.

1.2. Need for Nonlinear, Robust Methods of Modeling Peptide Binding to MHC

MHC class II–peptide recognition is known to be a more complex process to model than class I recognition. The theoretical and experimental studies to date have suggested strongly that the interaction of peptides with the MHC is nonlinear and complex. Interactions between amino acids as well as between the peptides and the protein play an important role in modulating binding affinity. Buus *(1)* reviewed a number of general approaches for MHC-binding affinity prediction and advocated strongly for the application of neural networks because they were better suited to recognizing complicated peptide patterns than binding motifs and other methods. In addition, it is clear that QSAR methods that rely on explicit determination of interaction terms are problematic. They require a subjective input from the modeler concerning which interactions are important. More importantly, the number of interactions that must be included in the model increases dramatically, potentially causing problems with deriving statically valid models that are not overfitted. For example, Gulukota et al. *(6)* encountered this problem when modeling peptide binding to the MHC class I HLA-A2.1 allele. The neural network architecture they used had 180 descriptors and 50 hidden layer neurodes resulting in used-over 9000 weights. As their training set was small ($\sim$200–300), and the number of adjustable weights high, the probability of overfitting is high also. Doytchinova, Blythe, and Flower's work provides another example where overfitting posed a problem. In their model, they included all neighbor linear interaction terms. As the number of possible terms and cross-terms was very large (6,180), they needed to employ a linear variable reduction method (partial least squares) to build a linear model of the peptide binding that avoided overfitting.

We have developed a robust, nonlinear, parsimonious, structure-property mapping methodology able to model relationships between chemical structure and a wide variety of properties. Using these methods, we have built predictive models of drug target activity *(9)*, ADME properties *(10)*, toxicity *(11)*, and phase II metabolism *(12)*, amongst other properties. We describe how this method can be applied to build predictive models of peptide binding to the MHC.

1.3. Advantages of Bayesian Regularized Neural Networks

Our methodology employs Bayesian regularized neural networks and several types of molecular descriptors to build predictive models of peptide binding *(13)*. Applying a Bayesian framework to a neural network provides a number of advantages over traditional backpropagation neural networks used in QSAR studies, including those modeling peptide binding to the MHC. Neural nets are "universal approximators" able to model continuous, complex, nonlinear response surfaces of arbitrary complexity, given sufficient training data. However, they can overfit data, be overtrained (compromising their ability to predict), require partitioning of sometimes scarce data into validation sets to determine when to stop training, and selecting the correct network architecture is difficult. Bayesian neural nets on the other hand are robust, difficult to overtrain, minimize the risk of overfitting, are tolerant of noisy or missing data, automatically find the least complex model that explains the data, and can automatically optimize their architecture ***(14)***.

1.4. Bayes' Theorem

Bayes' theorem employs conditional probability to make predictions. Bayes' theorem, sometimes called *the inverse probability law*, is an example of a powerful statistical inference method. Bayes' theorem is derived fairly simply. If two events are independent of each other, the probability of A and B occurring is the product of the individual probabilities, P(A and B) = P(A)P(B). The conditional probability (probability of B given A) is written as P(B|A). It is defined as

$$P(B|A) = P(A \text{ and } B)/P(A) \tag{1}$$

This can be rearranged to P(A and B) = P(A)P(B|A).
Similarly

$$P(A|B) = P(A \text{ and } B)/P(B) \tag{2}$$

Combining Eqs. 1 and 2 gives Bayes' theorem:

$$P(A|B) = P(B|A)P(A)/P(B)$$

We can use Bayes' theorem to find the conditional probability of event A given the conditional probability of event B and the independent probabilities of events A and B. Not everyone is comfortable with Bayes' theorem. Some find it difficult to accept that instead of using probability to predict the future, probability is used to make inferences about the past. Bayes' theorem can be used to optimally control regularization and make artificial neural networks (ANNs) more robust, parsimonious, and interpretable.

1.5. Bayesian Regularization of Neural Networks

Details of Bayesian regularization applied to backpropagation neural networks may be found in previous publications ***(13,14)***; so, only a brief summary is given here.

Regression can exhibit instability for some data sets, and regularization is used to improve the modeling robustness. For neural network-based regression, the aim is to modify the weights in the neural network to minimize the cost function $S(\mathbf{w})$.

$$S(\mathbf{w}) = \sum_{i=1}^{N_D} [\mathbf{y}_i - f(\mathbf{x}_i)]^2$$

It is difficult to find the optimum complexity for a regression model that balances the tendency for bias [model is too simple to capture the underlying (nonlinear) relationship] and variance (model is overly complex and explains the noise as well as the underlying relationship). To control this problem and minimize the risk of an overly complex model, the simple device of adding an extra *regularization* term λ (a weight penalty) to the cost function *regularizes* the solution. Regularization is used in many modeling methods such as ridge regression.

The cost function, $S(\mathbf{w})$, is subsequently minimized with respect to the weights.

$$S(\mathbf{w}) = \sum_{i=1}^{N_D} [\mathbf{y}_i - f(\mathbf{x}_i)]^2 + \lambda \sum_{j=1}^{N_p} w_j^2$$

This equation can be slightly rewritten in terms of hyperparameters α and β instead of λ.

$$S(\mathbf{w}) = \beta \sum_{i=1}^{N_D} [\mathbf{y}_i - f(\mathbf{x}_i)]^2 + \alpha \sum_{j=1}^{N_W} w_j^2$$

where N_W is the number of weights. Given initial values of the hyperparameters, α and β, the cost function, $S(\mathbf{w})$, is minimized with respect to the weights $\mathbf{w}$. A re-estimate of α and β is made by maximizing the Bayesian evidence for the model.

$$P(\alpha, \beta | D) = \frac{P(D|\alpha, \beta)\, P(\alpha, \beta)}{P(D)}.$$

Nabney *(15)* showed that only the evidence, $P(D|\alpha, \beta)$, needs to be maximized, and that the priors $P(D)$ and $P(\alpha, \beta)$ can be ignored. It can be shown *(16)* that the log of the evidence for α and β can be written as

$$\log P(D|\alpha, \beta) = -\alpha E_W^{\mathrm{MP}} - \beta E_D^{\mathrm{MP}} - \frac{1}{2}\ln(|\mathbf{C}|) - \frac{N_W}{2}\ln\alpha - \frac{N_D}{2}\ln\beta - \frac{k}{2}\log 2\pi$$

which is maximized with respect to α and β (N_W is the number of weights, and N_D is the number of data points).

Two optimizations are carried out: minimization with respect to the weights and maximization with respect to α and β until self-consistency is achieved. The new values of α and β are re-evaluated using

$$\alpha = \gamma/2E_W,$$

$$\beta = (N_D - \gamma)/2E_D, \text{ and}$$

$$\gamma = \sum_{i=1}^{N_W} \frac{\lambda_i}{\lambda_i + \alpha} = N_P - \alpha \text{ trace}\left(\mathbf{C}^{-1}\right).$$

γ is the effective number of parameters necessary for the model.

Applying a Bayesian framework to the neural net results in a probability distribution of weights, not a single set of weights, allowing error bars to be calculated for predictions.

2. Materials

We outline typical steps involved in building a Bayesian regularized neural network model of epitope binding to the MHC class II. We have published a complete descriptions of the use of Bayesian neural network methods to build QSAR models explaining MHC class II-binding activity of peptides to two HLA protein alleles, HLA-DRB1*0101 and HLA-DRB1*0301 *(17)*. This specific example is used to illustrate the methods. The primary steps are:

1. Collect a set of peptides of known sequences and binding affinities. The data set should be as large and diverse as possible (*see* **Note 2**). The peptides should have the recognition motif identified. For example, we employed a nine amino acid peptide-binding data from the MHCPEP database *(18)*. We used peptide-binding data for two alleles, the HLA-DRB1*0101 (data set 101; 1,408 peptides) and HLA-DRB1*0301 (data set 301; 349 peptides), to build binding models for each allele.
2. Partition the data set into a training set and test set. The partitioning can be random or based on a clustering method such as k-nearest neighbors (kNN). The proportions allocated to the training and test sets are typically 80 : 20, although in our example we used 70 : 30.

Table 1
Property descriptors for the amino acids

Number	Amino acid	Code	N_{accept}	N_{donor}	N_{hetero}	Mol. Volume	logP	Rot. bonds	Charge
1	Alanine	A	0	0	0	45	1.8	0	0
2	Cysteine	C	0	1	1	62	1.2	0	0
3	Aspartate	D	2	0	2	69	−1.5	1	−1
4	Glutamate	E	2	0	2	87	1.2	2	−1
5	Phenylalanine	F	0	0	0	115	3.3	1	0
6	Glycine	G	0	0	0	29	1.1	0	0
7	Histidine	H	2	2	2	97	1.0	1	0
8	Isoleucine	I	0	0	0	96	3.3	1	0
9	Lysine	K	0	1	1	108	−1.2	3	1
10	Leucine	L	0	0	0	96	3.3	1	0
11	Methionine	M	0	0	1	97	2.0	2	0
12	Asparagine	N	1	1	2	76	−1.5	1	0
13	Proline	P	0	0	0	76	2.9	0	0
14	Glutamine	Q	1	1	2	93	−1.0	2	0
15	Arginine	R	0	3	3	128	3.3	4	1
16	Serine	S	1	1	1	54	−0.2	0	0
17	Threonine	T	1	1	1	70	0.1	0	0
18	Valine	V	0	0	0	77	2.7	0	0
19	Tryptophan	W	0	1	1	145	2.6	1	0
20	Tyrosine	Y	1	1	1	124	2.7	1	0

N_{accept}, N_{donor}, and N_{hetero} are the numbers of hydrogen bond acceptor, hydrogen bond donor and heteroatoms, respectively. Mol. volume, logP, rot. bonds, and charge represent the molar volume, log of the octanol–water partition coefficient, number of rotatable bonds, and formal charge at physiological pH, respectively.

3. If the binding affinities to the MHC allele are continuous variables, they are usually converted to log values. If they are discrete or categorical values, they can be used without transformation if they correspond approximately to log affinity ranges. For example, categorical or discrete values may be 0 if the peptide does not bind and 1 if they do. They may also be integers chosen to represent no binding, weak binding, moderate binding, and strong binding (*see*, however, **Note 3**). In our example, we used an activity class of 1 = nil-binding activity (class N), 5 = low-binding activity (class L), 7 = moderate-binding activity (class M), and 9 = high-binding activity (class H).
4. The amino acids corresponding to each position in the sequence are converted into mathematical representations. This can be done in many ways (*see* **Notes 4** and **5**). Common types of representations (also known as descriptors) include a binary vector of length 20 which contains all zeros except for the position corresponding to the amino acid at that position in the sequence. It is also common to describe amino acids using a string of values that correspond to physicochemical properties of the amino acids (e.g., taken from the amino acid descriptor database) or related representations known as z descriptors ***(19)***. This provides a matrix of values used as input to the neural network. In our example, we used two types of peptide mathematical representation: binary vectors ($9 \times 20 = 180$ descriptors for each peptide); property descriptors based on seven physical properties of the 20 naturally occurring amino acids ($9 \times 7 = 63$ property descriptors for each peptide). The property descriptors for the 20 amino acids are summarized in Table 1.

3. Methods

The peptide classification problem is of substantially different character to the type of modeling to which Bayesian neural nets have been applied previously. However, results show that the method is capable of producing good, predictive models of MHC–peptide binding, due to its ability to deal with nonlinear response surfaces and interactions between components.

1. There Bayesian neural network has been implemented as a module (Mol.SAR) in a commercial software package, Know It All, distributed by the Bio-RAD Corporation (see http://www.knowitall.com). Information is available in the book by Nabney ***(15)***, which allows these neural networks to be implemented in a high-level language such as MATLAB. There is also a lot of useful information on implementing Bayesian regularized neural networks in the papers by Mackay ***(20)***.
2. Once the algorithm is implemented, the neural network architecture needs to be chosen. Neural networks typically consist of three layers: an input layer where the mathematical descriptions of the peptides are applied; the hidden layer that does the processing; the output layer, where the model makes its prediction and the error is derived and propagated back through the network. The mathematical representations of the amino acid sequence of the peptide are presented to the input layer. This layer is connected to the central hidden layer, which in turn is connected to the output layer, typically consisting of a single neurode. The neurodes

in one layer are usually fully connected to the neurodes of the next layer. The number of neurodes in the network determines the number of weights (number of fitted parameters in the model). In typical backpropagation neural networks, the architecture is commonly chosen so that the number of weights is approximately half the number of sequences in the training set. This avoids overfitting, which becomes more likely as the number of training sequences and number of weights approach each other. Bayesian neural networks are parsimonious, and the effective number of weights in the model is usually less than the number of weights, and so, overfitting is avoided. In our two-allele example, the number of neurodes in the hidden layer was one or two.

3. Other parameters of the Bayesian neural network must also be defined, typically the nature of the transfer function inside each neurode. Input neurodes usually contain linear transfer functions, hidden layer neurodes contain sigmoidal transfer functions, and output neurodes contain linear transfer functions for continuous data and sigmoidal transfer functions for discrete or categorical data. In our example, we employed a linear transfer function in a single output neurode.
4. Although Bayesian regularized neural networks are robust and generally train to almost identical models, it is possible for the training to locate a local minimum in the response surface. Training the network five times, starting from random initial weights, is usually sufficient to obtain an optimum model.
5. Training is carried out by presenting the mathematical descriptions of each peptide in turn to the input neurodes and using the network to determine the predicted biological response (e.g., log binding affinity or activity category). The predicted response is compared with the known (experimentally measured) response, generating an error (difference between the predicted and measured response). This error is propagated back through the network to adjust the weights. This is done many times for all sequences in the training set until a maximum in the evidence has been achieved, at which time training stops.
6. The residual error in predicting the biological response of the model is used to generate statistics for the model: the square of the correlation coefficient between the predicted and measured training set data [r^2 (train)]; the root mean square (RMS) error in the model (RMS deviation between predicted and measured response variables); and the standard error of estimation (SEE), which is essentially the RMS weighted by the number of parameters in the model. For discrete data, contingency tables and A_{ROC} are often used as the primary yardsticks of performance ***(21)***. An A_{ROC} value of 1 indicates a perfect model, and a value of 0.5 denotes a model no better than chance. The A_{ROC} measure removes biases due to differing numbers of binding and nonbinding peptides, and biases due to arbitrary defined decision thresholds ***(22)***. This measure is also not overly affected by the presence of a small number of outliers, some of which may result from classification ambiguities near the decision boundaries.
7. The biological response variables for the test set are then predicted by the model. This generates an analogous set of statistics to those of the training set: the square

of the correlation coefficient between the predicted and measured test set data [r^2 (test)]; the RMS error in the model (RMS deviation between predicted and measured test set response variables); and the standard error of prediction (SEP) for the test set. This gives a good indication of the prediction efficiency of the model for data not used to train the model.

8. Sometimes another independent test set is available that further assesses the ability of the model to generalize outside of the sequence space of the training set. In our example, we employed two independent external test sets (containing 30 and 343 peptides, respectively) for each of the two-allele models. These test set peptides consisted of sequences of length up to 25 amino acids (*see* **Note 6**). None of the nonamer motifs in the external test sets appeared in the training sets. For this example, the external test set compounds were classified into activity classes in a similar way to the training set data. Given the training set size was $\sim 10^3$ peptides and the number of possible peptides of nine amino acids is $\sim 10^{11}$, this is a stringent test of the predictive power of any peptide QSAR model (*see* **Note 2**). Our models were able to usefully predict the classifications of peptides not used in the training and internal test set procedures. As the nonamer motifs used in training do not appear in any of the external test sets, we have indication of the ability of models to extrapolate into unknown sequence space.
9. The model is then used to predict the biological response variables of additional real or virtual peptide sequences and may have considerable value in designing peptides with specific levels of activity or understanding the basis for the activity. Truth tables are also a useful, compact way to summarize how well QSAR models of peptide binding to MHC alleles perform.

   These models are able to make useful predictions of binding activity in a relatively large region of "peptide space," even when the sequences being predicted have not appeared in the training sets for the models. Models derived by these methods would be very useful in rapidly developing T-cell epitopes without the need to screen large libraries of peptides.

Notes

1. We make the assumption that peptide side chains bind independently. Gulukota et al. *(6)* proposed that the degree to which this assumption is true depends on the level of detail at which prediction is attempted. For binary classification (binding versus nonbinding), they found that the assumption appeared to be justified, but as the number of affinity classes increases this assumption will be less valid. One advantage of neural networks is that they can accommodate cross-terms between descriptors, achieving some degree of relaxation of the strict independent binding assumption.
2. Any realistic training set constitutes a very small selection of "peptide space" (1 in 10^8 in our example). Consequently, the diversity of the training set can have a major influence on the performance of predictive models derived from it. The few

examples of highly active peptides in the external test set of our example that are predicted to have no binding attest to these limitations in training set diversity.

3. It is clear that defining boundaries between classes are not sharp. The use of crisp sets to classify what is essentially continuous data results in ambiguities in class membership because some data lie close to the decision boundaries. This applies to both the training and test sets and results in some misclassification.
4. Clearly different representations of peptides will produce models with varying abilities to generalize well. Our results with the property-based descriptors show that it is possible to obtain models that generalize well using reduced descriptor sets chosen to capture peptide properties rather than simply indicate the presence or absence of a given amino acid at a certain position. The observation that models using these descriptors sometimes required more complex neural networks architectures suggests they require a more flexible modeling method to take into account a larger contribution from cross-terms or nonlinearity. Property-based descriptors may generalize better in "peptide space" than other models that learn the associations between motifs and activity. Limitations of descriptors in capturing all of the relevant properties (some of which will clearly depend on three-dimension structures adopted by the nonamer motifs) will cause misclassifications.
5. The quality of models depends on the efficacy and relevance of descriptors used. It is clear that the types of descriptors we used are relatively simple. The binary descriptors do no more than to identify each amino acid. The property-based descriptors attempt to incorporate molecular properties into the description. There are many ways amino acids could be described and some of these may produce better models. However, our work suggests that using a robust, model-free, nonlinear method to build models relating descriptors to activity can be surprisingly successful, even with relatively simple descriptors. Development of more efficient descriptors is an active area of research.
6. Peptides that are longer than nine residues used in building the models were dealt with by the following method. Peptide-binding affinity was predicted by scoring all possible overlapping nonamers in the peptide and choosing the one with highest affinity. In our example, it was clear that the same active nonamer motifs were occurring in different peptides with different activities. Consequently, the assumption that the activity of a nonamer motif was invariant with respect to the rest of the peptide in which it is embedded is not always true, although in the majority of cases it is approximately true. This would result in some misclassification.

References

1. Buus, S. (1999) Description and prediction of peptide-MHC binding: The 'Human MHC Project'. *Curr. Opin. Immunol.* **11**, 209–213.
2. Doytchinova, I.A. and Flower, D.R. (2001) Towards the quantitative prediction of T-cell epitopes: CoMFA and CoMSIA studies of peptides with affinity for the class I MHC molecule HLA-A*0201. *J. Med. Chem.* **44**, 3572–3581.

3. Doytchinova, I.A., Blythe, M.J., and Flower, D.R. (2002) Additive method for the prediction of protein-peptide binding affinity. Application to the MHC class 1 molecule HLA-A*0201. *J. Proteome Res.* **1**, 263–272.
4. Logean, A., Sette, A., and Rognen, D. (2000) Customized versus universal scoring functions: application to class I MHC-peptide binding free energy predictions. *Bioorg. Med. Chem. Lett.* **11**, 675–679.
5. Brusic, V., Bucci, K., Schönbach, C., Petrovsky, N., Zelezvikow, J., and Kazura, J.K. (2001) Efficient discovery of immune response targets by cyclical refinement of QSAR models of peptide binding. *J. Mol. Graph. Model.* **19**, 405–411.
6. Gulukota, K., Sidney, J., Sette, A., and DeLisi, C. (1997) Two complementary methods for predicting peptides binding major histocompatibility complex molecules. *J. Mol. Biol.* **267**, 1258–1267.
7. De Hann, E.C., Wauben, M.H.M., Grosfeld-Stulemeyer, M.C., Kruijtzer, J.A.W., Liskamp, R.M.J., and Moret, E.E. (2002) Major histocompatibility complex class II binding characteristics of peptoid-peptide hybrids. *Biorg. Med. Chem.* **10**, 1939–1945.
8. Bhasin, M. and Raghava, G.P.S. (2004) SVM-based method for predicting HLA-DRB1*0401 binding peptides in an antigen sequence. *Bioinformatics* **20**, 421–423.
9. Polley, M.J., Winkler, D.A., and Burden, F.R. (2004) Broad-based QSAR of farnesyltransferase inhibitors using a Bayesian regularized neural network. *J. Med. Chem.* **47**, 6230–6238.
10. Winkler, D.A. and Burden, F.R. (2004) Modelling blood brain barrier partitioning using Bayesian neural nets. *J. Mol. Graph. Model.* **22**, 499–508.
11. Burden, F.R. and Winkler, D.A. (2000) A QSAR model for the acute toxicity of substituted benzenes towards *Tetrahymena pyriformis* using Bayesian regularized neural networks. *Chem. Res. Toxicol.* **13**, 436–440.
12. Sorich, M.J., McKinnon, R.A., Winkler, D.A., Burden, F.R., Miners, J.O., and Smith, P.A. (2003) Comparison of linear and nonlinear classification algorithms: Prediction of drug metabolism by UDP-glucuronosyltransferase isoforms. *J. Chem. Inf. Comput. Sci.* **43**, 2019–2024.
13. Winkler, D.A. and Burden, F.R. (2000) Robust QSAR models from novel descriptors and Bayesian regularized neural networks. *Mol. Simul.* **24**. 243–258.
14. Burden, F.R. and Winkler, D.A. (1999) Robust QSAR models using Bayesian regularized artificial neural networks. *J. Med. Chem.* **42**, 3183–3187.
15. Nabney, I.T. (2002). *Netlab: Algorithms for Pattern Recognition.* Springer-Verlag, London.
16. Burden, F.R and Winkler, D.A. (2007) Bayesian Regularization of Neural Networks, in "Applications of Artificial Neural Networks in Chemistry and Biology", Livingston, D. (ed.) Humana Press.
17. Winkler, D.A. and Burden, F.R. (2005) Predictive Bayesian neural network models of MHC class II peptide binding. *J. Mol. Graph. Model.* **23**, 481–489.

18. Brusic, V., Rudy, G., and Harrison, L.C. (1998) MHCPEP, a database of MHC-binding peptides: update 1997. *Nucleic Acids Res.* **26**, 368–371.
19. Sandberg, M., Eriksson, L., Jonsson, J., Sjostrom, M., and Wold, S. (1998) New chemical descriptors relevant for the design of biologically active peptides. A multivariate characterization of 87 amino acids. *J. Med. Chem.* **41**, 2481–2491.
20. MacKay, D. J. C. (1992) A practical Bayesian framework for backpropagation networks. *Neural Comput.* **4**, 448–472.
21. Swets, J.A. (1988) Measuring the accuracy of diagnostic systems. *Science* **240**, 1285–1293.
22. Brusic, V., Rudy, G., Honeyman, M., Hammer, J., Harrison, L. (1998) Prediction of MHC Class II-binding peptides using an evolutionary algorithm and artificial neural network. *Bioinformatics* **14**, 121–130.

IV

Predicting other Properties of Immune Systems

28

TAPPred
Prediction of TAP-Binding Peptides in Antigens

Manoj Bhasin, Sneh Lata, and G. P. S. Raghava

Summary

The transporter associated with antigen processing (TAP) plays a crucial role in the transport of the peptide fragments of the proteolysed antigenic or self-altered proteins to the endoplasmic reticulum where the association between these peptides and the major histocompatibility complex (MHC) class I molecules takes place. Therefore, prediction of TAP-binding peptides is highly helpful in identifying the MHC class I-restricted T-cell epitopes and hence in the subunit vaccine designing. In this chapter, we describe a support vector machine (SVM)-based method TAPPred that allows users to predict TAP-binding affinity of peptides over web. The server allows user to predict TAP binders using a simple SVM model or cascade SVM model. The server also allows user to customize the display/output. It is freely available for academicians and noncommercial organization at the address http://www.imtech.res.in/raghava/tappred.

Key Words: TAP; MHC class I; T-cell epitopes; subunit vaccine; SVM; cascade SVM

1. Introduction

The processing of an endogenous antigen involves intracellular processes such as production of peptide fragments by proteasome and transport of peptides to endoplasmic reticulum (ER) through transporter associated with antigen processing (TAP) *(1)*. The understanding of these processes can help in filtering T-cell epitopes and reducing false-positive results. The binding of peptide to TAP is crucial for its translocation from cytoplasm to ER. Therefore, understanding about the binding of peptides to TAP transporter can play a vital role in improving the prediction accuracy of major histocompatibility complex (MHC) class I-restricted peptides.

From: *Methods in Molecular Biology, vol. 409: Immunoinformatics: Predicting Immunogenicity In Silico*
Edited by: D. R. Flower © Humana Press Inc., Totowa, NJ

TAP is a main channel for the transport of the antigenic fragments/peptides from cytosol to ER, where they bind to MHC molecules ***(2)***. This is a heterodimeric transporter belonging to the family of ABC transporters that uses the energy provided by ATP to translocate the peptides across the membrane ***(3–4)***. A TAP transporter can translocate peptides of 8–40 amino acids, with preference for peptides of length 8–11 amino acids ***(5–6)***. Beside length preference, the nature of peptides also influences the peptide selectivity. Because of extensive polymorphism in TAP2 subunit of rat transporter, distinct set of peptides bind and are translocated by TAP transporter with varying efficiency ***(7)***. The understanding of selectivity and specificity of TAP may contribute significantly in prediction of the MHC class I-restricted T-cell epitopes. Therefore, prediction of TAP-binding peptides is crucial in identifying the MHC class I-restricted T-cell epitopes and hence subunit vaccine designing. Only limited algorithms are developed till now to explore TAP-binding and translocation efficiency of peptides due to the lesser amount of data ***(8–9)***. The JenPep is the first publicly available compilation having ~400 TAP-binding peptides ***(10)***. The TAP-binding peptides are also included in version 3.1 of MHCBN ***(11)***. This chapter will provide an overview of the bioinformatics tool TAPPred developed for prediction of TAP-binding affinity and translocation efficiency of the peptide.

2. Materials

TAPPred is a user-friendly web server developed and launched on SUN server 420R under Solaris environment. Support vector machine (SVM) was implemented using the freely downloadable software, SVM_light. The web server was launched using public domain software package Apache. All web pages are written in hypertext markup language (HTML), and CGI scripts are written in PERL and JavaScript. ReadSeq (developed by Dr Don Gilbert) has been integrated in the server, which allows user to submit their sequence in any standard formats. This server is accessible from http://www.imtech.res.in/raghava/tappred/ or http://www.imtech.ac.in/raghava/tappred/. In addition, MHC2Pred has been mirrored at University of Arkansas for Medical Sciences, Little Rock, USA on SGI origin server under IRIX environment (http://bioinformatics.uams.edu/raghava/tappred/).

3. Method

The server TAPPred is freely available for academicians and noncommercial organization. Home page (Fig. 1) of the server has a menu list at the top, which has the following menu options:

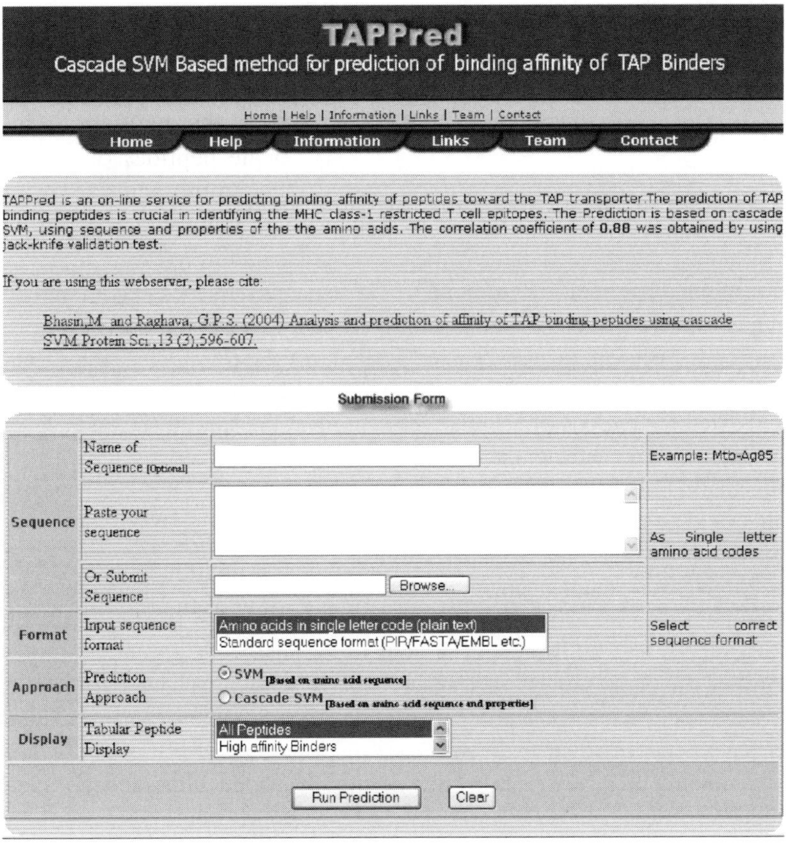

Fig. 1. A snapshot of home page of TAPPred server.

1. Home: Directs to the home page of the server. The home page itself is the submission form in this server.
2. Help: Provides step-by-step guidance to use the web server.
3. Information: This option is linked to the page that provides the detailed information about the TAP transport and the stepwise algorithm of the method.
4. Links: This link leads to a page having the links of relevant databases, prediction methods, and commercial epitope prediction methods.
5. Team: Directs to page providing the address and e-mail ID of the persons involved in developing the method.
6. Contact: By clicking this link one can access the address and e-mail ID of the person to be contacted in case of a query.

3.1. Instructions to Use the Server

The home page itself is the submission form of the server. A sequence submission form is a web interface wherein users can paste their query sequence, select among the choices provided, parameters of their choice, and submit it to the server that returns the result of this query. The fields that a user is required to fill in the submission form are as follows:

1. Name of antigenic sequence (optional): The name of sequence may have letters and number with the "-" or "_." Entry of all other characters would flash a warning. The sequence is assigned a default name "TAP1," if the field is left blank. It may be a problem with ä, ö, ü, or an empty space within the name of the sequence, which is not allowed for reasons of security. Also most of the special (i.e., nonalphabetical or nonnumerical) characters are not allowed.
2. Antigenic sequence**:** Protein sequence in single-letter amino acid code can be pasted in this field or can be uploaded from a local sequence file, in any of the standard formats. All the nonstandard characters such as [∗&∧%$@#!()_+∼=;′", <>?.\|} are ignored from the sequence. The minimum length of the submitted sequence should be nine; otherwise the server will show a warning message. A warning is also displayed if input from both or none of the sequences is detected.
3. Format of antigenic sequence**:** The server accepts both formatted and unformatted raw antigenic sequences. The server uses ReadSeq routine to parse the input. The user has to choose whether the sequence uploaded or pasted is plain or formatted before running prediction. The results of the prediction will be wrong if the format chosen is wrong.
4. Prediction approaches: The method predicts the binding affinity of peptides for TAP transporter. The server provides two options to predict the binding affinity of peptide on the basis of SVM:

   a. Simple SVM: In this the prediction is based on the sequential information of the peptides. The prediction by simple SVM is quick.
   b. Cascade SVM: In case of cascade SVM, the prediction is based on sequential and features of the amino acids. The prediction is done at two levels. In first level, the preliminary results are obtained by combining features of amino acids with sequential information. In second level, results of the first level are further filtered. The prediction by cascade SVM is slower, but it is more reliable as compared to simple SVM. User has to select a single approach of prediction in any one run of prediction. A warning will be displayed if none or both approaches of prediction were selected.

5. Run prediction: "Run Prediction" button has to be clicked in order to run the prediction method (*see* **Note 1**).

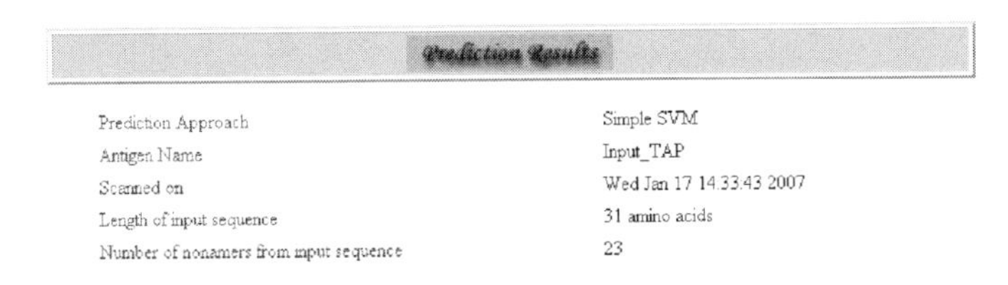

Fig. 2. Header of result display providing detailed information.

Tabular Display

Peptide Rank	Start Position	Sequence	Score	Predicted Affinity
1	22	AEWPRSNDC	6.984	High
2	18	GCSNAEWPR	4.634	Intermidate
3	4	KLAETDFLL	4.588	Intermidate
4	1	AMCKLAETD	3.953	Intermidate
5	6	AETDFLLAN	3.920	Intermidate
6	11	LLANDWRGC	3.733	Intermidate
7	12	LANDWRGCS	3.598	Intermidate
8	14	NDWRGCSNA	3.213	Intermidate
9	9	DFLLANDWR	3.179	Intermidate

Fig. 3. Result display in tabular format.

6. Result display: The result will be displayed in user-friendly format. The user can choose the type of peptides to be displayed in the result. The display, in a tabular form, provides four options.

   a. All peptides (indiscriminate of binding affinity)
   b. High-affinity binders only
   c. Intermediate-affinity binders only
   d. Low-affinity binders

The user can select only output display one at a time. The results of each prediction will display firstly an header, which will provide information about the length of peptide sequence, nonamers obtained from that sequence, and date of prediction as shown in Fig. 2.

The results will be displayed in two different formats. First format will show Fig. 3.

Acknowledgments

We acknowledge the financial support from the Council of Scientific and Industrial Research (CSIR) and Department of Biotechnology (DBT), Government of India.

Note

1. To avoid the misuse of the site, the services are available for the registered users only. Users who are interested to use these servers are required to register themselves

at http://www.imtech.res.in/errors/noauth.html. They need to fill up a registration form if they agree to the terms and conditions stated in the form. The user name and password is then sent by e-mail to the users.

References

1. Nussbaum, A.K., Kuttler, C., Tenzer, S., and Schild, H. 2003. Using the World Wide Web for predicting CTL epitopes. *Curr. Opin. Immunol.* **15:** 69–74.
2. Lankat-Buttgereit, B. and Tampe, R. 1999. The transporter associated with antigen processing TAP: Structure and function. *FEBS Lett.* **464:** 108–112.
3. Abele, R. and Tampe, R. 1999. Function of the transport complex TAP in cellular immune recognition. *Biochim. Biophys. Acta.* **1461:** 405–419.
4. van Endert, P.M., Saveanu, L., Hewitt, E.W., and Lehner, P. 2002. Powering the peptide pump: TAP crosstalk with energetic nucleotides. *Trends Biochem. Sci.* **27:** 454–461.
5. Heemels, M.T. and Ploegh, H.L. 1994. Substrate specificity of allelic variants of the TAP peptide transporter. *Immunity* **1:** 775.
6. Schumacher, T.N., Kantesaria, D.V., Heemels, M.T., Ashton-Rickardt, P.G., Shepherd, J.C., Fruh, K., Yang, Y., Peterson, P.A., Tonegawa, S., and Ploegh, H.L. 1994. Peptide length and sequence specificity of the mouse TAP1/TAP2 translocator. *J. Exp. Med.* **179:** 533–540.
7. Uebel, S. and Tampe, R. 1999. Specificity of the proteasome and the TAP transporter. *Curr. Opin. Immunol.* **11:** 203–208.
8. Doytchinova, I., Hemsley, S., Flower, D.R. 2004. Transporter associated with antigen processing preselection of peptides binding to the MHC: A bioinformatic evaluation *J. Immunol.* **173:** 6813–6819.
9. Bhasin, M. and Raghava, G.P.S. 2004. Analysis and prediction of affinity of TAP binding peptides using cascade SVM. *Protein Sci.* **13:** 596–607
10. Bhasin, M., Singh, H., and Raghava, G.P.S. 2003. MHCBN: A comprehensive database of MHC binding and non-binding peptides. *Bioinformatics* **19:** 666–667.
11. Blythe, M.J., Doytchinova, I.A., and Flower, D.R. 2002. JenPep: A database of quantitative functional peptide data for immunology. *Bioinformatics* **18:** 434–439.

29

Prediction Methods for B-cell Epitopes

Sudipto Saha and Gajendra P. S. Raghava

Summary

In this chapter, two prediction servers of linear B-cell epiotpes have been described; (i) BcePred, based on physico-chemical properties that include hydrophilicity, flexibility/mobility, accessibility, polarity, exposed surface, turns, and antigenicity and ii) ABCpred, based on recurrent neural network. Both of the servers assist in locating linear epitope regions in a protein.

Key Words: B-cell epitope; linear epitope; physico-chemical properties; flexibility; hydrophilicity; surface accessibility; turns; recurrent neural network; vaccine

1. Introduction

A crucial step in designing of peptide vaccines involves the identification of B-cell and T-cell epitopes. The experimental scanning of B-cell epitope active regions requires the synthesis of overlapping peptides, which span the entire sequence of a protein antigen. This is costly and labor-intense task. In silico techniques are the best alternative to find out which regions of a protein out of thousands possible candidates are most likely to evoke immune response. Most of the existing B-cell epitope prediction methods are based on physico-chemical properties of amino acids. Based on these scales, a web server BcePred (www.imtech.res.in/raghava/bcepred/) was developed to predict B-cell epitope regions in an antigen sequence *(1)*. BcePred can predict continuous B-cell epitopes, and physico-chemical scales used were hydrophilicity *(2)*, flexibility/mobility *(3)*, accessibility *(4)*, polarity *(5)*, exposed surface *(6)*, turns *(7)*, and antigenicity *(8)*. Quantification of these properties is determined by assigning a value to each of the 20 natural amino acids. Users can select

From: *Methods in Molecular Biology, vol. 409: Immunoinformatics: Predicting Immunogenicity In Silico*
Edited by: D. R. Flower © Humana Press Inc., Totowa, NJ

any physico-chemical properties or combination of two or more properties for epitopes prediction. It presents the results in graphical and tabular frame. In case of graphical frame, this server plots the residue properties along protein backbone, which assist the users in rapid visualization of B-cell epitope on protein. The peak of the amino acid residue segment above the threshold value (default is 2.38 in combined approach) is considered as predicted B-cell epitope. The tabular output is in the form of a table, which will give the normalized score of the selected properties with the corresponding amino acid residue of a protein along with the maximum, minimum and averages values of the combined methods, selected. Blythe and Flower ***(9)*** examined 484 amino acid propensity scales in prediction of B-cell epitopes and found that even the best set of scales and parameters performed only marginally batter than random.

ABCpred server (http://www.imtech.res.in/raghava/abcpred/) was developed for predicting continuous B-cell epitopes based on machine learning techniques. ABCpred has been trained on B-cell epitopes obtained from Bcipep database ***(10)***. This server can predict continuous (linear) B-cell epitopes. Users can select window length of 10, 12, 14, 16, and 20 as predicted epitope length. It presents the results in graphical and tabular frame. In case of graphical frame, this server plots the epitopes in blue color along protein backbone (black color), which assist the users in rapid visulaziation of B-cell epitope on protein. The tabular output is in the form of a table, which will provide the aminoacids length from N-terminal to C-terminal in a protein predicted by the server.

2. Materials and Methods

2.1. Usage of B-Cell Epitope Prediction Servers

The users are required to fill a request form available at http://www.imtech.res.in/errors/noauth.html for using web servers developed by raghava's group (http://www.imtech.res.in/raghava). The user name (e-mail ID) and password are provided through e-mail. The old users can directly access the database by providing the user name and password.

2.2. Description of BcePred Server

The web-based server allows prediction of linear B-cell epitope using physico-chemical properties of amino acids. The common gateway interface (CGI) script for these servers are written using PERL version 5.03. These servers are installed on a Sun Server (420E) under a UNIX (Solaris 7) environment. Users can enter the primary amino acid sequence for prediction using file uploading or cut-and-paste options. The results provide summarized information about the query sequence and prediction.

SUBMISSION FORM

Sequence name (optional) :

Paste your sequence below:
(Amino acid sequence in one lettercode. No header line)

Or Submit sequences from file : Browse...

Threshold [-3 to 3] :

Hydrophilicity:	2	**Flexibility:**	1.9
Accessibility:	2	**Turns:**	1.9
Exposed Surface:	2.4	**Polarity:**	2.3
Antegenic Propensity:	1.8	**Combined:**	1.9

Select physico-chemical properties to use:
For multiple selection use Ctrl Key

Hydrophilicity (Parker et al., Biochemistry, 25, 5425 (1986))
Flexibility (Karplus et al., Naturwissenschaften, 72, 212 (1985))
Accessibility (Emini et al., J.virol., 55, 836 (1985))
Turns (Pellequer et al., Immunol.Lett., 36, 8 3(1993))

Clear fields | Submit sequence

GRAPHICAL RESULT : : SEQ 1 to 60

Fig. 1. The display of BcePred server (**A**) submission form; (**B**) graphical output (sequences 1–60). Solid line represents "hydrophilicity," round dot line represents "turns," sqaure dot represents "surface," dashed line represents "flexibility," dash dotted line represents "polar," long dashed line represents "accessibility," long dash dotted line represents "antigenic," and long dash dot dotted line represents "combined."

The home page of BcePred server is available at http://www.imtech.res.in/raghava/bcepred/. The available menus are home, submission form, help page, output format, algorithm, and team. The submission form has been shown in Fig. 1A. The help page describes the usage of the server, and the output format explains the interpretation of the output format. The algorithms used in developing the methods based on physico-chemical properties were explained in detail in algorithm menu.

2.2.1. Steps to Follow for Using the Submission Form of BcePred Server

1. Enter a name for the sequence (optional).
2. Enter the sequence in the sequence window (with no header line) or give a file name (file uploading).
3. The sequence must be written using the one-letter amino acid code: "acdefghiklmnpqrstvwy" or "ACDEFGHIKLMNPQRSTVWY." Other letters will be converted to "X" and treated as unknown amino acids. Other characters, such as whitespace and numbers, will simply be ignored.
4. Change the threshold [–3 to +3]: Default thresholds for different parameters are selected based on the best sensitivity and specificity obtained (*see* **Note 1**).
5. Select the physio-chemical properties: Users can select any physico-chemical properties such as hydrophilicity or flexibility or accessibility or turns or exposed surface or polarity or antegenic propensity or combined methods (All). For multiple selection, use Ctrl key (*see* **Note 2**).
6. Press the "Submit sequence" button.
7. A WWW page will return the results when the prediction is ready. Response time depends on system load.

2.2.2. BcePred Server Output

It presents the results in graphical, tabular frame, and overlap display.

2.2.2.1. Graphical Format

An example of graphical output of BcePred has been shown in Fig. 1B. In case of graphical frame, server plots the residue properties along protein backbone, which assist the users in rapid visualization of B-cell epitope on protein. The peak of the amino acid residue segment above the threshold value is considered as predicted B-cell epitope.

2.2.2.2. Tabular Display

The tabular output is in the form of a table, which will give the normalized score of the selected properties with the corresponding amino acid residue of a protein along with the maximum, minimum, and average values of the combined methods selected.

2.2.2.3. Overlap Display

In overlap display, consensus prediction of two or more physico-chemical properties are displayed. Predicted epitope regions are shown in blue color.

2.3. Description of ABCpred Server

The web-based server allows prediction of linear B-cell epitope using artificial neural network. ABCpred has been trained on B-cell epitopes obtained from Bcipep database ***(10)*** and will therefore presumably have better performance for prediction of B-cell epitope of an antigen ***(11)***. This server can predict continuous (linear) B-cell epitopes. The CGI script for these servers are written using PERL version 5.03. These servers are installed on a Sun Server (420E) under a UNIX (Solaris 7) environment. Users can enter the primary amino acid sequence for prediction using file uploading or cut-and-paste options.

The home page of ABCpred server is available at http://www.imtech.res.in/raghava/abcpred/. The available menus are home, submission form, help page, method, and team. The submission page has been shown in Fig. 2 A. The help menu explains the usage of the submission form, and method menu describes the artificial neural network (recurrent neural network) ***(12)*** based on which the server was developed.

2.3.1. Steps to Follow for Using the Submission Form of BcePred Server

1. Enter a name for the sequence (optional).
2. Enter the sequence in the sequence window (with no header line) or give a file name (file uploading).
3. The sequence must be written using the one-letter amino acid code: "acdefghiklmnpqrstvwy" or "ACDEFGHIKLMNPQRSTVWY.". Other letters will be converted to "X" and treated as unknown amino acids. Other characters, such as whitespace and numbers, will simply be ignored.
4. Change the threshold [+0.1 to +1.0]: Users can select threhold value from 0.1 to 1. Default threshold is the optimum value (*see* **Note 3**).
5. Users can select the window length (10, 12, 14, 16, 18, 20); default length is 16 (*see* **Note 4**).
6. Users can can select overlapping filter on or off (*see* **Note 5**).
7. Press the "Submit sequence" button.
8. A WWW page will return the results when the prediction is ready. Response time depends on system load.

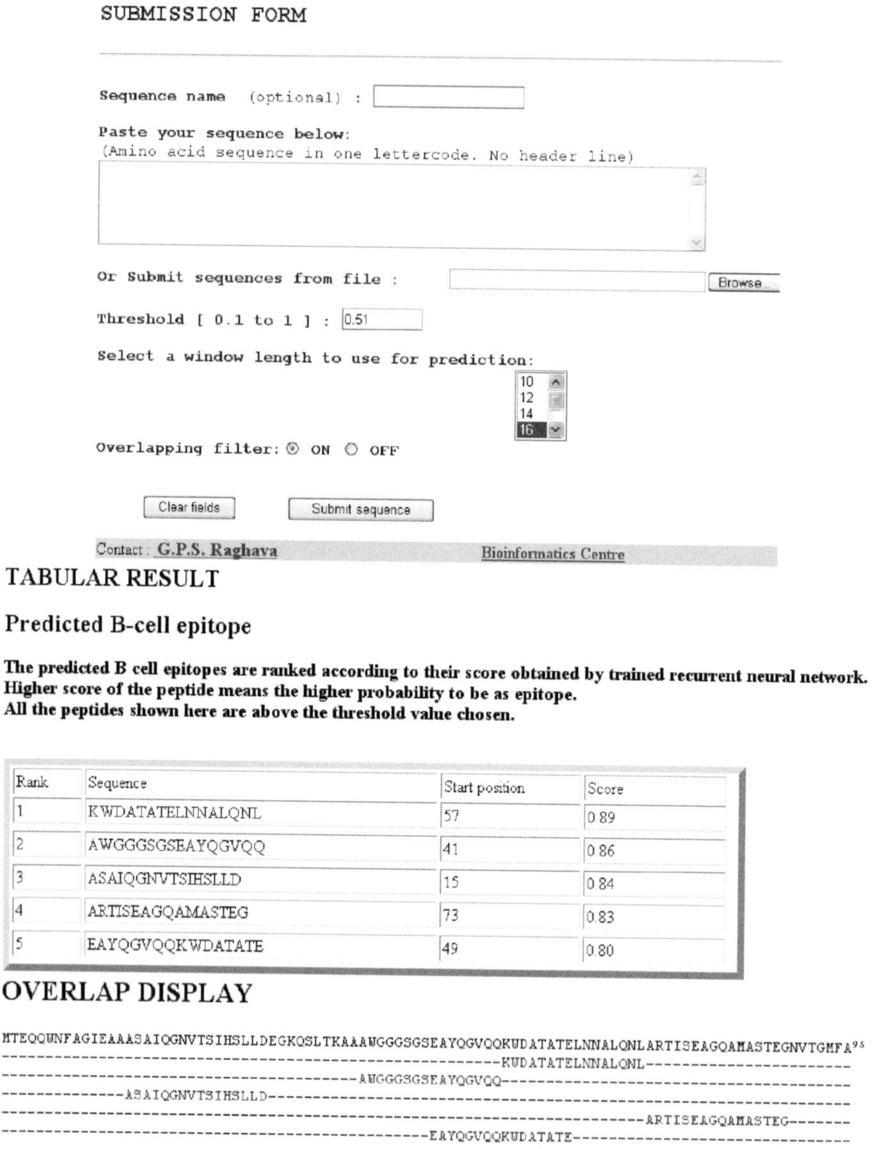

SUBMISSION FORM

Sequence name (optional) :

Paste your sequence below:
(Amino acid sequence in one lettercode. No header line)

Or Submit sequences from file : Browse...

Threshold [0.1 to 1] : 0.51

Select a window length to use for prediction: 10 12 14 16

Overlapping filter: ◉ ON ○ OFF

Clear fields Submit sequence

Contact : G.P.S. Raghava Bioinformatics Centre

TABULAR RESULT

Predicted B-cell epitope

The predicted B cell epitopes are ranked according to their score obtained by trained recurrent neural network. Higher score of the peptide means the higher probability to be as epitope. All the peptides shown here are above the threshold value chosen.

Rank	Sequence	Start position	Score
1	KWDATATELNNALQNL	57	0.89
2	AWGGGSGSEAYQGVQQ	41	0.86
3	ASAIQGNVTSIHSLLD	15	0.84
4	ARTISEAGQAMASTEG	73	0.83
5	EAYQGVQQKWDATATE	49	0.80

OVERLAP DISPLAY

```
MTEQQWNFAGIEAAASAIQGNVTSIHSLLDEGKQSLTKAAAWGGGSGSEAYQGVQQKWDATATELNNALQNLARTISEAGQAMASTEGNVTGMFA95
--------------------------------------------------------KWDATATELNNALQNL-----------------------
----------------------------------------AWGGGSGSEAYQGVQQ---------------------------------------
--------------ASAIQGNVTSIHSLLD-----------------------------------------------------------------
------------------------------------------------------------------------ARTISEAGQAMASTEG-------
------------------------------------------------EAYQGVQQKWDATATE-------------------------------
```

Fig. 2. Screenshot of ABCpred (**A**) submission form where user can input their antigen sequence and can select appropriate threshold and window length; (**B**) output in tabular format where epitope sequence (rankwise in first column), their start position, and score are shown in column 2 and 3 respectively. The predicted B-cell epitopes are ranked according to their score obtained by trained recurrent neural network. Higher score of the peptide means the higher probability to be as epitope. All the peptides shown here are above the threshold value chosen. (**C**) Output in graphical format.

2.3.2. ABCpred Server Output

It presents the result in tabular frame and overlap display.

2.3.2.1. Tabular Display

In case of tabular frame, the server ranked epitopes based on the score obtained from the trained recurrent neural network. An example of tabular frame output has been shown in Fig. 2B. The higher score values of the peptides indicates the higher probability to be as epitope.

2.3.2.2. Overlap Display

In overlap display, predicted epitopes were aligned (see Fig. 2C).

Acknowledgments

We acknowledge the financial support from the Council of Scientific and Industrial Research (CSIR) and Department of Biotechnology (DBT), Govt. of India.

Notes

1. The default values are the optmized value set at each physico-chemical properties. For hydrophilicity and accesibility thereshold value is 2, flexibility and turns threshold value is 1.9, polarity is 2.3, exposed surface is 2.4, antigenic propensity is 1.8, and combined (flexibility + hydrophilicity + polarity + exposed surface) is 2.38. Increase in the threshold results in better specificity, but worse sensitivity.
2. Users can select more than one physico-chemical properties by pressing Ctrl key. This option will allow users to select consensus epitope predicted by more than one physico-chemical scale.
3. Default threshold is 0.5. An increase in the threshold results in better specificity, but worse sensitivity. Only those peptides having score greater than threshold will be shown in the output result.
4. By using default length 16, high sensitivity was observed at low false positive.
5. Overlapping filter allows to minimize the output result of overlapping predicted peptides.

References

1. Saha, S. and Raghava, G.P.S. (2004) BcePred: prediction of continuous B-cell epitopes in antigenic sequences using physico-chemical properties. In Artificial Immune Systems, Nicosia, G., Cutello, V., Bentley, P.J., and Timis, J., (eds.) ICARIS, LNCS 3239, Springer. Publishers, pp. 197–204.
2. Hopp, T.P. and Woods, R.K. (1981) Predictions of protein antigenic determinants from amino acid sequences. *Proc. Natl. Acad. Sci. U.S.A.* **78**, 3824–3828.

3. Karplus, P.A. and Schulz, G.E. (1985) Prediction of chain flexibility in proteins: a tool for the selection of peptide antigen. *Naturwissenschaften* **72**, 212–213.
4. Emini, E.A., Hughes, J.V., Perlow, D.S. and Boger, J. (1985) Induction of hepatitis A virus-neutralizing antibody by a virus-specific synthetic peptide. *J. Virol.* **55**, 836–839.
5. Ponnuswamy, P.K., Prabhakaran, M. and Manavalan, P. (1980) Hydrophobic packing and spatial arrangements of amino acid residues in globular proteins. *Biochim. Biophys. Acta.* **623**, 301–316.
6. Janin, J. and Wodak, S. (1978) Conformation of amino acid side-chains in proteins. *J. Mol. Biol.* **125**, 357–386.
7. Pellequer, J.-L., Westhof, E. and Regenmortel, M.H.V. (1993) Correlation between the location of antigenic sites and the prediction of turns in proteins. *Immunol. Lett.* **36,** 83–99.
8. Kolaskar, A.S. and Tongaonkar, P.C. (1990) A semi-emperical method for prediction of antigenic determinants on protein antigens. *FEBS Lett.,* **276**, 172–174.
9. Blythe, M.J. and Flower, D.R. (2005) Benchmarking B cell epitope prediction: underperformance of existing methods. *Protein Sci.*, **14**, 246–248.
10. Saha, S., Bhasin, M. and Raghava, G.P.S. (2005) Bcipep: a database of B-cell epitopes. *BMC Genomics*, **6**, 79
11. Saha, S. and Raghava, G.P.S. (2006) Prediction of continuous B-cell epitopes in an antigen using recurrent neural network. PROTEINS: *Structure, Functions, and Bioinformatics* **65**, 40–48.
12. Zell, A. and Mamier, G. (1997) *Stuttgart Neural Network Simulator version 4.2.* University of Stuttgart, Stuttgart, Germany.

30

HistoCheck

Evaluating Structural and Functional MHC Similarities

David S. DeLuca and Rainer Blasczyk

Summary

The HistoCheck webtool provides clinicians and researchers with a way of visualizing and understanding the structural differences among related major histocompatibility complex (MHC) molecules. In the clinical setting, human leukocyte antigen (HLA) matching of hematopoietic stem cell donors and recipients is essential to minimize "graft versus host disease" (GvHD). Because exact HLA matching is often not possible, it is important to understand which alleles present the same structures (HLA–peptide complexes) to the T-cell receptor (TCR) despite having different amino acid sequences. HistoCheck provides a summary of amino acid mismatches, positions, and functions as well as 3-dimensional (3D) visualizations. In this chapter, we describe how HistoCheck is used and offer advice in interpreting the query results

Key Words: Histocheck; HLA; MHC; class I; class II; peptide; binding; GvHD; donor; stem cell transplantation; matching; T-cell receptor

1. Introduction

The collection of genes known as the major histocompatibility complex (MHC) was discovered during studies initiated by J. Dausset, R. Payne and J. J. van Rood, which attempted to describe a genetically inherited system of alloantigens (antigens resulting from genetic discrepancies during transplantation) in the 1950s ***(1–3)***. During the early 1960s, multi-transfused patients and parous women were shown to often have circulating antibodies against alloantigens, now known to be encoded by the human form of MHC—human leukocyte antigen (HLA). Consequently, anti-HLA antibody screening is a standard practice when matching organ donors and recipients. Later, it became

From: *Methods in Molecular Biology, vol. 409: Immunoinformatics: Predicting Immunogenicity In Silico*
Edited by: D. R. Flower © Humana Press Inc., Totowa, NJ

clear that MHC-derived proteins restrict the specificity of the antigen receptor expressed on the surface of T lymphocytes and thus play a major role in the regulation of the immune response ***(4)***.

In the context of organ transplantation between non–HLA-identical donors and recipients, the recipient's T cells identify the donor's HLA proteins as foreign and initialize an immune response against the transplant. Consequently, the survival rate among recipients of HLA-matched organs is significantly higher than when mismatches are present ***(5,6)***.

HLA matching for organ donors and recipients is complicated by HLA's high rate of polymorphism. The latest release of the IMGT/HLA database contains 2,088 alleles ***(7)***. Exact matching across multiple HLA loci (e.g., HLA-A, HLA-B, HLA-C, and HLA-DRB1) is very difficult. For kidney, heart, cornea, and pancreas transplantations, "low-resolution" matching is used—HLA alleles are only required to belong to the same serological group. For hematopoietic stem cell transplantations during leukemia therapy, "high-resolution" matching is required; patient and recipient alleles are required to produce the same protein sequence. After total body irradiation for eliminating malignant hematopoietic cells, leukemia patients need to receive a new hematopoietic and immune system through stem cell transplantation. From the perspective of the donor's immune cells, the recipient's entire body is foreign, which leads to the so-called *graft versus host disease* (GvHD).

The likelihood of finding a high-resolution match for stem cell transplantation is low, and therefore, clinicians often seek a "next-best" match. This requires an understanding of which amino acid differences are not expected to result in a functional change to the HLA protein. Here, the selective binding of HLA to short peptide sequences, as well as the T-cell receptor (TCR), is of the greatest interest. Amino acid differences in regions of the protein that do not play a role in peptide or TCR binding could be acceptable between stem cell donor and recipient.

The peptide binding groove is encoded by exons 2 and 3 for class I HLA and exon 2 for class II HLA. The binding groove is formed by a beta-sheet "floor" with two alpha-helical "walls." Peptides bind by squeezing in between the alpha helices, typically deeply anchored at the second amino acid from the N terminus, as well as the C-terminal position. The TCR contacts the binding groove from above, interacting with the surface amino acids of the alpha helices and peptide ***(8)***.

HistoCheck (http://www.histocheck.org) is an online tool which helps clinicians and researchers visualize the amino acid substitutions of HLA alleles so that they can make informed judgments about their functional similarity.

HistoCheck provides crystallography-based 3-dimensional (3D) visualizations of the allelic mismatches by highlighting amino acids substitution positions. The user is provided with dissimilarity scores (DSSs) for the amino acids involved as well as an over-all DSS for the two alleles ***(9)***.

2. Implementation

HistoCheck is written in Java, runs on a Tomcat application server, utilizes servlets, Java server pages, and a MySQL database. The HLA alleles and their sequences are updated regularly via the IMGT/HLA database: ftp://ftp.ebi.ac.uk/pub/databased/imgy/mhc/hla/ ***(7)***.

2.1. Three-Dimensional Visualization

GIF images of the HLA structures with highlighted mismatches are generated on a linux server using RasMol version 2.7.1.1. A description of RasMol script commands can be found in the University of Massachusetts web server http://www.umass.edu/microbio/rasmol/distrib/rasman.htm. Chime can be integrated into the HTML of a website using the EMBED tag. Here is an example:

<embed src="PDB_FILE_NAME.pdb" bgcolor=black display3d= cartoon color3d=chain height="590" width ="600" startspin="false" script="script SCRIPT_NAME.spt;">

Commands used in the *.spt file correspond largely with standard RasMol commands.

2.2. The DSS Algorithm

In addition to providing information on the specific amino acid substitutions involved between two HLA alleles, HistoCheck generates a DSS, which attempts to quantify the overall functional differences between the two alleles (*see* **Note 1**) ***(10)***. The score is based on the Risler substitution matrix as well as data on the function of specific amino acids positions (i.e., their role in peptide binding or TCR interaction) (*see* **Note 2**) ***(11)***. The score is generated by

1. summing the Risler scores across all mismatches,
2. dividing this score by 100,
3. adding a penalty of 1 for each mismatch that occurs on a position that either interacts with the TCR or the peptide, or both.

An example calculation is given in Table 1.

Table 1
Calculating the dissimilarity score for A*2402 and A*2304

Position	Mismatch	Function	Penalty	Risler score
144	Lysine → Glutamine	–		13
151	Histidine → Arginine	TCR	+1	64
156	Glutamine → Leucine	PEP	+1	27
166	Aspartic acid → Glutamic acid	TCR	+1	30
167	Glycine → Tryptophan	TCR+PEP	+1	87
		Total	4	221
		Divide Risler scores by 100	221/100 = 2.21	
		Dissimilarity score	4 + 2.21 = 6.21	

PEP, Peptide contact site; TCR, T-cell receptor.

The dissimilarity score is based on the Risler scores of mismatched amino acids combined with penalties for positions which interact with the TCR or peptide. Note that although position 157 is involved in both TCR contact and peptide binding, the penalty is only counted once.

3. Application

HistoCheck can be accessed online at http://histocheck.org using any javascript-enabled browser. Although HistoCheck is available free of charge, first-time users are required to register for a user name and password, because the developers are interested in what kinds of medical and research institutes find HistoCheck userful.

3.1. Comparing a Patient's HLA to Specific Donor HLA

After signing in to HistoCheck, the user is presented with a query form (Fig. 1). The first option is the type of display to be used in showing the 3D structure of HLA. Chime is a web-browser plug-in that presents molecules interactively in 3D, allowing the user to rotate the molecule and choose between a variety of display options. Alternatively, a still GIF image can be generated, which shows the alleles' 3D structure, but is not interactive.

Next, the user may select one of the following HLA loci: A, B, Cw, DRB1, DRB3, DRB4, DRB5, DQA1, DQB1, DPA1, and DPB1. The specific alleles for donor and recipient can then be specified. Two donors may be specified, for a side-by-side comparison.

The resulting webpage shows a list of amino acid mismatches between donor and recipient (Fig. 2). For each mismatch, the domain, exon, pocket, and amino acid position are displayed (*see* **Note 3**). To help understand the significance of each mismatch, additional information is given: the position's

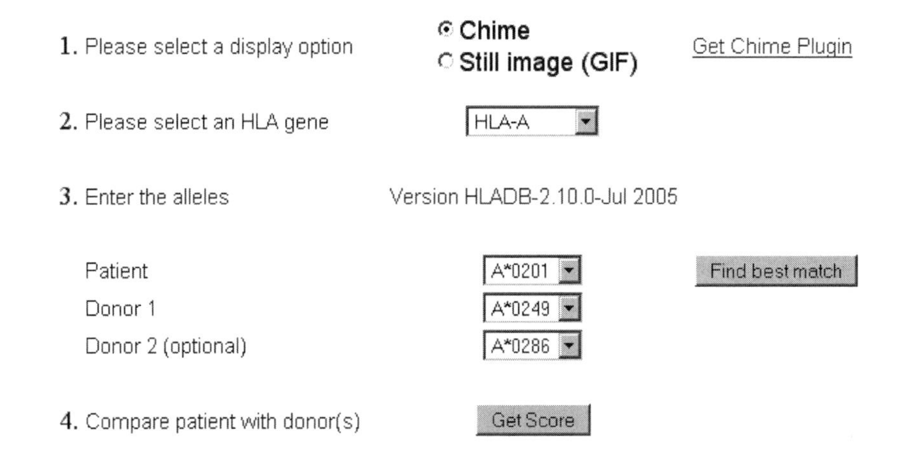

Fig. 1. The query page for the HistoCheck website. The user may choose display options and human leukocyte antigen (HLA) alleles for structural comparison. Patient alleles can be compared directly with donor candidates with the "Get Score" button. Alternatively, all the alleles of a locus can be ranked by similarity to the patient's allele by clicking the "Find best match" button.

role in binding the peptide and/or TCR, as well as the Risler score for the two amino acids involved (*see* **Note 4**). The combination of functional significance of the position (TCR binding/peptide binding), and the extent of biophysical dissimilarity between the amino acids, is the basis for the DSS (*see* **Note 5**). The summary table lists the total number of mismatches, the affected pockets, total number of mismatches that affect peptide binding, the total number of mismatched positions that interact with the TCR, and the overall DSS.

Underneath the mismatch tables, the HLA mismatches are displayed visually either as a GIF image or in an interactive Chime window. The mismatched positions are highlighted yellow. For class I HLA, the structure is based up HLA-A*0201 in complex with a decameric peptide from Hepatitis B nucleocapsid protein. The α_1, α_2, and α_3 domains are displayed in blue. The α_1 and α_2 domains form the peptide binding groove, which also interacts with the TCR.

The β_2-microglobulin domain is shown in green. A decamer peptide is shown bound to the protein in red. If class II alleles were selected, the 3D structure is based on crystallographic data from HLA-DRA with HLA-DRB1*0101. The β_1 and β_2 domains of DRB1 are shown in dark blue. The α_1 and α_2 domains of DRA are shown in turquoise. The bound 13-mer peptide is shown in green. The α_1 and β_1 domains form the peptide binding groove. Although class II HLA proteins are heterodimers, the user selects only one gene at a time, for simplicity. In this case, only the mismatches for the protein of the selected gene are displayed. Because HLA-DRA, encoding for the alpha chain of the various DR heterodimers, is not polymorphic, it is not offered in the list of genes.

If the Chime display option was selected, the user can rotate the molecule and zoom in on particularly interesting locations. Chime also provides various display options. The default option is "cartoons," which allows one to quickly orient and locate secondary, tertiary, and quaternary structures. Other options, such as wireframe, ball and stick, and space-fill can be used for more detail, once the major landmarks have been identified.

A large GIF image or Chime representation can be obtained by clicking the "Big GIF" or "Big Chime" links. The "RasMol Script" link provides an rsm file, which contains the atomic coordinate information from the standard pdb format, as well as commands which orient the HLA molecule and highlight the mismatches. The rsm files can be downloaded and viewed locally using the RasMol viewer, RasTop 2.0.

3.2. Ranking Alleles by their Similarity to a Patient's HLA

HistoCheck can also be used to find the most similar variants of an allele. The procedure is almost identical to that described in Section 3.1. However, after selecting the donor's allele on the query page, the user may also click the "Find Best Match" button instead of the "Get Score" button. In this case, all of the alleles of the given locus are considered and ranked by ascending DSS (i.e., the most similar alleles are at the top of the list). The ordered list of alleles appears in the right frame, and the mismatch result page for the best match is displayed in the center frame.

For example, if HLA-A*0201 is chosen as the donor's allele, a report comparing A*0201 with A*0209 appears in the center frame. Because A*0201 and A*0209 have no amino acid differences in the key domains (α_1 and α_2), the DSS is zero. These alleles are different at position 236 of the mature protein, but this position is part of the α_3 domain, which does not interact with the TCR or peptide. Although no mismatches are reported, the footnote "Additional differences found outside key domains" as well as the 3D image with the highlighted

Welcome to HistoCheck - an HLA Sequence Interpreter

Detailed Results

Donor 1 **New Query | Print**

A*0201 -- A*0210

Amino Acid Mismatch	Domain	Exon	Pocket	Position	Function	R Score
Phenylalanine--> Tyrosine	α1	2	BC	9	PEP	4
Tyrosine--> Phenylalanine	α2	3	ABD	99	PEP	4
Tryptophan--> Glycine	α2	3		107		87

Summary

Total Differences	Affected pockets	Total PEP	Total TCR	DSS Score
3	ABCD	2	0	2.95

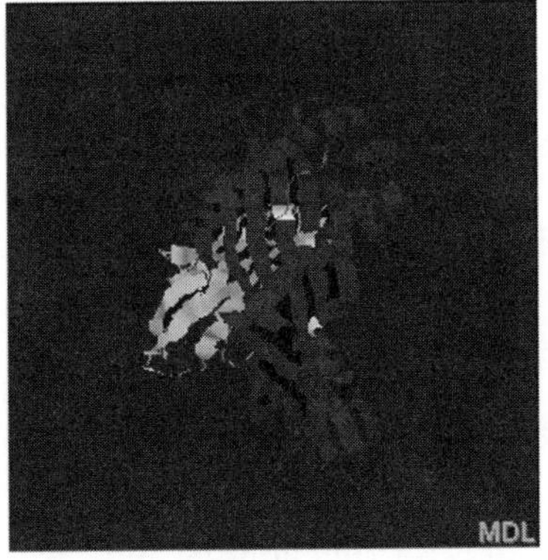

Legend:

DSS = Total dissimilarity between alleles.
PEP = Residue is likely to belong to the peptide binding site.
TCR = Residue is likely to have contact to the T-cell receptor.
R Score = Amino acid dissimilarity score according to **Risler et al.** See **Table 3.**
* = Peptide binding residue without pocket assignment
Total PEP = Total number of residues assigned to the peptide binding site
Total TCR = Total number of residues that are likely to have contact to the TCR

Big GIF
Big Chime
Rasmol Script

Fig. 2. The results page from a HistoCheck query. Here, the user has chosen to compare HLA-A*0201 with A*0210. Three amino acid differences were found at positions 9, 99, and 107. Positions 9 and 99 are involved in peptide binding. The SSM score quantifies the functional differences of these alleles. In the crystallographic structure of HLA bound to a peptide, the three mismatch positions are highlighted. Two mismatches can be seen on the beta-sheet, and one in a loop structure on the lower right.

mismatch appears. In the ranking of the most similar alleles to A*0201 on the right, one can see that A*0201 has a zero mismatch score with A*0209, A*0266, and A*0275. Clicking on the allele's name in this list brings up the detailed report for the comparison. Clicking on the fourth allele in the list, A*0268, one can see a single amino acid substitution: arginine to lysine. at position 157. Although this position is in the α_2 domain, it does not interact directly with the peptide or the TCR and is therefore of low significance. Visual inspection of the 3D structure shows that position 157 is part of the domain's alpha helix, but faces away from the peptide binding groove. Furthermore, arginine and lysine (both long and basic) are structurally very similar, as reflected by the very low Risler score (3). It can be concluded that despite a mismatch in the α_2 domain, A*0201 and A*0268 can be expected to bind the same peptides and appear identical to the TCR.

3.3. Interpretation of DSS

As described in Section 2.2, the DSS is based up the functional role of the mismatched positions, as well as the structural similarity of the amino acids involved. The example involving A*0201 mentioned above describes comparisons where it is quite clear that the amino acid differences are unlikely to affect HLA function. The best matches are of course those with DSS of zero, indicating that there are no differences in the key domains. Amino acid substitutions which are in the key domains, but which are not involved in peptide binding or contact with the TCR, are likely to be tolerable. Mismatches in peptide or TCR-binding regions could only be expected to be tolerable when the Risler score is very low (below 10). See (*see* **Notes 1–3**) for more information on interpreting the DSS.

3.4. Chime Installation

Interactive protein viewers are useful tools for understanding protein structure. Chime is a web-browser plug-in, allowing for integration into websites. Chime works with Internet Explorer, Netscape, and FireFox. Downloading Chime requires free registration at the MDL website. Good instructions on downloading and installing Chime can be found at the University of Massachusetts website http://www.umass.edu/microbio/chime/ getchime.htm..

Although the Chime installation is straightforward for all versions of Internet Explorer, problems may arise when installing for the newest Netscape and FireFox browsers. A trick for installing chime in these browsers is worth mentioning here. The instructions given below refer to MDL Chime version 6.2 SP6.

1. Install Chime normally for Internet Explorer.
2. Copy the npchime.dll file from the Internet Explorer plug-in folder (C:\Program Files\Internet Explorer\plugins\).
3. Paste the file into the plug-in folder of FireFox or Netscape. For FireFox the folder is likely to be C:\Program Files\Mozilla Firefox\plugins\.

Acknowledgments

We are grateful for the ongoing contributions of Holger-Andreas Elsner to the HistoCheck project. The development of HistoCheck was partially funded by the German José Carreras Leukemia-Foundation (grant DJCLS R 04/01).

Notes

1. This manuscript describes the functionality of HistoCheck at end of 2005. The next version of HistoCheck will involve several improvements. New crystallographic data are available, which have been re-analyzed to determine the functional roles of HLA amino acid positions. This analysis includes locus-specific definitions for TCR and peptide interactions. Furthermore, static correlations between certain HLA mismatches and GvHD have been identified. These "special mismatches" will be highlighted in HistoCheck's mismatch report, and the reference papers will be sighted.
2. Alternatives to the current DSS will be offered. The BLOSUM62 scoring matrix, for example, has delivered improvements in the area of sequence alignments. Whether this matrix is better than the Risler matrix for comparing HLA alleles has not been determined. This question is complicated by the fact that such matrices are based on the assumption that the rate of amino acid substitution among related proteins is proportional to amino acid similarity. The HLA binding groove is an exception to this rule because of the evolutionary pressure for diversity, driven by the need to respond to rapidly mutating pathogens. For this reason, a dissimilarity algorithm will be provided, which weighs the HLA positions according to the variability analysis provided by Reche et al. ***(13)***.
3. A refreshing aspect of HistoCheck in the age of black-box-bioinformatics (i.e., artificial neural networks and hidden Markov models) is that the primary biological data are provided to the user. These so-called "hard data" include the nucleic acid and protein sequences that have been validated by numerous work groups and are, in effect, irrefutable. The mismatched positions reported by HistoCheck are primary data, and the user is left with the freedom to interpret them. Other aspects of HistoCheck can be considered secondary data (also called "soft" or "semi-soft" data). The crystallographically determined structures of HLA are models, whose limitations should be recognized. In particular, the fluidity and elasticity of protein structures are not represented in these models. It can be expected that the conformation of loops, for example, differs greatly in aqueous versus crystal environments. That said, comparison of many crystallographic HLA structures shows that the protein backbone is remarkably conserved. Although "semi-soft," crystallographic models are extremely informative, concerning tertiary/quaternary protein structure, using this data to draw conclusions about TCR interactions and peptide binding can be considered secondary or even tertiary data.
4. Risler's similarity scores are also soft data. The scores are based on the rate of amino acid substitution among structurally similar proteins. HistoCheck's DSS is an attempt to summarize secondary data concerning amino acid substitutions. That this score is highly theoretical and removed from primary data is indisputable. In a preliminary analysis performed with more than 1,700 HLA class I mismatched transplant pairs from the hematopoietic stem cell transplant component of the 13th International Histocompatibility Workshop (Effie Petersdorf, Fred Hutchinson

Cancer Research Center, Seattle, WA), the DSS was not superior in predicting the severity of GvHD compared to just counting the number of HLA class I mismatches (unpublished data). Furthermore, a small preliminary study did not show a correlation between the DSS and T-cell alloreactivity in vitro ***(12)***. Because this study was performed in an allogeneic transplantation setting, in which non-HLA differences (i.e., minor histocompatibility antigens) affected alloreactivity, it is unclear to which extent non-HLA differences overshadowed HLA similarities. To clarify this point, further studies involving autologous cells, modified to express additional HLA proteins, are necessary.

5. HistoCheck's DSS is an elementary mathematical model that represents a first step in quantifying the structural differences between HLA alleles. HistoCheck users are encouraged to study the primary data that this website provides, such as number and location of amino acid substitutions, and to examine the 3D structures provided in order to make informed conclusions about the similarity/dissimilarity of HLA alleles.

References

1. Dausset, J. (1954). Leuco-agglutinins IV. Leuco-agglutinins and blood transfusion. *Vox Sang* 4, 190–8.
2. Payne, R. & Rolfs, M. R. (1958). Fetomaternal leukocyte incompatibility. *J Clin Invest* 37, 1756–63.
3. Van Rood, J., Eernisse, J. G. & van Leeuwen, A. (1958). Leukocyte antibodies in sera from pregnant women. *Nature* 181, 1735–6.
4. Zinkernagel, R. M. & Doherty, P. C. (1974). Restriction of in vitro T cell-mediated cytotoxicity in lymphocytic choriomeningitis within a syngeneic or semiallogeneic system. *Nature* 248, 701–2.
5. Saba, N. & Flaig, T. (2002). Bone marrow transplantation for nonmalignant diseases. *J Hematother Stem Cell Res* 11, 377–87.
6. Hansen, J. A., Gooley, T. A., Martin, P. J., Appelbaum, F., Chauncey, T. R., Clift, R. A., Petersdorf, E. W., Radich, J., Sanders, J. E., Storb, R. F., Sullivan, K. M. & Anasetti, C. (1998). Bone marrow transplants from unrelated donors for patients with chronic myeloid leukemia. *N Engl J Med* 338, 962–8.
7. Robinson, J., Waller, M. J., Parham, P., de Groot, N., Bontrop, R., Kennedy, L. J., Stoehr, P. & Marsh, S. G. (2003). IMGT/HLA and IMGT/MHC: sequence databases for the study of the major histocompatibility complex. *Nucleic Acids Res* 31, 311–4.
8. Saper, M. A., Bjorkman, P. J. & Wiley, D. C. (1991). Refined structure of the human histocompatibility antigen HLA-A2 at 2.6 A resolution. *J Mol Biol* 219, 277–319.
9. Elsner, H. A., DeLuca, D., Strub, J. & Blasczyk, R. (2004). HistoCheck: rating of HLA class I and II mismatches by an internet-based software tool. *Bone Marrow Transplant* 33, 165–9.

10. Elsner, H. A. & Blasczyk, R. (2002). Sequence similarity matching: proposal of a structure-based rating system for bone marrow transplantation. *Eur J Immunogenet* 29, 229–36.
11. Risler, J. L., Delorme, M. O., Delacroix, H. & Henaut, A. (1988). Amino acid substitutions in structurally related proteins. A pattern recognition approach. Determination of a new and efficient scoring matrix. *J Mol Biol* 204, 1019–29.
12. Heemskerk, M. B., Doxiadis, I. I., Roelen, D. L., Claas, F. H. & Oudshoorn, M. (2005). Letter: the HistoCheck algorithm does not predict T-cell alloreactivity in vitro. *Bone Marrow Transplant* [Epub ahead of print, Sep. 5], with 36, 927–8.
13. Reche, P. A. & Reinherz, E. L. (2003). Sequence variability analysis of human class I and class II MHC molecules: functional and structural correlates of amino acid polymorphisms. *J Mol Biol* 331, 623–41.

31

Predicting Virulence Factors of Immunological Interest

Sudipto Saha and Gajendra P. S. Raghava

Summary

In this chapter, three prediction servers used for predicting virulence factors, bacterial toxins, and neurotoxins have been described. VICMpred server predicts the functional proteins of gram-negative bacteria that include virulence factors, information molecule, cellular process, and metabolism molecule. BTXpred server allows users to predict bacterial toxins, its release, and further classification of exotoxins. NTXpred server allows prediction of neurotoxins and further classifying them based on their function and source.

Key Words: Virulence factors; bacterial toxins; exotoxins; endotoxins; toxoid; toxin-neutralizing antibodies; neurotoxins; vaccine

1. Introduction

Most of the proteins in an organism involve in cellular process, metabolism, and information storage, the remaining can be classified under virulence factors that allow the germs to establish themselves in the host. Virulence factors include adhesions, toxins, and hemolytic molecules. VICMpred server predicts the functional proteins of Gram-negative bacteria using amino acid patterns and composition. The ability of the toxoid vaccine to induce toxin-neutralizing antibodies has provided the basis for the use of therapeutic antitoxins and immunoglobulins for the prophylaxis and treatment of diseases caused by bacterial toxin. The discovery of an effective method to detoxify tetanus and diphtheria toxins by formaldehyde treatment allowed the introduction of mass immunization that led to almost complete elimination of both diseases from developed countries. BTXpred server predicts bacterial toxins and classifying

From: *Methods in Molecular Biology, vol. 409: Immunoinformatics: Predicting Immunogenicity In Silico*
Edited by: D. R. Flower © Humana Press Inc., Totowa, NJ

them based on release (exotoxins and endotoxins) and function as (i) activate adenylate cyclase, (ii) activate guanylate cyclase, (iii) food poisioning, (iv) neurotoxins, (v) macrophage cytotoxin, (vi) vacuolating cytotoxin, (vii) thiol activated, and (viii) hemolysin. The knowledge of neurotoxins is very important for the development of drugs against pain and epilepsy. A number of pharma companies are working on the use of these neurotoxins for the development of potent drugs. NTXpred server predicts neurotoxins and classifies them based on source as (i) eubacteria (produced by genus *Clostridium*), (ii) cnidarians (where cnidoblast organelles store and deliver toxins), (iii) molluscans (cone), (iv) arthropods (mainly scorpion and spider), (v) chordates (snake) and on function as (i) ion channel blockers, (ii) blockers of acetylcholine receptors, (iii) inhibitors of neurotransmitter release through metalloproteolytic activity, (iv) inhibitors of acetylcholine release with phospholipase A_2 activity, and (v) facilitators of acetylcholine release. Thus, identification of virulence factors is crucial for vaccine and drug development.

2. Materials and Methods

2.1. Usage of Web Servers

The users are required to fill a request form available at http://www.imtech.res.in/errors/noauth.html for using web servers developed by raghava's group (http://www.imtech.res.in/raghava/). The user name (e-mail ID) and password are provided through e-mail. The old users can directly access the database by providing the user name and password.

2.2. Description of VICMpred

The web-based server allows prediction of broad function of a protein (e.g., virulence factors, information molecule, cellular process, and metabolism molecule) from its amino acid sequences. The common gateway interface (CGI) script for the server has been written using PERL version 5.03. The server has been installed on a Sun Server (420E) under a UNIX (Solaris 7) environment. Users can enter the primary amino acid sequence for prediction using file uploading or cut-and-paste options. The server accepts the protein sequences in any standard format such as EMBL, GCG, and FASTA or in plain text format. Web servers use the readseq program to read the input sequences. The results provide summarized information about the query sequence and prediction.

VICMperd is freely available at http:www.imtecg.res.in/raghava/vicmpred/ and mirror site available at http://bioinformatic.uams.edu/mirror/vicmpred/. The available menus in VICMpred server are help page, submission, algorithm, references, developers, and contact. The help page describes the general information and stepwise help to submit sequence in the submission page. The submission menu links to the server submission form, shown in Fig. 1A.

2.2.1. Types of Prediction

The server allows the prediction on the basis of two different approaches:

1. Pattern based
2. Pattern based combined with amino acid composition and dipeptide composition (*see* **Note 1**).

2.2.2. About VICMpred

The detailed information on methods used in developing the server is available at algorithm menu. The output format of the server has been shown in Fig. 1B (*see* **Note 2**). It is important in drug and vaccine point of view to select virulence proteins from the pool of proteins or the proteome of an organism.

2.3. Description of BTXpred Server

The aim of BTXpred server is to predict bacterial toxins and its function from primary amino acid sequence using SVM, HMM, and PSI-Blast. The CGI script for the server has been written using PERL version 5.03. The server has been installed on a Sun Server (420E) under a UNIX (Solaris 7) environment. Users can enter the primary amino acid sequence for prediction using file uploading or cut-and-paste options. The server accepts the protein sequences in any standard format such as EMBL, GCG, and FASTA or in plain text format. Web servers use the readseq program to read the input sequences. The results provide summarized information about the query sequence and prediction.

BTXpred server and related information is available from http://www.imtech.res.in/raghava/btxpred. The mirror site of BTXpred server is accessible from http:bioinformatics.uams.edu/mirror/btxpred/. The server allows users to predict bacterial toxins, its release, and further classification of exotoxins. The server provides the option of predicting toxins either on the basis of amino acid or dipeptide composition-based SVM method *(1)* or PSI-BLAST *(2)* and classifies exotoxins using HMM *(3)* and PSI-BLAST. The server predicts bacterial toxins, classifies bacterial toxins into exotoxins and endotoxins, and further classifies exotoxins into seven different functions depending on their molecular targets (i) activate adenylate cyclase, (ii) activate guanylate cyclase, (iii) food poisoning, (iv) neurotoxins, (v) macrophage

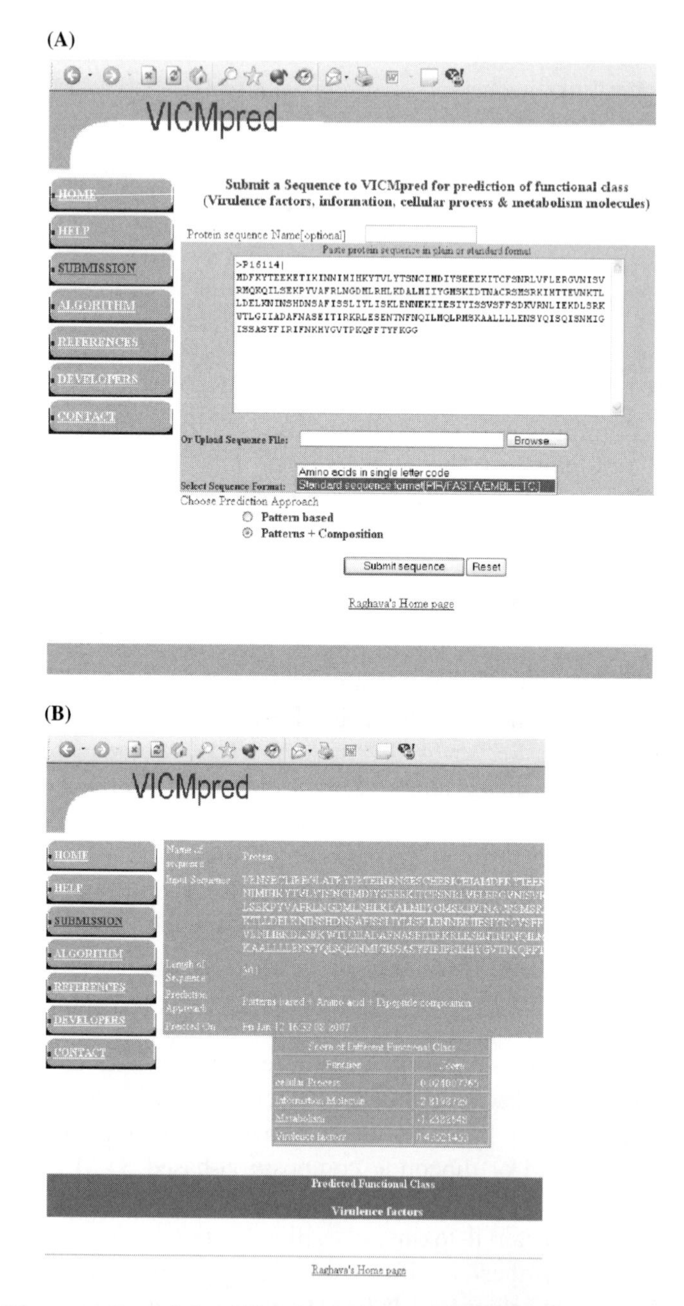

Fig. 1. The snapshot of the (**A**) submission and (**B**) output format of VICMpred server.

cytotoxin, (vi) vacuolating cytotoxin, and (vii) thiol-activated cytotoxin. The available menus in the server are submission form, help page, supplementary, epitope prediction, and developers link. The submission menu links to the submission form of the server as shown in Fig. 2A.

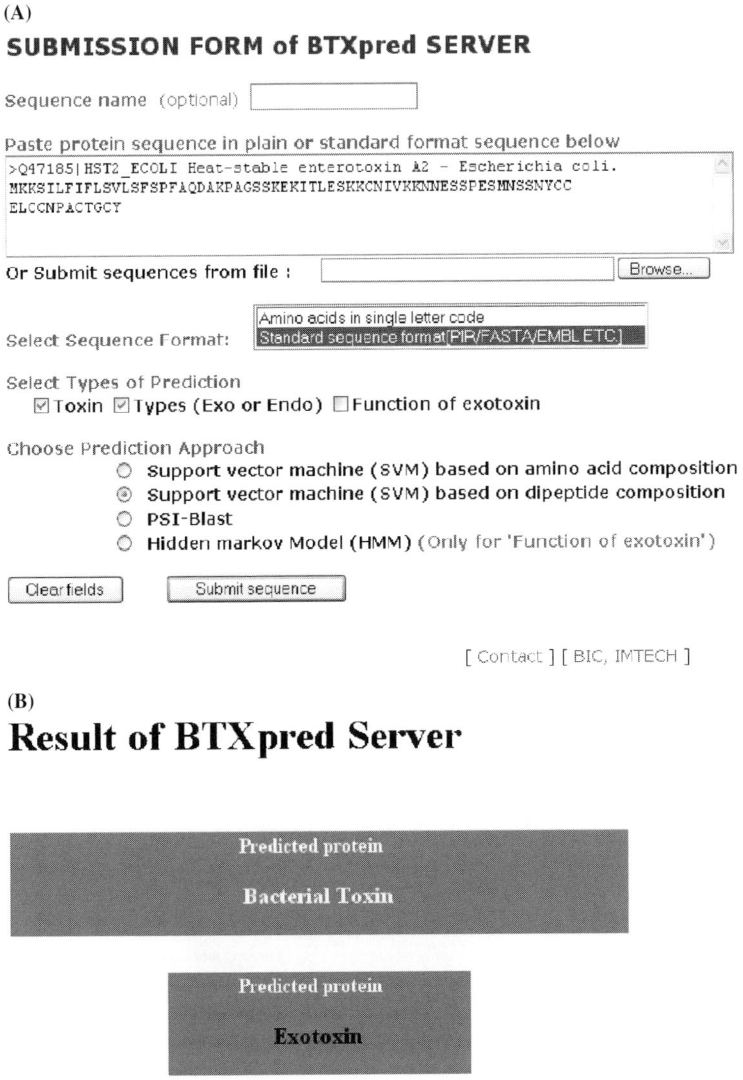

Fig. 2. The snapshot of the (**A**) submission and (**B**) output format of BTXpred server.

(A)

Submission

Protein sequence Name[optional]

Paste protein sequence in plain or standard forma

>P81755
MRCFPVFIILLLLIASAPCFDARTKTDDDVPLSSLRDNLKRTIRTRLNIRECCEDGWCC
TAAPLTGR

Or Upload Sequence File: Browse...

Select Sequence Format: Amino acids in single letter code / Standard sequence format[PIR/FASTA/EMBL ETC]

Select Types of Prediction
☑Neurotoxin ☑Source ☑Function ☑ion channels inhibitors

Choose Prediction Approach
- ○ Amino acid Composition
- ○ Amino acid Composition + Length
- ○ Dipeptide
- ◉ Dipeptide + Length
- ○ PSI-BLAST

Submit sequence | Reset

HOME | Submission | Help | BIC, IMTECH | Contact

(B)

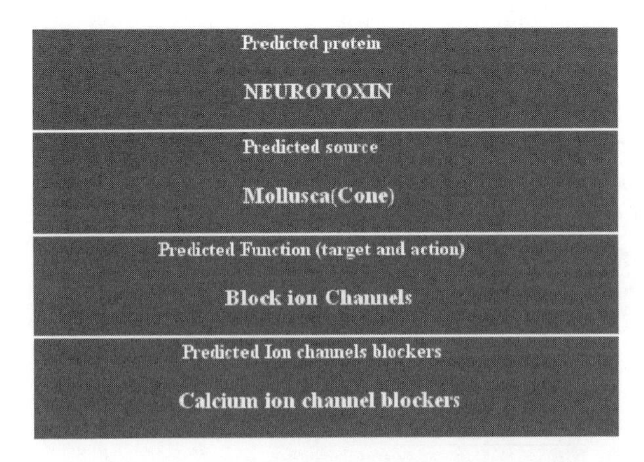

NTXpred Result
Predicted protein
NEUROTOXIN
Predicted source
Mollusca(Cone)
Predicted Function (target and action)
Block ion Channels
Predicted Ion channels blockers
Calcium ion channel blockers

2.3.1. Types of Prediction

The server allows three types of prediction (*see* **Note 3**).

1. Bacterial toxin or non-toxin.
2. Types of toxin—endotoxin or exotoxin.
3. Function of exotoxins.

2.3.2. Different Approaches Provided by BTXpred

The server allows the prediction on the basis of three different approaches.

1. SVM (for toxin and types of toxin).
2. PSI-Blast (for toxin, types of toxin, and functions of exotoxins).
3. HMM (only for function of exotoxins).

2.3.3. About BTXpred Server

The help page describes the general information and stepwise help to submit sequence in submission page. Additional information of this server is linked to supplementary menu. The epitope prediction menu links to Bcepred server *(4)* for prediction of B-cell epitope in the bacterial toxin protein. This will help the users interested in generating antibodies against the toxin. The output of the server provides summarized information about the query sequence and the prediction. The snapshot of the output format has been shown in Fig. 2B.

2.4. Description of NTXpred Server

The aim of NTXpred server is to predict neurotoxins and its source and probable function from primary amino acid sequence using SVM based on composition and PSI-Blast. The CGI script for the server has been written using PERL version 5.03. The server has been installed on a Sun Server (420E) under a UNIX (Solaris 7) environment. Users can enter the primary amino acid sequence for prediction using file uploading or cut-and-paste options. The server accepts the protein sequences in any standard format such as EMBL, GCG, and FASTA or in plain text format. Web servers use the readseq program to read the input sequences. The results provide summarized information about the query sequence and prediction.

The server and related information is available at http://www.imtech.res.in/raghava/ntxpred and mirror site at http://bioinformatics.uams.edu/mirror/

Fig. 3. The snapshot of the (**A**) submission and (**B**) output format of NTXpred server.

ntxpred/. The server predicts neurotoxins, its source mainly eubacteria, cnidaria (sea anemone), mollusca (cone), arthropoda (scorpion and spider) and chordata (snake), probable function mainly the ion channel blockers, blockers of acetylcholine receptors, inhibitors acetylcholine release throughmetalloproteolytic activity or through phospholipase A2 activity and facilitators acetylcholine release and further sub-classification of ion channels blockers into calcium, sodium, potassium, and chloride ion channels inhibitors. The available menus in the server are submission form, help page, data set, algorithm, B-cell epitope prediction, supplementary information, developers, and contact information. The submission menu links to the submission form of the server as shown in Fig. 3A.

2.4.1. Types of Prediction

The server allows four types of prediction.

1. Neurotoxins or non-toxin
2. Source of the neurotoxin
3. Function of the neurotoxin
4. Sub-classification of ion channel inhibitors

2.4.2. Different Approaches Provided by NTXpred

The server provides the prediction on the basis of five different approaches (*see* **Note 4**).

1. SVM module based on amino acid composition.
2. SVM module based on amino acid composition and length.
3. SVM module based on dipeptide.
4. SVM module based on dipeptide and length.
5. PSI-Blast.

2.4.3. About NTXpred Server

The help page describes the general information and stepwise help to submit sequence in the submission page. Additional information of this server is linked to supplementary menu. The epitope prediction menu links to Bcepred server ***(4)*** for prediction of B-cell epitope in the bacterial toxin protein. The results provide summarized information about the query sequence and prediction. The snapshot of the submission and output format is shown in Fig. 3B.

Acknowledgments

We acknowledge the financial support from the Council of Scientific and Industrial Research (CSIR) and Department of Biotechnology (DBT), Govt. of India.

Notes

1. Combined approach of pattern based and composition of amino acid give higher accuracy than pattern based alone.
2. Scores of four different classes are given in tabular form. The highest score achieved by individual class is the predicted functional class.
3. For choosing types of prediction by BTXpred, SVM does not allow predicting function of exotoxins, and PSI-BLAST allows all the three, but HMM allows only prediction of function of exotoxins.
4. SVM module based on amino acid composition and length give higher accuracy in predicting neurotoxins, source, and function.

References

1. Joachims, T. 1999. Making large-scale SVM learning particle. In Scholkopf, B., Burges, C., and Smola, A. (eds), Advances in Kernal Methods Support Vector Learning. MIT Press, Cambridge, MA and London, pp. 42–56.
2. Altschul, S.F., Madden, T.L., Schaffer, A.A., Zhang, J., Zhang, Z., Miller, W., and Lipman, D.J. 1997. Gapped BLAST and PSI-BLAST: a new generation of protein database search programs. *Nucleic Acids Res.* **25**: 3389–3402.
3. Eddy, S.R. 1998. Profile hidden Markov models. *Bioinformatics* **14**: 755–763.
4. Saha, S. and Raghava, G.P.S. 2004. BcePred: prediction of continuous B-cell epitopes in antigenic sequences using physico-chemical properties. In Artificial Immune Systems, Nicosia, G., Cutello, V., Bentley, P.J., and Timis, J. (eds.) ICARIS, LNCS 3239, pp. 197–204.

Index